Burkhard Büdel, Thomas Friedl (Eds.)
Life at Rock Surfaces
Life in Extreme Environments

Life in Extreme Environments

Series Editor
Dirk Wagner

Volume 9

Life at Rock Surfaces

Challenged by Extreme Light, Temperature and Hydration Fluctuations

Edited by
Burkhard Büdel and Thomas Friedl

DE GRUYTER

Editors
Prof. Dr. Burkhard Büdel
University of Kaiserslautern
Biology
Erwin-Schrödinger-Strasse 13
67663 Kaiserslautern, Germany
buedel@bio.uni-kl.de

Prof. Dr. Thomas Friedl
Experimental Phycology and Culture Collection of Algae (EPSAG)
University of Göttingen
Nikolausberger Weg 18
37073 Göttingen, Germany
tfriedl@uni-goettingen.de

Series Editor
Prof. Dr. Dirk Wagner
GFZ German Research Centre for Geosciences, Helmholtz Centre Potsdam
Section Geomicrobiology
Telegrafenberg
14473 Potsdam, Germany
dirk.wagner@gfz-potsdam.de

ISBN 978-3-11-064261-2
e-ISBN (PDF) 978-3-11-064646-7
e-ISBN (EPUB) 978-3-11-064264-3
ISSN 2197-9227

Library of Congress Control Number 2021934662

Bibliographic information published by the Deutsche Nationalbibliothek
The Deutsche Nationalbibliothek lists this publication in the Deutsche Nationalbibliografie; detailed bibliographic data are available in the Internet at http://dnb.dnb.de.

Cover image: Burkhard Büdel
Typesetting: Compuscript Ltd., Shannon, Ireland
Printing and binding: CPI books GmbH, Leck

www.degruyter.com

Preface

Planet Earth has a total surface area of 510,000,000 km², of which 360,570,000 km² (70.7%) belong to the water expanse and 149,430,000 km² (29.3%) refer to the land surface of the planet. Although often a prominent part of the land surface, the proportion of geological rock formations on the land surface area is not really known. Even the amount of bedrock is not clear. Although it is undoubtedly possible to estimate mountain regions' base area, it seems almost impossible to include all types of rock surfaces, such as inselbergs, whose dimensions range from a few square meters to square kilometers. The entire rock surface of mountains is magnitudes larger than their base area because of altitudinal elongation and distinct relief formation by strong erosive forces driven by inclination and climate. Therefore, rock surface is an enormously extensive and important habitat for life on Earth. However, rock surfaces provide a challenging habitat for a broad diversity of micro- and small-sized organisms. They interact with each other, forming complex communities, and with their substrate, causing biodeterioration of the rock. Extreme fluctuations in light, temperature, and hydration are the main factors that determine the rock surface habitats. The rock surface habitat includes various more differentiated microhabitats. Epilithic organisms thrive on the surface without penetrating the rock. Endolithic organisms live just beneath the surface using a thin layer of the rock surface for protection against the environment's adverse conditions (e.g., light protection and storage of water). Chasmoendolithic organisms use fractures of the rock surface for a more habitable environment. In contrast to soils, rock surfaces are solid and last much longer, depending on rock type and climate. Thus, in all climatic regions on Earth, rock surfaces are a substratum challenging for all organisms. One of the significant limitations on rock surfaces of all climate zones for organisms is the absence of soil as a water buffer. Consequently, for any life on rock surface, water availability is the main limiting factor. Rock surfaces, therefore, may be the domain of poikilohydric organisms. They already live for the moment of water availability, and the total amount of water limits their presence on the rock surface. The water availability regime varies within a vegetation period, whatever a vegetation period may be in the Earth's different climatic zones. Poikilohydry is the strategy of microorganisms, including lichens, to adapt to the extremely water-limited habitat. In vascular plants, there is even a secondary reversion to poikilohydry. Vascular plants on rock are mostly limited to shale depressions or rock pools of any size where water accumulates after precipitation events. The erosive activity of microorganisms resulting in a rough rock surface provides the ground for vascular plant species to colonize even steep rock surfaces. Also, biogenic rock weathering mediated by microorganisms further facilitates the establishment of plants on rock surfaces. Therefore, the rock surface habitat draws an illustrative example for the coexistence and collaboration of very diverse organisms, which results in a mutually beneficial strategy for adaption to an extreme habitat.

https://doi.org/10.1515/9783110646467-202

Other limiting factors for life on rock are rock stability, rock chemistry, and rock surface temperature and insolation. From the tropics to the boreal zone, fully exposed rock surfaces' temperature can reach up to 60°C. Even in Antarctica's polar climate, rock surface temperatures may reach up to 23°C. Photosynthetic active light on exposed rock surfaces ranges from 2,700 µmol photons $\cdot m^{-2} \cdot s^{-1}$ in the tropics to 2,200 µmol photons $\cdot m^{-2} \cdot s^{-1}$ in the Dry Valleys of Antarctica. These are the landmarks for life on rock surfaces and illustrate the challenges organisms have to cope with when living on rocks. It explains why exposed rock surfaces are extreme habitats for life.

The book will provide an overview of the various organismal groups, from microorganisms to vascular plants, and survey organism-mediated interactions with the rock surface.

Burkhard Büdel and Thomas Friedl

Note: Most of the book chapters were written during the COVID-19 pandemic time. Because of the impact on their families, two authors could not finish their chapters. We deeply regret that and hope that their families will fully recover.

Volume published in the series

Volume 1
Jens Kallmeyer, Dirk Wagner (Eds.)
Microbial Life of the Deep Biosphere
ISBN 978-3-11-030009-3

Volume 2
Corien Bakermans (Ed.)
Microbial Evolution under Extreme Conditions
ISBN 978-3-11-033506-4

Volume 3
Annette Summers Engel (Ed.)
Microbial Life of Cave Systems
ISBN 978-3-11-033499-9

Volume 4
Blaire Steven (Ed.)
The Biology of Arid Soils
ISBN 978-3-11-041998-6

Volume 5
Jens Kallmeyer (Ed.)
Life at Vents and Seeps
ISBN 978-3-11-049475-4

Volume 6
Natuschka Lee (Ed.)
Biotechnological Applications of Extremophilic Microorganisms
ISBN 978-3-11-042773-8

Volume 7
Susanne Liebner, Lars Ganzert (Eds.)
Microbial Life in the Cryosphere and its Feedback on Global Change
ISBN 978-3-11-049645-1

Volume 8
Étienne Yergeau
Advanced Techniques for Studying Microorganisms in Extreme Environments
ISBN 978-3-11-052464-2

Contents

Contributing authors

Burkhard Büdel
Department of Biology
University of Kaiserslautern
Erwin-Schrödinger-Strasse 13
67663 Kaiserslautern, Germany
buedel@bio.uni-kl.de

Claudia Coleine
Department of Ecological and Biological Sciences
University of Tuscia
Viterbo, Italy
coleine@unitus.it

Don Cowan
Centre for Microbial Ecology and Genomics
Department of Genetics
University of Pretoria
Pretoria 0028, South Africa
don.cowan@up.ac.za

Tatyana Darienko
Experimental Phycology and Culture Collection of Algae (EPSAG)
University of Göttingen
Nikolausberger Weg 18
37073 Göttingen, Germany
tdarien@gwdg.de

Asunción de los Ríos
Museo Nacional de Ciencias Naturales
CSIC, Serrano 115 dpdo
28006 Madrid, Spain
arios@ccma.csic.es

Thomas Friedl
Experimental Phycology and Culture Collection of Algae (EPSAG)
University of Göttingen
Nikolausberger Weg 18
37073 Göttingen, Germany
tfriedl@uni-goettingen.de

Martin Grube
Institute of Plant Sciences
University of Graz
8010 Graz, Austria
martin.grube@uni-graz.at

Tomáš Hauer
Faculty of Sciences
University of South Bohemia
Branišovská 1645/31a
37005 České Budějovice, Czech Republic

Patrick Jung
University of Applied Sciences Kaiserslautern
Carl-Schurz-Str. 10-16
66953 Pirmasens, Germany
patrick_jung90@web.de

Jan Kaštovský
Faculty of Sciences
University of South Bohemia
Branišovská 1645/31a
37005 České Budějovice, Czech Republic
hanys@prf.jcu.cz

Jiří Komárek
Faculty of Sciences
University of South Bohemia
Branišovská 1645/31a
37005 České Budějovice, Czech Republic
komarek@butbn.cas.cz

Pedro H. Lebre
Centre for Microbial Ecology and Genomics
Department of Genetics
University of Pretoria
Pretoria 0028, South Africa
pedro.bixiraonetomarinholebre@up.ac.za

Thulani P. Makhalanyane
Centre for Microbial Ecology and Genomics,
Department of Genetics
University of Pretoria
Pretoria 0028, South Africa
thulani.makhalanyane@up.ac.za

Lucia Muggia
Department of Life Sciences
Università degli Studi di Trieste
Via Valerio 12/2
34128 Trieste, Italy
lucia_muggia@hotmail.com

Stefan Porembski
Department of Botany and Botanical Garden
Institute of Biosciences
University of Rostock
Wismarsche Strasse 44/45
18051 Rostock, Germany
stefan.porembski@uni-rostock.de

Christoph Scheidegger
Biodiveristy and Conservation Biology
Swiss Federal Institute WSL
Birmensdorf, Switzerland
christoph.scheidegger@wsl.ch

Laura Selbmann
Department of Ecological and Biological Sciences
University of Tuscia
Viterbo, Italy
Italian National Antarctic Museum (MNA), Mycological Section
Genoa, Italy
selbmann@unitus.it

Virginia Souza-Egipsy Sánchez
Instituto Estructura de la Materia, Spanish National Research Council
C/ Serrano 113 bis-119-121-123
28006 Madrid, Spain
Virginia.souza-egipsy@csic.es

1 Functionality

Asunción de los Ríos and Virginia Souza-Egipsy

1.1 The interface of rocks and microorganisms

Microorganisms inhabiting rock substrates, called lithobionts, have the ability to attach to rock surfaces but also to exploit protected microhabitats within or beneath rocks. The core of the chapter focuses on reviewing appropriate research strategies and techniques to characterize these different rock-microorganism interfaces using microscopy images and the combination with analytical and molecular biology techniques. Light microscopy techniques, including laser confocal, scanning, and transmission electron microscopy as well as atomic force microscopy, have all been considered, as well as emerging techniques such as compositional mapping, in which microscopy is combined with spectroscopy. These combined approaches allow to visualize lithobiontic communities as mixed polymicrobial aggregates embedded in a biopolymer matrix where microbial cells and minerals interplay. Hence, the main microbial-mineral interactions such as mineral precipitation and mineral weathering are also described here. This chapter summarizes the framework for deciding on suitable approaches to *in situ* characterization of lithobiontic communities.

1.1.1 The lithobiontic niche

Rocks are widely distributed and are a very good substrate for microbial colonization. In fact, microorganisms colonize most of the rock substrates around the world, acting as arbitrators in interactions that occur at the interfaces between the lithosphere, on the one hand, and the hydrosphere and atmosphere on the other [1]. Well attached to external surfaces or lodged inside hard mineral substrates, microorganisms have had the ability to exploit a wide range of lithic microhabitats since their appearance on Earth (Fig. 1.1).

Microorganisms inhabiting rock substrates in terrestrial, marine, and freshwater environments are called lithobionts. The term was coined by Golubic et al. [2] to include rock dwelling organisms present on and within rock substrates, epiliths, and endoliths, respectively. However, hypoliths, which are organisms and microorganisms occupying the interface between rock and soil, have also been considered later as lithobionts [3, 4] because of their close relationship with lithic substrates [5]. The lithobiontic niche is not restricted to rocks (Fig. 1.1A–E); lithobionts are also present in pedogenic soil crusts (Fig. 1.1F) and biogenic calcareous substrates. Depending on

https://doi.org/10.1515/9783110646467-001

Fig. 1.1: (A) University Valley at McMurdo Dry Valleys (Continental Antarctica). (B) Rock boulders at Namib Desert (Namibia). (C) Epilithic lichen community at Livingston Island (Maritime Antarctica). (D) Epilithic community dominated by cyanobacteria from Spanish heritage assets (Santiago de Compostela, Spain). (E) Endolithic lichen (*Bagglietoa marmorea*) inducing biopitting in calcareous rock (Cuenca, Spain). (F) Gypsum crust from Namib Desert in cross-sectional view showing endolithic growth. (G) Upturned quartz pebble from Namib Desert showing hypolithic colonization.

their relation to the substrate, we can therefore establish the following classification (Fig. 1.2):

- Epilithic microorganisms colonizing the external surface of rocks or pedogenetic soil crusts. They are associations of fungi, algae, cyanobacteria, and heterotrophic bacteria (Figs. 1.1C, D and 1.3A).
- Hypolithic microorganisms inhabiting the ventral surfaces of rocks, especially beneath translucent pebbles (Figs. 1.1G and 1.3B). Microorganisms dominate most of hypolithic communities, but they also coexist with organisms such as mosses in these locations.
- Endolithic microorganisms occupying internal microhabitats of rocks and soil crusts. They are divided according to their ecological niche into chasmoendoliths, cryptoendoliths, euendoliths, and hypoendoliths [2, 6, 7].
  - Chasmoendoliths (arrows in Fig. 1.3C) refer to microorganisms that inhabit fissures and cracks within rocks.
  - Cryptoendoliths (arrows in Fig. 1.3D) occupy preexisting structural cavities and pore spaces within the rock or soil crusts.
  - Euendoliths (Fig. 1.3E) actively penetrate into the interior of the hard mineral substrates, creating the endolithic microhabitat themselves and forming tunnels that conform to the shape of the microbial cells.
  - Hypoendoliths settle in the underside of soil crusts, but within them [6].

Nonetheless, the different types of lithobiontic colonization are not always strictly delimited. Some microorganisms are partially epilithic and partially endolithic, such as some lichen and cyanobacteria species; others penetrate carbonate substrates as euendoliths but also colonize preexisting structural cavities within rocks; hypoliths can also be endoliths and colonize fissures of the rock above them [1, 5, 8].

The ability of rocks to become colonized, or bioreceptivity [9, 10], notably conditions the ecological niche played and the rock-microbial interfaces generated. While fissures and cracks of siliceous rocks harbor chasmoendolithic microbial communities, pores of carbonate rocks favor the formation of cryptoendolithic communities [11]. Euendoliths are present mostly in relatively soluble substrates such as carbonates and phosphates [2]. However, we cannot discard that species-specific abilities facilitate the development of certain ecological niches [12].

Lithobiontic communities can be found in a wide range of settings and environmental conditions, from polar deserts to subtropic and tropic regions, following gradients of critical environmental determinants, such as water supply and light [1]. In fact, they inhabit lithic substrates in almost every terrestrial, marine, and freshwater environment on the surface of the Earth. Lithobiontic communities are widespread in milder climatic conditions where they compete with vascular vegetation [8, 13]. However, their highest ecological importance is reached in extreme environments as they are the main colonizers there [3, 11, 14, 15, 16]. In addition, endolithic microbial communities have been detected on the seafloor and subseafloor of the upper oceanic

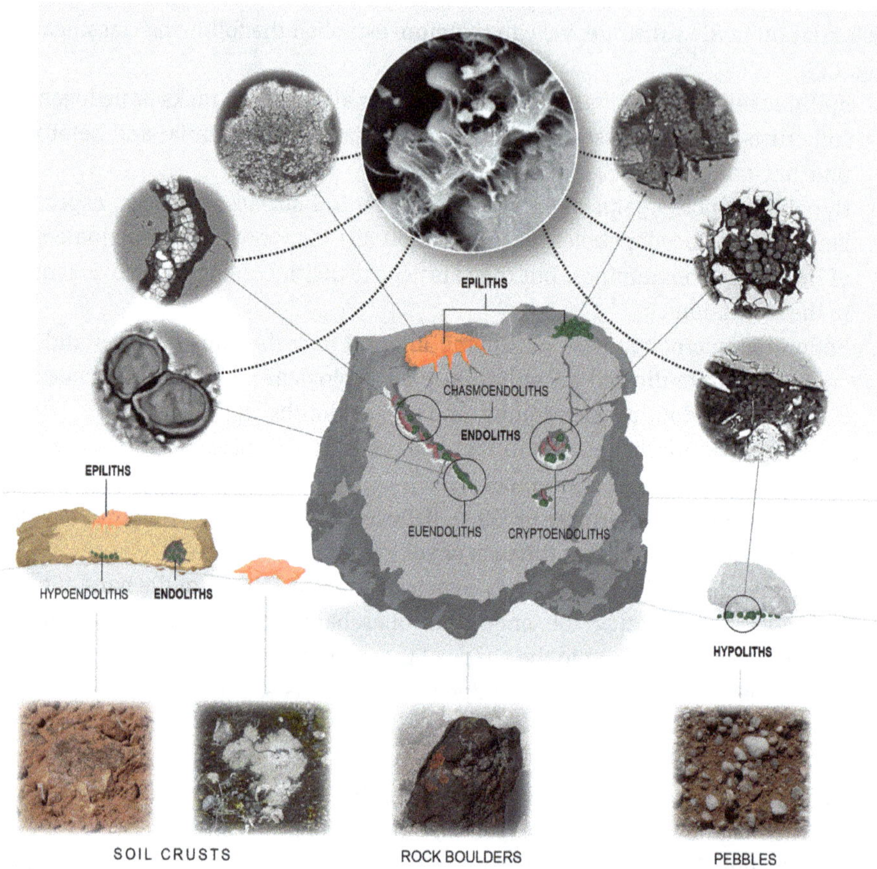

Fig. 1.2: Schematic representation illustrating the main lithic microhabitats occupied by lithobiontic communities: epilithic and endolithic microhabitats in rock boulders, epilithic and endolithic microhabitats in pedogenic soil crusts, and hypolithic microhabitats under translucent pebbles. Small circles at the side show scanning electron microscopy (SEM) images representing the different lithobiontic ecological niches (from top to bottom and first of the right: epilithic lichen– and epilithic cyanobacteria–dominated community, chasmoendolithic cyanobacteria community and cryptoendolithic lichen association, and euendolithic cyanobacteria and hypolithic cyanobacteria). The biggest circle located at the center above contains a low-temperature scanning microscopy (LTSEM) image showing the biofilm organization shared by the different types of lithobiontic communities.

crust at the spreading ridges and hydrothermal areas [17] and associated with rock deposits at the continental subsurface [18, 19].

1.1.2 Research strategies for the study of the rock-microorganism interface

In the lithobiontic habitat, microorganisms live, thrive, and create microenvironments that are difficult to characterize. Multiscale and correlative approaches are needed to

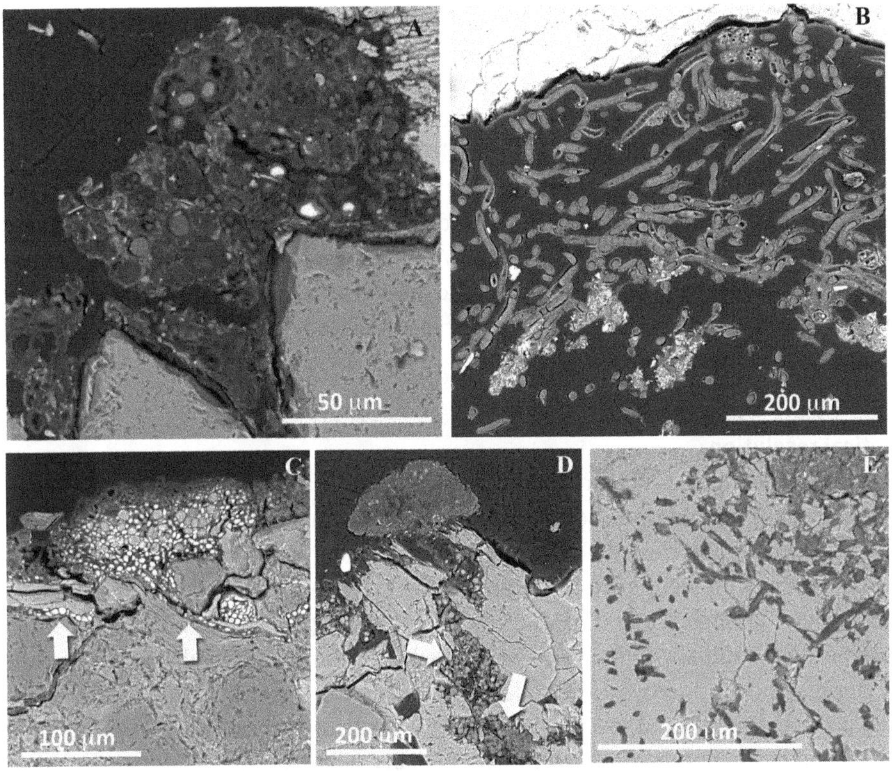

Fig. 1.3: (A) SEM-BSE image of an epilithic microbial community harboring algae, cyanobacteria, and heterotrophic bacteria on sandstones (Dry Valleys, Antarctica). (B) SEM-BSE image of a hypolithic community dominated by filamentous cyanobacteria under a quartz pebble at Namib Desert. (C) SEM-BSE image of fungal chasmoendolithic colonization (arrows) under an epilithic lichen thallus in granite from Livingston Island (Antarctica). (D) SEM-BSE image of a cryptoendolithic growth (arrows) under a lichen thallus in soil gypsum crust from Namib Desert (Namibia). (E) SEM-BSE image of euendolithic fungal growth located under an epilithic lichen thallus in a soil crust rich in carbonates from Namib Desert.

describe the rock-microorganism interface and the processes occurring therein. The precise description of microbial communities in the lithobiontic niche requires combining the output from several visualization and analytical techniques. Researchers using a combination of techniques will be able to answer questions about diversity and distribution of organisms within a community and what they are doing there.

Visualization techniques

The first identification of the presence of lithobiontic communities is made using the optical microscopy approach. In the study of the interface of microorganisms and rocks or sediments, petrographic polished thin sections must be prepared for observation with transmitted and reflected light microscopy. To achieve the necessary degree of preservation of the microbial cells, samples should be aldehyde fixed in buffered solutions, as simple drying of the specimens destroys microbial structures. After fixation, samples are gradually dehydrated by increasing concentrations of ethanol and embedded with a low viscosity resin. Finally, the resin is polymerized, and thin sections of these hardened specimens can be made following conventional procedures for preparing rocks for microscopy studies. The use of fluorescence stains based on intracellular binding sites of DNA or RNA promotes marked fluorescence enhancement for epifluorescence microscopy of the microbial communities. The objective of this approach is to discriminate the organic material from the mineral particles and to locate areas of interaction.

Optical resolution

Microbial communities of more than 10 μm thickness usually cannot be handled with common light microscopes because material above and below the focal plane will scatter light and interfere with the direct measurement. Such biofilms can be investigated using wide-field deconvolution epifluorescence microscopy [20] or confocal laser scanning microscopy (CLSM) [21]. CLSM allows optical sectioning of the structure of the microbial communities associated with the mineral substrate by means of fluorescence induced by glutaraldehyde fixation of organic material [22], enabling a three-dimensional reconstruction of the undisturbed sample [23]. Combining CLSM and *in situ* fluorescence hybridization using 16S rRNA targeted oligonucleotide probes (FISH) enables the detection and identification of specific groups of bacteria in biofilm and microbial mats [24]. To increase the sensibility, the fluorescence signals can be amplified by catalyzed reporter deposition–fluorescence *in situ* hybridization (CARD-FISH) [25]. In the case of microbial communities inside carbonates, the samples should be decalcified to avoid the false binding to positively charged mineral surfaces [26]. The use of specific probes labeled with fluorophores of different emission wavelengths allows researchers to visualize colocalization and can be useful in determining the architecture of the microbial community [27]. In addition, staining with live/dead fluorescence markers allows the characterization of the physiological state of microorganisms in lithobiontic communities [21, 28]. Other fluorescence stains can provide information about microenvironments such as pH measurements around the microbial cells [23, 29]. The use of fluorescent lectin-binding probes allows to detect the extracellular polymeric substance (EPS) placed around the cells, yielding information about different structural domains related to mineral nucleation [30, 31].

In the case of photosynthetic organisms, one additional parameter that can be analyzed is the fluorescent emissions at 680 to 690 nm after proper excitation, which is due to chlorophyll molecules in reaction centers of photosystem II (PS II) of *in vivo* oxygenic phototrophs (Fig. 1.4A). Wavelengths that excite PS II fluorescence depend on associated accessory antenna pigments, which vary in different groups of oxygenic phototrophs [32]. This technique has been used to study the composition and structure of microbial communities from stromatolites [33]. The lambda scan confocal microscopy has been used to record a series of individual images obtained using a defined emission fluorescence wavelength range. This procedure has been successfully used to assess, at the single-cell level, the physiological state of endolithic photosynthetic cyanobacteria [34].

High-resolution techniques

The analysis of the fine interaction of microorganisms with the minerals requires study via high-resolution techniques, such as SEM using secondary electron microscopy and backscattered electron (BSE) modes on flat surfaces [35]. Wierzchos and Ascaso [36] included postfixation with osmium tetroxide during the preparation of the sample, allowing to visualize the cytological aspect of the microorganisms present in close contact with the mineral substrate [37]. The analysis of finely polished sample cross sections reveals the contact between the organisms and the surrounding substrate [38] and allows to identify biochemical and biophysical weathering induced on the minerals [39]. This approach (Fig. 1.3A–E) has been used, among others, to describe the endolithic habitat of lichens [40, 41] and cyanobacteria [42] in Antarctic granites, epilithic lichens on gypsum crystals and soil crust in arid and semiarid habitats [6, 43, 44], and endolithic communities in halite and ignimbrite rocks from Atacama Desert [45, 46]. These techniques have also been used in describing endolithic microbial activity in oceanic basalts [47, 48].

LTSEM is another interesting technique to analyze lithobiontic communities (Fig. 1.4B) because it has proved to be valuable for visualizing water at the interface between microorganisms and their substrate [49] and the spatial structure of EPS matrix [6, 42], with implications in lithification in microbial mats [50]. The distribution of water in lichen-dominated soil crust was also described with this technique [51, 52], showing areas of hydrophobicity on the surface of soil crust lichens as one of the key factors controlling the variability in runoff response in semiarid habitats.

The characterization of the biogeochemical activity of microorganisms on the mineral surfaces requires the use of microscopy with higher resolution such as transmission electron microscopy (TEM) (Fig. 1.4C). In fact, to study physicochemical mineral transformations, it is advisable to use a high-resolution transmission electron microscope (HRTEM), which works at atomic resolution scale [53]. HRTEM provides simultaneous structural information at the nanometer scale via selected area electron diffraction (SAED) and chemical data through energy-dispersive X-ray spectroscopy (EDX) or by electron energy loss spectroscopy (EELS) [54–56]. TEM has been used extensively to define the architecture of the cell envelope, the location and nature of the interactions with metal ions, and the structure and composition of associated minerals [57]. Using this technique, it has been demonstrated that different constituents of cell walls from gram-positive and gram-negative bacteria are capable of binding metal ions by means of exposure of anionic charges [58]. Difficulties inherent to producing sufficiently thin samples to allow electrons to pass through them while the organic-mineral interface is preserved intact have limited its application. However, nowadays samples can be prepared using a focused ion beam (FIB) to thin a region of interest selected under SEM observations for analysis under TEM [59, 60]. Other techniques such as atomic force microscopy (AFM) do not need to process the samples and allow the observation of the interactions of microorganisms in their native state [61, 62]. Taylor and Lower [63] have determined the thickness and surface density of layers of EPS around the cells using AFM. New instrumentation and methodologies of liquid-phase TEM and advances in cryogenic TEM are also transforming our understanding of the physical and chemical mechanisms underlying the formation of biominerals [64]. Cryo-TEM has been applied to study the dynamic processes of nucleation, self-assembly, crystal growth, and coarsening of biominerals [65, 66].

Up to this point, we have been revising different methodologies used to study microbial communities following a sequential approach, by increasing the scale of resolution from light microscopy to high-resolution TEM and AFM. Samples fixed and embedded in resin can be polished to be used for light, fluorescent microscopy, and SEM. Selected areas under SEM can be extracted by FIB and studied by TEM. This strategy of visualizing the same sample with different microscopy techniques is known as correlative microscopy. However, not all techniques are easy to combine. For instance, fluorescent probes are incompatible with classic EM sample preparation. Fluorescence reduction caused by treatment with high concentrations of osmium and by complete dehydration may be prevented by cryofixation of the samples [67]. By using Cryo-TEM, samples are preserved in a frozen hydrated state, which does not require chemical fixation or embedding, and allows for fluorescence preservation. Sample transfer between microscopes offers the possibility to use fluorescent visualization before electron microscopy visualization, enabling analyses of microbe-mineral interactions at different scales and microenvironments.

Analytical techniques

An important feature of microbial communities is the spatial heterogeneity of their extracellular chemistry. The combination of visualization and chemical analysis allows us to spatially resolve these existing chemical patterns. In addition, the study of older sediments can provide us with information regarding the fossilization and diagenetic processes acting on the previously live microbial communities [22]. There are several chemical imaging techniques that can be non-invasive, such as EDX or electron probe micro-analyzer (EPMA) and Raman spectroscopy. In other techniques, the atoms or molecules are sputtered away from the sample surface, such as X-ray photoelectron spectroscopy (XPS) and secondary ion mass spectroscopy (SIMS). Selecting among techniques depends on the resolution needed and the information sought. In some cases, additional techniques such as microsensors can be used to describe the changes at the micron scale along the daily cycle of the endolithic communities [68] or in different areas of microbial mats [69].

At the optical level, analytical imaging techniques such as Raman spectroscopy are very useful to characterize mineral-microbial interfaces. The frequency of light scattered from a molecule depends on the structural characteristics of the chemical bands, such that each peak corresponds to a given Raman shift (from the incident radiation energy) related to a specific molecular vibration and represents a particular chemical constituent of the sample. Raman requires minimal specimen preparation and can be used on the rock chips or on standard uncovered geological thin sections without the use of high vacuum conditions. The achievable spatial resolution is down to 20–50 nm when coupled with confocal microscopy, near-field optical techniques, or super resolution imaging techniques. A good example of this technique was shown on the study of endolithic phototrophic microbial communities from Atacama Desert [70, 71]. Raman imaging revealed that pigment composition within the microbial aggregates was useful to understand their adaptation strategy to survive in extreme environments [72]. This technique was also used to describe the presence of melanin pigments in the cell walls of fungi colonizing the interior of gypsum crusts [73].

As higher resolution is needed, the use of scanning transmission X-ray microscopy (STXM), a synchrotron-based X-ray spectromicroscopy that can both image and reveal oxidation states, is of great help for studying microbially associated minerals [74, 75]. FIB milling in combination with STXM analyses has been used to measure the accelerated weathering of a sheet silicate (biotite) by ectomycorrhizal fungi [76]. STXM, which uses near-edge X-ray absorption fine structure (NEXAFS) as its contrast mechanism, can be used to analyze hydrated biological samples. NEXAFS image sequences can be converted into quantitative maps of the chemical species present in biofilms [77]. With these techniques, it is possible to visualize bacteria and associated mineral phases in fully hydrated biofilms [78, 79]. The use of proton-induced X-ray emission (PIXE), a non-destructive elemental analysis technique, was used to study the differences in major element concentrations among the endolithic colonized areas in granitic sandstones [80]. Recent approaches using micro–X-ray absorption near-edge structures have revealed the distribution of elements such as P, Cl, and S in cross sections of biological soil crusts from arid and semiarid habitats [81]. This approach revealed apatite hotspots related to the presence of microorganisms and the accumulation of S and Cl containing compounds within green algae and on their outer surface. These methods, in combination with TEM, provide remarkably clear chemical state–specific images of the areas of the microorganism-mineral interface. In addition, STXM and photoemission electron microscopy (PEEM) may provide the ability to identify biogenic signatures in natural samples [59, 82, 83]. Until now, the highest precision is achieved using the NanoSIMS technique, which, using a Cs^+ ion beam with a size as small as 50 nm scanning over sample surfaces, is able to obtain high spatial resolution images of up to 7 isotopic species simultaneously [84, 85].

In vitro studies involving the purification and characterization of molecules that drive biogeochemical processes can also be very relevant to understand the mechanisms of mineral-microbial interactions. These experimental studies are important to determine which forces are operational during the adhesion of microorganisms to minerals and which macromolecules contribute to metal bioreduction [86]. Mineral-respiring bacteria use a process called direct extracellular electron transfer (DEET) to route their respiratory electron transport chain to insoluble electron acceptors on the exterior of the cell [87]. *In vivo* experiments can complement the *in vitro* studies to delve into the metabolic strategies that have been evolutionarily optimized to exploit a particular ecological niche [88]. Following these approaches, the relationships between the microbial mechanisms of energy metabolism and the physicochemical features of the environments can be established.

Molecular biology techniques for the detection and identification of microorganisms

The identification of lithobiontic microorganisms using molecular biology techniques is a hard task given the difficulties of obtaining high-quality DNA, in terms of both purity and concentration, from the rocks (especially from endolithic microhabitats). To describe microbial communities at rock interfaces, it is especially relevant to combine the identification of microbial components by means of molecular biology techniques with morphological and ultrastructural characterization via microscopy without disturbing the lithic microhabitats [42, 89, 90] because with this strategy, it is possible to localize, describe, and simultaneously identify them.

The isolation and the culturing of lithobiontic microorganisms allow to easily extract DNA for polymerase chain reaction (PCR) or genomic analysis and to perform subsequent phylogenetic analysis to classify and identify them [91, 92]. One of the best methods for the isolation of lithobiontic microorganisms is crushing the part of the rock of interest and plating the rock powder on a media plate after suspension [93]. However, only a small fraction of environmental microorganisms has been cultivated because they have special growth requirements such as specific nutrients, growth signals, or dependence on other microorganisms [94].

In simple microbial communities, single taxa can be identified by DNA barcoding. The collection of small samples from the rocks allows extracting DNA and later amplify via PCR specifically targeted DNA sequences that are taxonomically informative. Sequencing of PCR products and comparison with reference databases allow identification, but lithic habitats are generally underexplored and close relatives are frequently not found in reference databases. The extension of cryptic colonization, such as endolithic lichens, can be characterized by comparing the sequences of endolithic growths with those of identifiable epilithic forms or recognizable external fruiting bodies [40]. However, most of the microbial lithobiontic communities are complex, and this strategy cannot be followed. At present, the development of high-throughput DNA sequencing technologies allows us to simultaneously obtain information at a large taxonomic scale about all the components of the communities. The method more broadly used on lithobiontic communities is metabarcoding [89, 90, 95–97], in which short regions of one or a few genes (called DNA barcodes) from environmental DNA are mass amplified by PCR using universal primers followed by high-throughput sequencing taxa, which are later classified using reference databases. Metabarcoding is providing important new insights into the cryptic diversity of lithobiontic communities, but like all PCR-based methods, it is PCR biased. At present, metagenomic studies begin to be also applied to lithobiontic communities [98–100]. Metagenomics is based on shotgun high-throughput sequencing and avoids the PCR bias, and it also gains insight into functional diversity but leads to more complex bioinformatic work.

1.1.3 Composition and spatial organization of lithobiontic communities

Composition

Most of the lithobiontic communities harbor microorganisms from different taxonomic groups (Fig. 1.3A). Fungi, algae, and bacteria are the most conspicuous microbial components of lithobiontic communities at ground surface [3, 90, 101, 102]. Archaea have been only recently acknowledged in different lithic habitats at ground surface, predominantly from extreme environments [19, 103–105]. However, archaea are commonly found coexisting with heterotrophic bacteria on the seafloor and subseafloor of the upper oceanic crust in spreading ridges and hydrothermal areas [17, 47, 106]. Viruses are also probably important components of different lithobiontic communities, but as far as we know, they have been only analyzed in hypolithic communities [107, 108].

The lithobiontic association of heterotrophic bacteria and cyanobacteria is one of the most widely spread across different lithic substrates; they share rock microhabitats and establish intimate associations between them [23, 28, 109, 110]. These mixed bacterial communities are present at epilithic (Fig. 1.3A), endolithic (Fig. 1.4B and C), and hypolithic positions (Fig. 1.3B), forming very successful associations in extreme environments. These communities were dominated by cyanobacteria and, among others, also harbor Acidobacteria, Actinobacteria, Bacteroidetes, and Proteobacteria [111–114]. Closely affiliated microbial lineages are often shared across rocks in different geographical regions, but some endemic taxa are also frequently found [114, 115].

Lichens are another successful lithobiontic association (Figs. 1.3C, D and 1.4D), which are basically the symbiosis between specialized fungi (mycobionts), mostly Ascomycetes, and unicellular green algae and/or cyanobacteria (photobionts). In addition to the main symbiotic partners, lichen harbor diverse communities of prokaryotes and fungi as cohabitants [116]. They are primary colonists of rocks and occupy approximately 6–8% of the Earth's land surface, forming part of epilithic communities around the world [117], also being key components of endolithic communities from extreme and temperate environments [40, 118–120].

Non-lichenized black fungi exhibiting meristematic growth are also frequent colonizers of rock surfaces in extreme environments [121, 122]. These melanized fungi belong, for the most part, to the fungal classes Dothideomycetes and Eurotiomycetes. Dothideomycetes were more frequently found in cold deserts, whereas Eurotiomycetes were more abundant in hot drylands [123]. Black fungi belonging mainly to the Dothideomycetes class have been found in lichen-dominated endolithic communities from Antarctica [124], as well as endolithic yeast from the Basidiomycota [125].

Spatial organization

The structural organization of lithobiontic communities is determined not only by their microbial composition but also by the spatial configuration of the occupied lithic microhabitats [3, 11, 102]. In addition, specific microenvironments are created at these microhabitats where microorganisms, which favor or are favored by these conditions, coexist [23]. The assembly of the lithobiontic communities at rock interfaces is consequently clearly limited by microenvironmental features differing on a small three-dimensional spatial scale [46, 101, 126, 127].

The distribution of certain microorganisms in the lithic substrate is also determined by their own growth requirements, such as light, moisture, and mineral nutrients, especially in endolithic communities [11]. Cryptoendolithic microbial colonization normally occurs as different colored bands running parallel to the rock surface, which respond to a steep light gradient [46, 118, 128]. The most extensive colonization by chasmoendolithic cyanobacteria and algae has been found in near-to-surface areas of the fissures and under the more translucent minerals [42, 101]. Hypoendolithic colonization has been detected only in crusts with a microporous matrix allowing light transmittance [6, 104]. The establishment of hypolithic communities is also associated with the arrival of sunlight to the ventral area of the rocks because this colonization is not detected in fully nontranslucent rocks [129]. In fact, in opaque rocks from Polar regions, hypolithic growth has been reported only as a result of light penetration around the edges of rocks that were sorted by periglacial action [14]. In addition, cryptic rock microhabitats, such as endolithic and hypolithic ones, favor water retention in them after rainfall or fog events and allow microbial cell hydration, which is especially relevant to the persistence of microbial communities in drylands [6, 13, 46, 129, 130]. Lithic microhabitats can also be considered nutrient reservoirs. Some nutrients required for microbial growth can be obtained by *in situ* weathering of the host microhabitat, but other essential nutrients are more likely derived from allochthonous sediments such as dust, which can also become accumulated in specific lithic microhabitats [115, 131, 132].

Microorganisms colonizing the lithic substrate are not dispersed but rather form mixed polymicrobial aggregates embedded in an organic matrix resulting mostly from the accumulation of EPS (polysaccharides, proteins, and nucleic acids), produced and secreted by the resident microbial cells [15, 23, 133]. The EPS adheres to the microbial aggregates of external (epilithic and hypolithic) and internal (endolithic) rock surfaces and configures a biofilm structure (Figs. 1.2 and 1.4B). Microbial cells are kept in proximity within the EPS matrix, thus allowing intense interactions, including cell-cell communication (Fig. 1.4C) and the formation of synergistic microconsortia [28, 134]. Other functional roles have also been attributed to the EPS; they favor water retention and sorption of organic compounds and ions and provide protection to microbial cells against freezing [28]. Minerals are also embedded within the EPS matrix, which seem closely associated with the EPS surrounding microbial cells, some of them corresponding to detached mineral fragments from the lithic substrate (Fig. 1.4B, asterisk) but also biogenic (Fig. 1.4B, white arrows) and allochthonous minerals are associated. An important chemical feature of the EPS matrix is the cation-binding ability of the negative functional groups, which plays a twofold role, either inhibiting or promoting mineral formation [135, 136]. Acidic EPS can also generate low pH in the proximity of the microbial aggregates [23].

Multiple interrelations occur within these biofilms, which involve not only the establishment of different relationships between biological components but also close interactions between the microbial cells and the immediate mineral environment in which they are found [54, 55, 137]. These microbial-mineral interactions, through their physical and chemical effects, contribute together with the EPS matrix to the formation of specific microenvironments in the biofilms. These areas have conditions different from those of the bulk, and the different microorganisms are organized depending on their physiological requirements [11, 23].

The spatial biofilm organization notably contributes to the establishment and functioning of the lithobiontic community because the arrangement of microbial communities facilitates the establishment of the close relationships among the different components and the configuration of specialized lithobiontic microconsortia [23, 28]. They are basically composed of cyanobacteria and algae as primary producers; fungi, which may be regarded as consumers; and heterotrophic bacteria as decomposers. Live, dead, or dormant algal, fungal, and bacterial cells can appear intermixed in several zones of the lithic substrate (Fig. 1.4A), favoring nutrient recyclability [21, 28]. The proportion of dead and dormant cells can be high in lithobiontic communities because they live close to the limits of their physiological potential, especially in communities of extreme environments [11].

Epilithic communities

Epilithic subaerial biofilms are frequently formed on external lithic surfaces in terrestrial, marine, and aquatic ecosystems [15, 138]. Depending on the development stage and the presence of pigmented microorganisms in the epilithic communities, some aggregates form visible colored biofilms detectable by the naked eye on the lithic substrate (Fig. 1.1D), whereas others are undistinguishable. Photosynthetic microorganisms lead the formation of green or greenish-blackish patinas on the rock in natural and man-built lithic substrates [139, 140]. Black fungi form compact, melanized colonies on bare rock surfaces in extreme and temperate climates, generating distinguishable black colorations on the rock [121, 122, 141]. The establishment of lichen symbiotic associations on rock surfaces is very common from milder to the most extreme environments, where the main symbiotic partners usually form a compact and macroscopically recognizable structure known as the lichen thallus (Fig. 1C). Mineral fragments and other microorganisms are frequently embedded in the saxicolous lichen thallus structure (Fig. 1.4D). Depending on their bioweathering capacity and the susceptibility of the substrate to biological attack, mineral fragments can be more or less relevant [142]. Calcium oxalate crystals are also frequent in epilithic and soil crust lichen, where they appear intermixed with substrate mineral fragments [40, 143].

Hypolithic communities

Hypolithic communities are organized in biofilms situated under translucent rocks (Fig. 1.1F), which can be dominated by different organisms such as mosses, bacteria, and fungi, and harbor mineral fragments of soil origin trapped within the EPS matrix [144, 145]. There are two main types: cyanobacteria-dominated and moss-dominated hypolithic communities. Their spatial structure was mostly determined by the dominant biological component [5]. In moss-dominated communities, the organization of the community was determined by the structure of the moss plantlets, including both live and dead tissues. In cyanobacteria-dominated hypolithic communities, the filamentous cyanobacterial cells and the surrounding EPS are the principal structural elements of a multilayered structure (Fig. 1.3B). These communities are arranged in two different layers. On the ventral rock surface, a layer harboring a higher density of mainly filamentous cyanobacteria cells is formed. Overlaying the latter, a second layer of dispersed cyanobacteria cells extending toward the soil, and trapping a higher amount of soil particles, is formed [5].

Endolithic communities

Microorganisms hidden within lithic microhabitats avoid high radiation, extreme temperatures, lack of water supply, and attack by grazing organisms, making this type of colonization very successful in extreme environments [3, 5, 13, 146]. Physicochemical features of the lithic substrate, especially textural properties, considerably limit the settlement of endolithic communities and determine their extension and organization [147, 148]. Microbial aggregates are accommodated in the empty spaces within the rock created during rock formation (primary porosity) or after rock formation (secondary porosity). An extensive porosity system and the preexistence of a weathering structure facilitate endolithic colonization [23, 149]. The porosity system (mainly extent and type) notably conditions the spatial organization of cryptoendolithic (Fig. 1.3D) and chasmoendolithic (Fig. 1.3C) communities. Epilithic growth can also induce the formation of fissures that are colonized by chasmoendolithic microorganisms [8]. Euendolithic colonization involves an active process of dissolution,

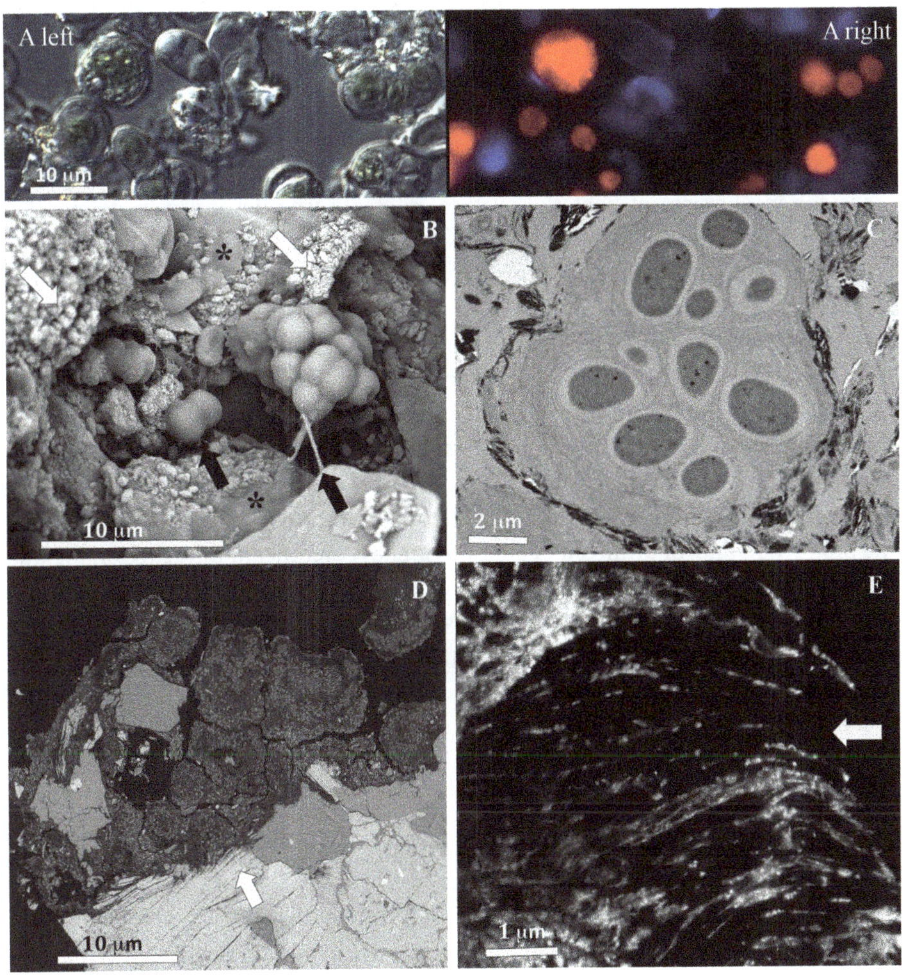

Fig. 1.4: (A) Correlative light microscopy (left) and UV fluorescent microscopy (right) images of living (red in panel A, right) and dead (blue in panel A, right) endolithic algae from Antarctic granite at Miers Valley (Antarctica). (B) LTSEM image of an endolithic biofilm formed by cyanobacteria embedded in an EPS matrix (black arrows) harboring mineral fragments (asterisks) and fine biomineral crystals (white arrows) from Dry Valleys (Antarctica). (C) TEM image of endolithic cyanobacteria in gypsum from Sorbas (Almería, Spain). (D) SEM-BSE image of epilithic *Caloplaca* sp. thallus in granite from Dry Valleys (Antarctica), inducing mica separation and exfoliation on the granite surface (arrow). (E) CLSM image of glutaraldehyde fixed endolithic fungal hyphae (white) colonizing interlayer spaces of mica (white arrow).

which allows boring into the substrate and the formation of tunnels [150]. Hence, the boring abilities of euendolithic microorganisms determine their distribution and organization. In intertidal marine carbonates (calcareous substrates formed by living organisms and those formed by geological processes) or carbonate crusts (Fig. 1.3E), euendolithic cyanobacteria and fungi occupy broad layers under the surface [151, 152]. Endolithic lichen with symbionts showing euendolithic ecological niches can also become established in marine intertidal environments [12].

1.1.4 Microbial-mineral interactions

Since the mechanism of bacterial surface reactivity was described in detail [153], it was established that the close spatial association of active cells with minerals creates chemical gradients that drive the formation of minerals through recrystallization, solid-state transformation, and/or heterogeneous nucleation [154], including reactions on mineral surfaces and cell components that serve as templates [82, 155, 156]. The action of microorganisms on minerals may be attained by slow dissolution processes mediated by biogenic organic acids or complexants [157], or by chemical transfer between cell surface components and mineral [158]. The specific environmental conditions and adaptation mechanisms may induce changes in cell surface chemistry, modifying the adsorption properties of the cell walls. These changes in the production and/or arrangement of EPS can have significant influence on the fractions of complexing elements present in the microenvironment surrounding the cells [159]. The lithobiontic community growing on the rock surface potentially functions initially as a weathering agent (through the production of organic acids or ligands that facilitate mineral dissolution) and later as reactive surfaces for the nucleation of secondary clay phases [22].

Mineral precipitation

The surfaces of the microbial cells are interfaces where the precipitation of metal ions and the development of fine-grained minerals take place [160, 161]. Although the chemical structure of these minerals may be identical to that of minerals produced by geochemical mechanisms, their biological origin can be traced by their isotope concentrations [162, 163]. The formation of mineral precipitates in microbial communities can be either biologically induced as a result of microbial activities or biologically influenced if cells and EPS act as passive nucleation templates [135]. On the other hand, organomineralization, as the mineralization process mediated by organic molecules or particles independent of the living organisms that produce them [164], occurs also in natural environments. Biologically induced and influenced mineral precipitation and organomineralization reactions often occur at the same time in natural environments and are difficult to discriminate. Furthermore, it was stated that the production of EPS and decay of microbial biofilms are critical for the development of localized conditions that favor mineral precipitation.

Silicates

Field studies have led to recognize that microorganisms can mediate the formation of clay minerals. The inorganic phases develop in a predictable manner, beginning with the adsorption of cationic ions to the anionic cellular surfaces by supersaturation of the proximal fluid. Then, the nucleation and precipitation of a precursor hydroxide phase on the cell surface, is followed by reaction with dissolved silica and aluminum, and eventually the growth of an amorphous claylike phase. Ferris et al. [155, 165] initially described complex (Fe, Al)-silicates on bacterial cells growing in metal-contaminated lake sediments. Colloidal species of (Fe, Al)-silicate composition that are either present or form by weathering initially in the water column may react directly with the outermost cellular layers. These precipitates ranged from amorphous gel-like, poorly ordered capsular material to fine-grained (nm size particles) authigenic mineral as complex (Fe, Al) silicates of variable composition [166]. Some clays form as replacement products from the alteration of primary minerals as was described in cryptoendolithic lichen communities in sandstones from the Antarctica McMurdo Dry Valleys. Cells became mineralized by epicellular deposition of these primary minerals [41]. Léveillé et al. [167] similarly documented the formation of kerolite (Mg-silicates) in association with biofilms colonizing the surfaces of walls and ceilings within basaltic sea caves on the northern coast of Kauai, Hawaii. Similar associations of Mg-silicates to cell walls of algae and cyanobacteria were described in other cryptoendolithic habitats such as Mono Lake pinnacles [168], Yellowstone geothermal environments [115], biofilms in acidic waters from Rio Tinto [169], cryptoendolithic communities in gypsum crust from Atacama Desert [170], and biosedimentary structures in lakes [85]. Handley et al. [171] described silicification of biofilm EPS in hot springs in Champagne Pool, New Zealand, where EPS provided direct templates for silica precipitation.

Carbonates

Calcite precipitation often starts on cell walls and sheath harboring EPS. This mode of calcite precipitation is related to crystal nucleation on cell sheath material [172] or to increased saturation induced by photosynthesis [173, 174]. The EPS have negatively charged radicals, such as carboxyl groups, that can absorb calcium ions [135]. Other studies have also shown inhibition of calcite precipitation by EPS [175–177]. This inhibition was related to differences in the biochemical composition of the EPS of different microorganisms or different physiological states. Kawaguchi and Decho [178] found that EPS from the unlithified layers of stromatolites had higher uronic acid and carbohydrate content than the EPS from lithified layers.

In addition to calcite, other carbonates can precipitate as a result of microbial activity. Vasconcelos et al. [179] showed the precipitation of dolomite at low temperatures in the presence of sulfate-reducing bacteria (SRB). Warthmann et al. [180] also isolated a halotolerant SRB bacterium, which mediates dolomite formation. Other natural environments, under both anoxic and oxic conditions, have been described as promoting dolomite formation [181]. García del Cura et al. [182] described calcite and dolomite precipitation in close association with EPS of coccoid microbial cells and filamentous cyanobacteria in springs. Visscher et al. [183] described the precipitation of aragonitic needles associated with the activity of SRB within Bahamian stromatolites. EPS produced by *Desulfovibrio* have calcium ions binding and buffering capacities that influence carbonate mineral precipitation [184]. In addition, other laboratory studies [185] reported the precipitation of vaterite by the soil bacterium *Myxococcus xanthus*. When minerals precipitate surrounding microbial cells, they can preserve remnants of organic matter and microbial cast, long after the cells have been degraded [186]. Furthermore, calcite precipitated in the presence of organic compounds often has distinct shapes [187], although abiotic conditions can also form the same shapes.

The precipitation of phosphate minerals such as apatite or struvite has been reported in batch culture studies [188, 189]. Phosphate biomineralization has been reported in association with EPS in natural environments [190–192]. Calcium phosphate precipitation was also described in sedimentary phosphatic stromatolites associated with microbial mats [193]. In older sediments, the initial precipitation of metastable carbonates and phosphates (vaterite, aragonite, and high-Mg calcite) tends to be an overgrowth of crystals of more stable phases (low-Mg calcite and dolomite), followed by the disappearance of the initial structure.

Iron and sulfur precipitation

The formation of intracellular iron sulfide minerals in magnetotactic bacteria is one of a few occurrences of a genetically mediated microbial-mineral formation [194]. Other biogenic iron oxides can form as a consequence of microbial metabolism, when bacteria oxidize Fe (II) to Fe (III) under oxic conditions. Ferris et al. [155] proposed that the surface of microorganisms may provide templates for the binding of Fe (II) and for the subsequent nucleation and growth of iron sulfide minerals, Fe(III) oxyhydroxides, in the presence of oxygen or Fe(II) sulfide minerals under anoxic conditions [58, 195]. In addition, the presence of iron sulfide minerals forming at the surface of microbial cells has been described in several experimental studies [196]. Iron oxides have been observed to form in bacterial EPS, as well as directly on cell walls as amorphous or nanocrystalline [197, 198]. Tazaki et al. [199] analyzed the structures and mineralogy of pisoliths from a hot spring and found that there were Fe-rich layers precipitated in the EPS of bacterial cells. Chan et al. [82] found that mine waters contained iron oxide precipitates at the center of which were elongated akagenite pseudocrystals. It was shown that the akagenite formed around extracellular polymeric fibers extruded by the microbes. Similar precipitations associated with Mn oxides have also been described [200].

Sedimentary metal sulfide mineral formation takes place in low-temperature environments and requires the production of free sulfide by sulfate-reducing microorganisms under anoxic conditions [201, 202]. Pyrite formation in the laboratory does not seem to occur readily in the presence of SRB. However, the metastable solid phases mackinawite (FeS) and greigite (Fe_3S_4) are rarely preserved in the sedimentary rock record but are abundant in culture experiments. It has been suggested that the presence of organic substances under culture conditions could inhibit pyrite formation and produce greigite [203]. Moreau et al. [204] studied biofilms of SRB and found that they contained spheroidal biogenic ZnS nanocrystals. Cellular encrustations and associations between EPS and minerals present in dissolution can lead to the preservation of microbial organic compounds and casts of the organisms with iron hydroxides [169, 205].

Mineral weathering

The detection of bioweathering processes is not always easy because weathering induced by abiotic agents is prevalent in all habitats. Hence, it is necessary to precisely characterize microsites at the interface between living organisms and their substrate [56]. The bioweathering activity of microorganisms such as cyanobacteria, free-living algae, and fungi and lichen symbionts on rocks has been reported across a wide range of microenvironments and climatic regimes [206]. The distribution and extension of induced bioweathering processes is extremely complex, being a multifactorial process strongly controlled by microenvironmental conditions. A common bioweathering-related depositional feature in arid regions is rock varnish related to bacteria [207]. Bioweathering in rock boulders and soils crusts can contribute to pedogenic processes and play a key role in biogeochemical cycles through the liberation of nutrients [208]. However, bioweathering induced by lithobionts colonizing stone in cultural heritage generates negative effects, biodeterioration, because it contributes to the deterioration of the cultural assets [49, 209].

The bioweathering effects of lithobionts on their mineral substrates can be attributed to biogeophysical and biogeochemical processes [210]. Biogeophysical action is reflected by the mechanical disruption of rocks caused by different processes. Adhesion, penetration, and growth, especially of filamentous microorganisms, generate enlargement of the pores and widening of fissures (Fig. 1.3C), inducing the loss of mineral particles from the substrate and the incorporation of mineral fragments into the biofilms (Fig. 1.4B and D). Expansion and contraction of biofilms in response to hydration/desiccation cycles, as well as changes in volume due to freezing processes, also induce mechanical damages in the substrate [49]. As regards biogeochemical action, metabolic products and other substances produced by the microbial communities can generate chemical effects on the colonized substrate, including dissolution and chelating processes. The dissolution of rock minerals can be achieved by organic acids released from cells, chemical impact of respiratory CO_2, H^+, and redox reactions with metal ions. The chemical effects comprise the solubilization and mobilization of elements from the minerals resulting in the weathering of minerals (silicates, phosphates, carbonates, sulfides, and oxides) and the formation of new deposits (carbonates, oxides, sulfides, and phosphates).

Silicates

Experimental studies of bioweathering processes have been very valuable to understand the mechanisms and the role of microorganisms in lithic substrate degradation. Fungi are generally more effective in releasing elements from silicates than bacteria or archaea [211]. Weed et al. [212] experimentally showed the alteration of mica to vermiculite by fungi. The ability of specific lichen substances to chelate Al, Fe, Ca, and Mg from granitic rocks and their minerals was investigated early by Ascaso and Galván [213]. When the lichen-rock interface was studied with infrared spectroscopy (IR) [214], it was observed that bands that were present in the rock disappear in the interface, whereas other new bands appear. Ascaso and Wierzchos [39] and Barker et al. [215] showed that mineral surfaces covered with EPS, or where cells were attached, appeared more extensively etched than the surrounding uncolonized surfaces. Vandevivere et al. [216] showed that the presence of gluconic acid from glucose utilization appeared to correlate to an increase in bacteria-mediated mineral dissolution. Organic acids, such as oxalic acid, were also related to the weathering of minerals by lichen [217] and has been reported as a strong solubilizing agent [218]. Metabolites of the citric acid cycle are excreted in addition to the production of CO_2 by respiration by fungi [219]. This acidolysis by fungi is a well-known mechanism [220], and quartz etching can be explained by biochemical weathering through organic acid secretion. Organic acids are also strong chelators of trivalent metals as Al^{3+} and Fe^{3+} [221], and by metal removal by chelation, the dissolution of the mineral may be promoted [222]. This implies that microorganisms promoted dissolution predominantly in minerals that contain interlayer ions [223] and in minerals with redox-sensitive metals. The separation and exfoliation of mica layers (Fig. 1.4D and E, white arrows) as a result of these dissolution processes is a common process in granite bioweathering [40, 137]. A different bioweathering pattern was observed in endolithic cyanobacteria in a tropical climate. In this case, the alkalization of the microenvironment around the living photosynthetic cells was reported as responsible for the dissolution of the binding material and the presence of etch marks on the quartz grains [68]. Bundeleva et al. [224] also showed forsterite dissolution to be promoted by the complexation of Mg^{2+} ions by cyanobacterial EPS. The bioweathering of phyllosilicates was also described in a coculture of alkaliphilic anaerobic bacteria [225]. In addition, olivine dissolution and hydrous Mg carbonate and silicate precipitation have been related to the presence of a consortium of photo-autotrophic and heterotrophic bacteria [226]. In the mineral weathering by bacteria, it is unclear whether this interaction is simply the coincidental result of microbial metabolism or a competitive ecological advantage. Bennet et al. [227]

indicated that, in the subsurface, silicate weathering by bacteria is sometimes driven by the nutrient requirements of the microbial consortium.

Carbonates

The clearest signal of bioweathering in carbonate rocks is the development of biopitting on rock surfaces (Fig. 1.1E) associated with the colonization and/or activity of endolithic and epilithic microorganisms. This process is especially relevant on monumental stone because it induces the disaggregation of the surface and biodeterioration processes, which can affect the preservation of the heritage asset [162, 228–230]. The porosity of carbonate rocks facilitates the microbial colonization (Fig. 1.3D) and the development of close mineral-microbial interactions [142]. The dissolution of sedimentary $CaCO_3$ minerals is driven in oxic environments by microbially mediated processes associated with the oxidation of organic material [153]. Euendolithic microorganisms such as fungi, cyanobacteria, and microalgae penetrate carbonates (Fig. 1.3E) through active processes of dissolution [2, 231]. It has even been recently demonstrated that euendolithic cyanobacteria are able to fix carbon liberated during mineral dissolution [163]. Photosynthetic activity of endolithic lichens also creates a distinctive exfoliating weathering pattern of sandstones [118] by the dissolution of the carbonate cementing material. The respiration process of endolithic lichen was also related to the bioweathering of limestones [232]. Cámara et al. [147] associated the disaggregation of dolostones with fungal colonization in a sequential process. First, fungi colonize the intercrystalline porosity, and second, they penetrate the crystals by means of physical and chemical activities. The formation of calcium oxalate patinas as a consequence of microbial colonization of calcareous rocks has been reported by Krumbein [233] as the main bioweathering action. Although there is indisputable evidence of mineral disintegration and chemical modification of carbonates by lithobionts, other studies have attributed bioprotective activities to them. It has been even recently proposed that the growth of microbial biofilms and various biomineralization phenomena can lead to the formation of layers that can stabilize surfaces and protect them from further weathering [234].

Sulfides and metals oxides

Studies on the dissolution of sulfides come under the general topic of bioleaching, which is increasingly used by the mining industry. Pyrite alteration is often mediated by the activity of iron-oxidizing bacteria and archaea, which are able to oxidize Fe^{2+} [235]. The presence of EPS has been observed to enhance bioleaching, either in an industrial context during ore processing or in natural settings during the production of acid rock drainage. The complexation and concentration of metal ions in the EPS layer is thought to aid the oxidative dissolution of the rock matrix by concentrating adsorbed Fe^{3+} ions within the EPS layer [236]. In other cases, the presence of EPS inhibited the dissolution rate by irreversibly binding to mineral surfaces [237]. In addition to bacteria, fungi bioaccumulate heavy metals in their fruiting bodies due to enzymes or organic acids chelating metals ions and actively transporting them into cells [238]. Siderophores, which are secondary metabolites that scavenge iron and other elements from the environment and deliver it to cells via specific receptors [239], are often synthesized and exuded by fungi and bacteria under iron-limiting conditions favoring mineral etching and dissolution [220]. In addition, an increasing number of microorganisms have been isolated, which are capable of using transition metal ions such as Fe and Mn as electron acceptors or donors. In their most oxidized form, Fe(III) and Mn(IV) exist predominantly as metal oxides with low solubility in neutral pH environments. The reduced forms of Fe and Mn can also exist as mineral solids, and some neutrophilic Fe-oxidizing microorganisms can use some of these phases as electron donors. DEET is a form of microbial respiration that transfers electrons from cytochromes localized on the exterior face of the outer membrane to metal ions on the surface of minerals [240]. Other microorganisms are able to form highly conductive networks of filaments that transfer electrons along their length to the minerals [241]. These mechanisms of microbial electron transfer to and from minerals will affect biogeochemical processes that control element cycling and the formation and weathering of geological materials.

1.1.5 Concluding remarks

The characterization of the interfaces between rocks and microorganisms needs multiscale approaches based on the combination of different techniques of molecular biology and high-resolution microscopy and analytical techniques. The objectives are to identify not only the microorganisms involved but also the specific microbial-mineral interactions occurring in the systems and their spatial distribution. Geomicrobiology is now a broad discipline with a great potential to solve previous open questions by means of the application of the most advanced techniques developed in the last decades. At present, proper multiscale approaches can be performed since microanalytical and microscopy techniques are now working at nanoscale resolution. These new techniques allow delving into the mechanisms of physicochemical mineral transformations induced by microorganisms and their consequences at different scales. In addition, recent progress in molecular biology techniques enables the simultaneous detection of a wider range of microorganisms and, together with the possibility of labelling soluble molecules, allows the retrieval of phylogenetic and metabolic information. High-resolution analysis of spatial structures and microbial-mineral interactions can complete the characterization. Only researchers using a combination of complementary techniques can properly map the presence, activity, and consequences of lithobiontic colonization and, consequently, the processes occurring at the interface between rock and microorganisms.

The close association existing between the lithobionts and the lithic substrate induces numerous and relevant microbial-mineral interactions, which drive their contributions to global biogeochemical cycles. These interactions play key roles in the settlement and development of lithobiontic colonization but also have a beneficial influence on the establishment of other biological communities in the ecosystem. Lithobiontic colonization can have a great influence on the diversity and functioning of edaphic communities, as well the development of soil crusts. However, the bioweathering processes that positively contribute to soil formation and the liberation of relevant mineral nutrients to the surroundings in natural environments also generate negative effects when colonization takes place in rocks or stones from our cultural heritage because they favor their deterioration. To design and evaluate treatments targeted to eliminate, reduce, or avoid this harmful lithobiontic colonization, it is necessary to properly characterize the interface between microorganisms and the colonized lithic substrate and the changes induced by the treatments in this interface.

To conclude, the phenomena driven by biologic interactions at the interface of rock and microorganisms are significant to major environmental and geoscience research areas, including rock weathering and soil formation; environmental mineralogy; biogeochemical cycling of elements, biotransformation of organic and inorganic contaminants, and environmental sustainability; biodeterioration and conservation of cultural heritage.

Acknowledgments

The authors dedicate this chapter to their colleagues, Professor Carmen Ascaso and Dr. Jacek Wierzchos, who introduced them to geomicrobiology and exerted a positive influence in their careers. This work was supported by AGE-MICINN grant no. PID2019-105469RB-C22. The authors also thank Esther Rodríguez (MNCN) for her technical assistance, Carlos Aguilera for the illustration work, and Omar Tarzi for English language editorial assistance.

References

[1] Golubic S, Schneider J. Microbial endoliths as internal biofilms. In: Krumbein WE, Paterson DM, Zavarzin GA, editors. Fossil and Recent Biofilms. Dordrecht: Springer; 2003. p. 249–63.

[2] Golubic S, Friedmann I, Schneider J. The lithobiontic ecological niche, with special reference to microorganisms. *J Sediment Petrol.* 1981; 51: 475–8.

[3] Wierzchos J, de los Ríos A, Ascaso C. Microorganisms in desert rocks: the edge of life on Earth. *Int Microbiol.* 2012; 15: 171–81.

[4] Makhalanyane TP, Pointing SB, Cowan DA. Lithobionts: cryptic and refuge niches. In: Cowan D, editor. Antarctic Terrestrial Microbiology. Berlin: Springer; 2014. p. 163–79.

[5] De los Ríos A, Cary C, Cowan D. The spatial structures of hypolithic communities in the Dry Valleys of East Antarctica. *Polar Biol.* 2014; 37: 1823–33.

[6] Wierzchos J, Cámara-Gallego B, de los Ríos A, et al. Microbial colonization of Ca-sulfate crusts in the hyperarid core of the Atacama Desert: implications for the search for life on Mars. *Geobiology*. 2011: 9: 44–60.

[7] Friedmann EI, Lipkin Y, Ocampo-Paus R. Desert algae of the Negev (Israel). Phycologia. 1967; 6: 185–200.

[8] De los Ríos, A., Wierzchos J, Ascaso C. Microhabitats and chemical microenvironments under saxicolous lichens growing on granite. *Microb Ecol*. 2002; 43: 181–8.

[9] Guillitte O. Bioreceptivity: a new concept for building ecology studies. *Sci Total Environ*. 1995; 215–20.

[10] Miller AZ, Sanmartin P, Pereira-Pardo L, et al. Bioreceptivity of building stones: a review. *Sci Total Environ*. 2012; 426: 1–12.

[11] De los Ríos A, Wierzchos J, Ascaso C. The lithic microbial ecosystems of Antarctica's McMurdo Dry Valleys. *Antarct Sci*. 2014; 26: 459–77.

[12] Pérez-Ortega S, Garrido-Benavent I, Grube M, Olmo R, de Los Ríos A. Hidden diversity of marine borderline lichens and a new order of fungi: Collemopsidiales (Dothideomyceta). *Fungal Divers*. 2016; 80: 285–300.

[13] Weber B, Büdel B. Endoliths. In: Reitner J, Thiel V, editors. Encyclopedia of Geobiology. Heidelberg: Springer; 2011. p. 348–54.

[14] Cockell CS, Stokes MD. Widespread colonization by polar hypoliths. *Nature*. 2004; 431: 414.

[15] Gorbushina A. Life on the rocks. *Environ Microbiol*. 2007; 9: 1613–31.

[16] Wierzchos J, Casero MC, Artieda O, Ascaso C. Endolithic microbial habitats as refuges for life in polyextreme environment of the Atacama Desert. *Curr Opin Microbiol*. 2018; 43: 124–31.

[17] Giovannoni SJ, Fisk MR, Mullins TD, Furnes H. Genetic evidence for endolithic microbial life colonizing basalic glass/seawater interfaces. Proceedings of the Ocean Drilling Program Scientific Results. 1996; 148: 207–14.

[18] Escudero C, Vera M, Oggerin M, Amils R. Active microbial biofilms in deep poor porous continental subsurface rocks. *Sci Rep*. 2018; 8: 1538.

[19] Dutta A, Sar P, Sarkar J, et al. Archaeal communities in deep terrestrial subsurface underneath the Deccan Traps, India. *Front Microbiol*. 2019; 10: 1362.

[20] Manz W, Arp G, Schumann-Kindel G, Szewzyk U, Reitner J. Widefield deconvolution epifluorescence microscopy combined with fluorescence *in situ* hybridization reveals the spatial arrangement of bacteria in sponge tissue. *J Microbiol Methods*. 2000; 40: 125–34.

[21] Wierzchos J, de los Ríos A, Sancho LG, Ascaso C. Viability of endolithic microorganisms in rocks from the McMurdo Dry Valleys of Antarctica established by confocal and fluorescence microscopy. *J Microsc*. 2004; 216: 57–61.

[22] Wierzchos J, Ascaso C. Life, decay and fossilisation of endolithic microorganisms from the Ross Desert, Antarctica: suggestions for *in situ* further research. *Polar Biol*. 2001; 24: 863–8.

[23] De los Ríos A, Wierzchos J, Sancho LG, Ascaso C. Acid microenvironments in microbial biofilms of Antarctic endolithic microecosystems. *Environ Microbiol*. 2003; 5: 231–7.

[24] Amann RI, Ludwig W, Schleifer KH. Phylogenetic identification and *in situ* detection of individual microbial cells without cultivation. *Microbiol Rev*. 1995; 59: 143–69.

[25] Amann R, Fuchs BM. Single-cell identification in microbial communities by improved fluorescence in situ hybridization techniques. *Nat Rev Microbiol*. 2008; 6: 339–48.

[26] Shiraishi F, Zippel B, Neu TR, Arp G. In situ detection of bacteria in calcified biofilms using FISH and CARD-FISH. *J Microbiol Methods*. 2008; 75: 103–8.

[27] Herrera A, Cockell CS, Self S, et al. Bacterial colonization and weathering of terrestrial obsidian in Iceland. *Geomicrobiol J*. 2008; 25: 25–37.

[28] De los Ríos A, Wierzchos J, Sancho LG, Ascaso C. Exploring the physiological state of continental Antarctic endolithic microorganisms by microscopy. *FEMS Microbiol Ecol*. 2004; 50: 143–52.

[29] Hunter RC, Hitchcock AP, Dynes JJ, Obst M, Beveridge TJ. Mapping the speciation of iron in *Pseudomonas aeruginosa* biofilms using scanning transmission X-ray microscopy. *Environ Sci Technol*. 2008; 42: 8766–72.

[30] Kawaguchi T, Decho AW. In situ analysis of carboxyl and sulfhydryl groups of extracellular polymeric secretions by confocal laser scanning microscopy. *Anal Biochem*. 2002; 304: 266–7.

[31] Zippel B, Neu TR. Characterization of glycoconjugates of extracellular polymeric substances in tufa-associated biofilms by using fluorescence lectin-binding analysis. *Appl Environ Microbiol*. 2010; 77: 505–16.

[32] Beutler M, Wiltshire KH, Meyer B, et al. A fluorometric method for the differentiation of algal populations *in vivo* and *in situ*. *Photosynth Res*. 2002; 72: 39–53.

[33] Papineau D, Walker JJ, Mojzsis SJ, Pace NR. Composition and structure of microbial communities from stromatolites of Hamelin Pool in Shark Bay, Western Australia. *Appl Environ Microbiol*. 2005; 71: 4822–32.

[34] Roldan M, Ascaso C, Wierzchos J. Fluorescent fingerprints of endolithic phototrophic cyanobacteria living within halite rocks in the Atacama Desert. *Appl Environ Microbiol*. 2014; 80: 2998–3006.

[35] Ascaso C, Wierzchos J. Structural aspects of the lichen–rock interface using backscattered electron imaging. *Bot Acta*. 1994; 107: 251–6.

[36] Wierzchos J, Ascaso C. Application of back-scattered electron imaging to the study of the lichen rock interface. *J Microscopy*. 1994; 175: 54–9.

[37] Ascaso C, Wierzchos J, de los Ríos A. Cytological investigations of lithobiontic microorganisms in granitic rocks. *Bot Acta*. 1995,108: 474–81.

[38] Sanders WB, Ascaso C, Wierzchos J. Physical interactions of two rhizomorph-forming lichens with their rock substrate. *Bot Acta*. 1994; 107: 432–9.

[39] Ascaso C, Wierzchos J. Study of the biodeterioration zone between the lichen thallus and the substrate. *Cryptogam Bot*. 1995; 5: 270–81.

[40] De los Ríos A, Sancho LG, Grube M, Wierzchos J, Ascaso C. Endolithic growth of two *Lecidea* lichens in granite from continental Antarctica detected by molecular and microscopy techniques. *New Phytol*. 2005; 165: 181–90.

[41] Wierzchos J, Sancho LG, Ascaso C. Biomineralization of endolithic microbes in rocks from the McMurdo Dry Valleys of Antarctica: implications for microbial fossil formation and their detection. *Environ Microbiol*. 2005; 7: 566–75.

[42] De los Ríos A, Grube M, Sancho LG, Ascaso C. Ultrastructural and genetic characteristics of endolithic cyanobacterial biofilms colonizing Antarctic granite rocks. *FEMS Microbiol Ecol*. 2007; 59: 386–95.

[43] Souza-Egipsy V, Wierzchos J, Ascaso C. Ultrastructural and biogeochemical features of microbiotic soil crusts and their implications in a semiarid habitat. *Geomicrobiol J*. 2002; 19: 567–80.

[44] Souza-Egipsy V, Wierzchos J, Sancho C, Belmonte A, Ascaso C. Role of biological soil crust cover in bioweathering and protection of sandstones in a semi-arid landscape (Torrollones de Gabarda, Huesca, Spain). *Earth Surf Process Landf*. 2004; 29: 1651–61.

[45] Wierzchos J, Ascaso C, McKay C. P. Endolithic cyanobacteria in halite rocks from the hyperarid core of the Atacama Desert. *Astrobiology*. 2006; 6: 415–22.

[46] Wierzchos J, Davila A, Artieda O, et al. Ignimbrite as a substrate for endolithic life in the hyper-arid Atacama Desert: implications for the search for life on Mars. *Icarus*. 2013; 224: 334–46.

[47] Torsvik T, Furnes H, Muehlenbachs K, Thorseth IH, Tumyr O. Evidence for microbial activity at the glass–alteration interface in oceanic basalts. *Earth Planet Sci Lett*. 1998; 162: 165–76.

[48] Thorseth IH, Pedersen RB, Christie DM. Microbial alteration of 0–30–Ma seafloor and sub-seafloor basaltic glasses from the Australian Antarctic discordance. *Earth Planet Sci Lett*. 2003; 215: 237–47.

[49] Ascaso C, Wierzchos J, Souza-Egipsy V, de los Ríos A, Delgado Rodrigues J. *In situ* evaluation of the biodeteriorating action of microorganisms and the effects of biocides on carbonate rock of the Jerónimos Monastery (Lisbon). *Int Biodeterior Biodegrad.* 2002; 49: 1–12.

[50] Wierzchos J, Berlanga M, Ascaso C, Guerrero R. Micromorphological characterization and lithification of microbial mats from the Ebro Delta (Spain). *Int Microbiol* 2006; 9: 289–95.

[51] Souza-Egipsy V, Valladares F, Ascaso C. Water distribution in foliose lichen species: interactions between method of hydration, lichen substances and thallus anatomy. *Ann Bot.* 2000; 86: 595–601.

[52] Souza-Egipsy V, Ascaso C, Sancho LG. Water distribution within terricolous lichens revealed by scanning electron microscopy and its relevance in soil crust ecology. *Mycol Res.* 2002; 106: 1367–74.

[53] Brown DA, Beveridge TJ, Keevil CW, Sheriff BL. Evaluation of microscopic techniques to observe iron precipitation in natural microbial biofilm. *FEMS Microbiol Ecol.* 1998; 26: 297–310.

[54] Barker WW, Banfield JF. Biologically versus inorganically mediated weathering reactions: relationship between minerals and extracellular microbial polymers in lithobiontic communities. *Chem Geol.* 1996; 132: 55–69.

[55] Barker WW, Banfield JF. Zones of chemical and physical interaction at interfaces between microbial communities and minerals. A model. *Geomicrobiol J.* 1998; 15: 223–44.

[56] Banfield JF, Barker WW, Welch SA, Taunton A. Biological impact on mineral dissolution: application of the lichen model to understanding mineral weathering in the rhizosphere. *Proc Natl Acad Sci USA.* 1999; 96: 3404–11.

[57] McLoughlin N, Wacey D, Kruber C, Kilburn MR, Thorseth IH, Pedersen RB. A combined TEM and NanoSIMS study of endolithic microfossils in altered seafloor basalt. *Chem Geol.* 2011; 289: 154–62.

[58] Beveridge TJ. Role of cellular design in bacterial metal accumulation and mineralization. *Annu Rev Microbiol.* 1989; 43: 147–71.

[59] Benzerara K, Menguy N, Guyot F, Vanni C, Gillet P. TEM study of a silicate–carbonate–microbe interface prepared by focused ion beam milling. *Geochim Cosmochim Acta.* 2005; 69: 1413–22.

[60] Obst M, Gasser P, Mavrocordatos D, Dittrich M. TEM–specimen preparation of cell/mineral interfaces by focused ion beam milling. *Am Minert.* 2005; 90: 1270–7.

[61] Obst M, Dittrich M. Living under an atomic force microscope an optimized approach for in vivo investigations on surface alterations towards biomineral nucleation on cyanobacterial cells. *Geobiology.* 2005; 3: 179–93.

[62] Teschke O. Volume of extracellular polymeric substance coverage of individual *Acidithiobacillus ferrooxidans* bacterium measured by atomic force microscopy. *Microsc Res Tech.* 2005; 67: 312–6.

[63] Taylor ES, Lower SK. Thickness and surface density of extracellular polymers on *Acidithiobacillus ferrooxidans*. *Appl Environ Microbiol.* 2008; 74: 309–11.

[64] Pouget EM, Bomans PHH, Goos J, Frederik PM, de With G, Sommerdijk N. The initial stages of template-controlled $CaCO_3$ formation revealed by Cryo-TEM. *Science.* 2009; 323: 1455–8.

[65] Comolli LR, Luef B, Chan CS. High-resolution 2D and 3D cryo-TEM reveals structural adaptations of two stalk-forming bacteria to an Fe-oxidizing lifestyle. *Environ Microbiol.* 2011; 13: 2915–29.

[66] Miot J, Maclellan K, Benzerara K, Boisset N. Preservation of protein globules and peptidoglycan in the mineralized cell wall of nitrate-reducing, iron(II)-oxidizing bacteria: a cryo-electron microscopy study. *Geobiology.* 2011; 9: 459–70.

[67] Miot J, Benzerara K, Kappler A. Investigating microbe–mineral interactions: recent advances in X-ray and electron microscopy and redox-sensitive methods. *Annu Rev Earth Planet Sci.* 2014; 42: 271–89.

[68] Büdel B, Weber B, Kühl M, Pfanz H, Sültemeyer D, Wessels DC. Reshaping of sandstone surfaces by cryptoendolithic cyanobacteria: bioalkalinization causes chemical weathering in arid landscapes. *Geobiology.* 2004; 2: 261–8.

[69] Nielsen M, Revsbech NP, Kuehl M. Microsensor measurements of hydrogen gas dynamics in cyanobacterial microbial mats. *Front Microbiol.* 2015; 6: 726.
[70] Vitek P, Edwards HGM, Jehlicka J, et al. Microbial colonization of halite from the hyper-arid Atacama Desert studied by Raman spectroscopy. *Philos Trans A Math Phys Eng Sci.* 2010; 368: 3205–21.
[71] Vitek P, Ascaso C, Artieda O, Casero MC, Wierzchos J. Raman imaging of microbial colonization in rock—some analytical aspects. *Anal Bioanal Chem.* 2020; 412: 3717–26.
[72] Vitek P, Ascaso C, Artieda O, Cristina Casero M, Wierzchos J. Discovery of carotenoid red-shift in endolithic cyanobacteria from the Atacama Desert. *Sci Rep.* 2017; 7: 11116.
[73] Culka A, Jehlicka J, Ascaso C, Artieda O, Casero CM, Wierzchos J. Raman microspectrometric study of pigments in melanized fungi from the hyperarid Atacama Desert gypsum crust. *J Raman Spectrosc.* 2017; 4: 1487–93.
[74] Geesey GG, Borch T, Reardon CL. Resolving biogeochemical phenomena at high spatial resolution through electron microscopy. *Geobiology.* 2008; 6: 263–9.
[75] Obst M, Dynes JJ, Lawrence JR, et al. Precipitation of amorphous $CaCO_3$ (aragonite-like) by cyanobacteria: a STXM study of the influence of EPS on the nucleation process. *Geochim Cosmochim Acta.* 2009; 73: 4180–98.
[76] Bonneville S, Morgan DJ, Schmalenberger A, et al. Tree–mycorrhiza symbiosis accelerates mineral weathering: evidences from nanometer-scale elemental fluxes at the hypha–mineral interface. *Geochim Cosmochim Acta.* 2011; 75: 6988–7005.
[77] Lawrence JR, Swerhone GDW, Leppard GG, et al. Scanning transmission X-ray, laser scanning, and transmission electron microscopy mapping of the exopolymeric matrix of microbial biofilms. *Appl Environ Microbiol.* 2003; 69: 5543–54.
[78] Toner B, Manceau A, Marcus MA, Millet DB, Sposito G. Zinc sorption by a bacterial biofilm. *Environ Sci Technol.* 2005; 39: 8288–94.
[79] Dynes JJ, Tyliszczak T, Araki T, et al. Speciation and quantitative mapping of metal species in microbial biofilms using scanning transmission X-ray microscopy. *Environ Sci Technol.* 2006; 40: 1556–65.
[80] Wierzchos J, Ascaso C, Ager FJ, García-Orellana I, Carmona A, Respaldiza MA. Identifying elements in rocks from the Dry Valleys desert (Antarctica) by ion beam proton induced X-ray emission. *Nucl Instrum Methods Phys Res B.* 2006; 249: 571–4.
[81] Baumann K, Jung P, Samolov E, et al. Biological soil crusts along a climatic gradient in Chile: richness and imprints of phototrophic microorganisms in phosphorus biogeochemical cycling. *Soil Biol Biochem.* 2018; 127: 286–300.
[82] Chan CS, De Stasio G, Welch SA, Girasole M, Frazer BH, Nesterova MV. Microbial polysaccharides template assembly of nanocrystal fibers. *Science.* 2004; 303: 1656–8.
[83] Gilbert P, Abrecht M, Frazer BH. The organic–mineral interface in biominerals. In: Banfield JE, Cervini Silva J, Nealson KH, editors. Reviews in Mineralogy and Geochemistry. Molecular Geomicrobiology. Washington (DC): Mineralogical Society of America and Geochemical Society. 2005; 59. p. 157–85.
[84] Kilburn MR, Wacey D. Nanoscale secondary ion mass spectrometry (NanoSIMS) as an analytical tool in geosciences. In: Grice K, editor. Principles and Practice of Analytical Techniques in Geosciences. United Kingdom: The Royal Society of Chemistry. 2015; 4. p. 1–34.
[85] Wacey D, Urosevic L, Saunders M, George AD. Mineralisation of filamentous cyanobacteria in Lake Thetis stromatolites, Western Australia. *Geobiology.* 2018; 16: 203–15.
[86] Korenevsky A, Beveridge TJ. The surface physicochemistry and adhesiveness of *Shewanella* are affected by their surface polysaccharides. *Microbiology.* 2007; 153: 1872–83.
[87] Shi L, Dong H, Reguera G, et al. Extracellular electron transfer mechanisms between microorganisms and minerals. *Nat Rev Microbiol.* 2016; 14: 651–62.

[88] Mahadevan R, Palsson BO, Lovley DR. In situ to in silico and back: elucidating the physiology and ecology of *Geobacter* spp. using genome-scale modelling. *Nat Rev Microbiol.* 2011; 9: 39–50.
[89] Wierzchos J, DiRuggiero J, Vitek P, et al. Adaptation strategies of endolithic chlorophototrophs to survive the hyperarid and extreme solar radiation environment of the Atacama Desert. *Front Microbiol.* 2015; 6: 934.
[90] Palinska KA, Abed RMM, Vogt JC, Radtke G, Golubic S. Microbial endoliths on east Adriatic limestone coast: morphological vs. molecular diversity. *Geomicrobiol J.* 2017; 34: 913–15.
[91] Sigler WV, Bachofen R, Zeyer J. Molecular characterization of endolithic cyanobacteria inhabiting exposed dolomite in central Switzerland. *Environ Microbiol.* 2003; 5: 618–27.
[92] Khan I, Zada S, Rafiq M, Jamil SUU, Shah AA, Hasan F. Morphological, biochemical and molecular characterization of culturable epilithic and endolithic bacteria from rocks of Ayubia (Murree), lower Himalaya, Pakistan. *Appl Ecol Environ Res.* 2017; 16: 563–77.
[93] Hirsch P, Eckhardt FEW, Palmer RJJ. Fungi active in weathering of rock and stone monuments. *Can J Botany.* 1995; 73: 1384–90.
[94] Pulschen AA, Bendia AG, Fricker AD, Pellizari VH, Galante D, Rodrigues F. Isolation of uncultured bacteria from Antarctica using long incubation periods and low nutritional media. *Front Microbiol.* 2017; 8: 1346.
[95] Van Goethem MW, Makhalanyane TP, Valverde A, Cary SC, Cowan DA. Characterization of bacterial communities in lithobionts and soil niches from Victoria Valley, Antarctica. *FEMS Microbiol Ecol.* 2016; 92, flv051.
[96] Lee J, Cho J, Cho Y-J, et al. The latitudinal gradient in rock inhabiting bacterial community compositions in Victoria Land, Antarctica. *Sci Total Environ.* 2019; 657: 731–38.
[97] Garrido-Benavent I, Pérez-Ortega S, Durán J, et al. Differential colonization and succession for microbial communities in rock and soil substrates along a maritime Antarctic glacier forefield. *Front Microbiol.* 2020; 11: 126.
[98] Crits-Christoph A, Robinson CK, Ma B, et al. Phylogenetic and functional substrate specificity for endolithic microbial communities in hyper-arid environments. *Front Microbiol.* 2016; 7: 301.
[99] Gomez-Silva B, Vilo-Muñoz C, Galetovic A, et al. Metagenomics of Atacama lithobiontic extremophile life unveils highlights on fungal communities, biogeochemical cycles and carbohydrate-active enzymes. *Microorganisms.* 2019; 7: 619.
[100] Coleine C, Albanese D, Onofri S, et al. Metagenomes in the borderline ecosystems of the Antarctic cryptoendolithic communities. *Microbiol Resour Announc.* 2020; 9, e01599–19.
[101] De los Ríos A, Wierzchos J, Sancho LG, Green A, Ascaso C. Ecology of endolithic lichens colonizing granite in continental Antarctica. *Lichenologist.* 2005; 37: 383–95.
[102] Choe Y-H, Kim M, Woo J, et al. Comparing rock-inhabiting microbial communities in different rock types from a high arctic polar desert. *FEMS Microbiol Ecol.* 2018; 94, fiy070.
[103] Bjelland T, Grube M, Hoem S, et al. Microbial metacommunities in the lichen–rock habitat. *Environ Microbiol.* Rep 2011; 3: 434–42.
[104] Khan N, TuYn M, StaVord W, et al. Hypolithic microbial communities of quartz rocks from Miers Valley, McMurdo Dry Valleys, Antarctica. *Polar Biol.* 2011; 34: 1657–68.
[105] Crits-Christoph A, Gelsinger DR, Ma B, et al. Functional interactions of archaea, bacteria and viruses in a hypersaline endolithic community. *Environ Microbiol.* 2016; 18: 2064–77.
[106] Thorseth IH, Torsvik T, Furnes H, Muehlenbachs K. Microbes play an important role in the alteration of oceanic crust. *Chem Geol.* 1995; 126: 137–46.
[107] Zablocki O, Van Zyl L, Adriaenssens EM, et al. High-level diversity of tailed phages, eukaryote-associated viruses, and virophage-like elements in the metaviromes of Antarctic soils. *Appl Environ Microbiol.* 2014; 80: 6888–97.
[108] Adriaenssens EM, Van Zyl L, De Maayer P, et al. Metagenomic analysis of the viral community in Namib Desert hypoliths. *Environ Microbiol.* 2015; 17: 480–95.

[109] Dong H, Rech JA, Jiang H, Sun H, Buck BJ. Endolithic cyanobacteria in soil gypsum: occurrences in Atacama (Chile), Mojave (United states), and Al-Jafr Basin (Jordan) deserts. *J Geophys Res.* 2007; 112, G02030.

[110] Horath T, Bachofen R. Molecular characterization of an endolithic microbial community in dolomite rock in the central Alps (Switzerland). *Microbiol Ecol.* 2009; 58: 290–306.

[111] McNamara CJ, Bearce K, Perry IV TD, Mitchell R. Microbial biofilms inside historic stones. In: Fort R, Alvarez de Buergo M, Gomez Heras M, Vazquez-Calvo C, editors. Heritage, Weathering and Conservation. London: Taylor & Francis; 2006; 1: 317–22.

[112] Tang Y, Lian B, Dong H, Liu D, Hou W. Endolithic bacterial communities in dolomite and limestone rocks from the Nanjiang Canyon in Guizhou Karst Area (China). *Geomicrobiol J.* 2012; 29: 213–25.

[113] Yung CCM, Chan Y, Lacap-Bugler DC, et al. Characterization of chasmoendolithic community in Miers Valley, McMurdo Dry Valleys, Antarctica. *Microb Ecol.* 2014; 68: 351–9.

[114] Lacap-Bugler DC, Lee KK, Archer S, et al. Global diversity of desert hypolithic cyanobacteria. *Front Microbiol.* 2017; 8: 867.

[115] Walker JJ, Pace NR. Phylogenetic composition of Rocky Mountain endolithic microbial ecosystems. *Appl Environ Microbiol.* 2007; 73: 3497–504.

[116] Grube M, Wedin M. Lichenized fungi and the evolution of symbiotic organization. In: Heitman J, Howlett B, Crous P, Stukenbrock E, James T, Gow N, editors. The Fungal Kingdom. Washington: ASM Press. 2016. p. 749–65.

[117] Nash TH. Lichen Biology. Cambridge, Cambridge University Press. 2008.

[118] Friedmann EI. Endolithic microorganisms in the Antarctic cold desert. *Science.* 1982; 215: 1045–53.

[119] Matthes-Sears U, Gerrath JA, Larson DW. Abundance, biomass, and productivity of endolithic and epilithic lower plants on the temperate-zone cliffs of the Niagara Escarpment, Canada. *Int J Plant Sci.* 1997; 158: 451–60.

[120] Selbmann L, Onofri S, Coleine C, Buzzini P, Canini F, Zucconi L. Effect of environmental parameters on biodiversity of the fungal component in lithic Antarctic communities. *Extremophiles.* 2017; 21: 1069–80.

[121] Selbmann L, de Hoog GS, Mazzaglia A, Friedmann EI, Onofri S. Fungi at the edge of life: cryptoendolithic black fungi from Antarctic Desert. *Stud Mycol.* 2005; 51: 1–32.

[122] Ruibal C, Gueidan C, Selbmann L, et al. Phylogeny of rock-inhabiting fungi related to *Dothideomycetes. Stud Mycol.* 2009; 64: 123–33.

[123] Coleine C, Stajich JE, de los Ríos A, Selbmann L. Beyond the extremes: rocks as ultimate refuge for fungi in drylands. Mycologia. 2020: 1–26. doi: 10.1080/00275514.2020.1816761.

[124] Onofri S, Zucconi L, Selbmann L, et al. Fungal associations at the cold edge of life. In: Seckbach J, editor. Algae and Cyanobacteria in Extreme Environments. Netherlands: Springer. 2007. p. 35–57.

[125] Selbmann L, Zucconi L, Onofri S, et al. Taxonomic and phenotypic characterization of yeasts isolated from worldwide cold rock-associated habitats. *Fungal Ecol.* 2014; 118: 61–71.

[126] Matthes U, Turner SJ, Larson DW. Light attenuation by limestone rock and its constraint on the depth distribution of endolithic algae and cyanobacteria. *Int J Plant Sci.* 2001; 162: 263–70.

[127] Coleine C, Gevi F, Fanelli G, Onofri S, Timperio AM, Selbmann L. Specific adaptations are selected in opposite sun exposed Antarctic cryptoendolithic communities as revealed by untargeted metabolomics. *PLoS One.* 2020; 15, e0233805.

[128] Nienow JA, McKay CP, Friedmann EI. The cryptoendolithic microbial environment in the Ross Desert of Antarctica: light in the photosynthetically active region. *Microb Ecol.* 1988; 16: 271–89.

[129] Warren-Rhodes KA, McKay CP, Boyle LN, et al. Physical ecology of hypolithic communities in the central Namib Desert: the role of fog, rain, rock habitat and light. *J Geophys Res.* 2013; 118: 1451–60.

[130] Ekwealor JTB, Fisher KM. Life under quartz: hypolithic mosses in the Mojave Desert. *PLoS One*. 2020; 15, e0235928.
[131] Siebert J, Hirsch P, Hoffmann B, Gliesche CG. Cryptoendolithic microorganisms from Antarctic sandstone of Linnaeus Terrace (Asgard Range): diversity, properties and interactions. *Biodivers Conserv*. 1996; 5: 1337–63.
[132] Omelon CR. Endolithic microorganisms and their habitats. In: Hurst C, editor. Advances in Environmental Microbiology, Volume 1: Their World: A Diversity of Microbial Environments. Cham: Springer. 2016. p. 171–201.
[133] Flemming H-C, Wingender J. The biofilm matrix. *Nat Rev Microbiol*. 2010; 8: 623–33.
[134] Flemming HC, Wuertz S. Bacteria and archaea on Earth and their abundance in biofilms. *Nat Rev Microbiol*. 2019; 17: 247–60.
[135] Dupraz C, Reid RP, Braissant O, Decho AW, Norman RS, Visscher PT. Processes of carbonate precipitation in modern microbial mats. *Earth Sci Rev*. 2009; 96: 141–62.
[136] Tornos F, Oggerin M, de los Ríos A, et al. Do microbes control the formation of giant copper deposits? *Geology*. 2019; 47: 143–6.
[137] Wierzchos J, Ascaso C. Morphological and chemical features of bioweathered granitic biotite induced by lichen activity. *Clay Clay Min*. 1996; 44: 652–7.
[138] Cutler NA, Oliver AE, Viles HA, Ahmad S, Whiteley AS. The characterisation of eukaryotic microbial communities on sandstone buildings in Belfast, UK, using TRFLP and 454 pyrosequencing. *Int Biodeterior Biodegrad*. 2013; 82: 124–33.
[139] Caneva G, Nugari MP, Salvadori O. Plan Biology for Cultural Heritage: Biodeterioration and Conservation. Los Angeles: The Getty Conservation Institute. 2009.
[140] Cutler N, Viles H. Eukaryotic Microorganisms and Stone Biodeterioration. *Geomicrobiol J*. 2010; 27: 630–46.
[141] Ruibal C, Platas G, Bills G. Isolation and characterization of melanized fungi from limestone formations in Mallorca. *Mycol Prog*. 2005; 4: 23–38.
[142] De los Ríos A, Cámara B, Garcia del Cura MA, Rico V, Galvan V, Ascaso C. Deteriorating effects of lichen and microbial colonization of carbonate building rocks in the Romanesque churches of Segovia (Spain). *Sci Total Environ*. 2009; 407: 1123–34.
[143] Souza-Egipsy V, Wierzchos J, García-Ramos J, Ascaso C. Chemical and ultrastructural features of the lichen–volcanic/sedimentary rock interface in a semiarid region (Almeria, Spain). Lichenologist 2002; 34: 155–67.
[144] Cowan DA, Khan N, Pointing SB, Cary C. Diverse hypolithic refuge communities in the McMurdo Dry Valleys. *Antarct Sci*. 2010; 22: 714–20.
[145] Lacap DC, Warren-Rhodes KA, McKay C, Pointing SB. Cyanobacteria and chloroflexi-dominated hypolithic colonization of quartz at the hyper-arid core of the Atacama Desert, Chile. *Extremophiles*. 2011; 15: 31–8.
[146] Kugler A, Dong H. Phyllosilicates as protective habitats of filamentous cyanobacteria *Leptolyngbya* against ultraviolet radiation. *PLoS One*. 2019; 14, e0219616.
[147] Cámara B, de los Ríos A, García del Cura MA, Galván V, Ascaso C. Dolostone bioreceptivity to fungal colonization. Mater Constr 2008; 58: 113–24.
[148] Cámara B, Suzuki S, Nealson KH, et al. Ignimbrite textural properties as determinants of endolithic colonization patterns from hyper-arid Atacama Desert. *Int Microbiol* 2014; 17: 235–47.
[149] Mergelov N, Mueller CW, Prater I, et al. Alteration of rocks by endolithic organisms is one of the pathways for the beginning of soils on Earth. *Sci Rep*. 2018; 8: 1–15.
[150] Golubic S, Radtke G, Le Campion-Alsumard. Endolithic fungi in marine ecosystems. *Trends Microbiol*. 2005; 13: 229–35.
[151] Gleason FH, Gadd GM, Pitt JI, Larkum AWD. The roles of endolithic fungi in bioerosion and disease in marine ecosystems. I. General concepts. *Mycology*. 2017; 8: 205–15.

[152] Roush D, García-Pichel F. Succession and colonization dynamics of endolithic phototrophs within intertidal carbonates. *Microorganisms*. 2020; 8: 214.
[153] Ehrlich HL. How microbes influence mineral growth and dissolution. *Chem Geol.* 1996; 132: 5–9.
[154] Fredrickson JK, Zachara JM. Electron transfer at the microbe–mineral interface: a grand challenge in biogeochemistry. *Geobiology*. 2008; 6: 245–53.
[155] Ferris FG, Fyfe WS, Beveridge TJ. Bacteria as nucleation sites for authigenic minerals in a metal-contaminated lake sediment. *Chem Geol.* 1987; 63: 225–32.
[156] Ferris FG, Wiese RG, Fyfe WS. Precipitation of carbonate minerals by microorganisms—implications for silicate weathering and the global carbon-dioxide budget. *Geomicrobiol J.* 1994; 12: 1–13.
[157] Taillefert M, Beckler JS, Carey E, Burns JL, Fennessey CM, DiChristina TJ. *Shewanella putrefaciens* produces an Fe(III)-solubilizing organic ligand during anaerobic respiration on insoluble Fe(III) oxides. *J Inorg Biochem*. 2007; 101: 1760–7.
[158] Lovley DR, Ueki T, Zhang T, et al. Geobacter: the microbe electric's physiology, ecology, and practical applications. *Adv Microb Physiol.* 2011; 59: 1–100.
[159] Langley S, Beveridge TJ. Effect of O-side-chain-lipopolysaccharide chemistry on metal binding. *Appl Environ Microbiol.* 1999; 65: 489–98.
[160] Bazylinski DA, Frankel RB. Biologically controlled mineralization in prokaryotes. In: Dove PM, DeYoreo JJ, Weiner S, editors. Reviews in Mineralogy and Geochemistry. Washington (DC): Mineralogical Society of America and Geochemical Society. 2003; 54. p. 217–47.
[161] Frankel RB, Bazylinski DA. Biologically induced mineralization by bacteria. In: Dove PM, DeYoreo JJ, Weiner S, editors. Biomineralization. Reviews in Mineralogy and Geochemistry. Washington (DC): Mineralogical Society of America and Geochemical Society. 2003; 54. p. 95–114.
[162] Ferris FG, Lowson EA. Ultrastructure and geochemistry of endolithic microorganisms in limestone of the Niagara Escarpment. *Can J Microbiol.* 1997; 43: 211–9.
[163] Guida BS, Bose M, Garcia-Pichel F. Carbon fixation from mineral carbonates. *Nat Commun.* 2017; 8: 1025.
[164] Trichet J, Defarge C. Non-biologically supported organomineralization. *Bulletin de l'Institut Oceanographique Numero Special (Monaco)* 1995; 14: 203–36.
[165] Ferris FG, Beveridge TJ, Fyfe WS. Iron–silica crystallite nucleation by bacteria in a geothermal sediment. *Nature*. 1986; 320: 609–11.
[166] Konhauser KO, Urrutia MM. Bacterial clay authigenesis: a common biogeochemical process. *Chem Geol.* 1999; 161: 399–413.
[167] Léveillé RJ, Fyfe WS, Longstaffe FJ. Geomicrobiology of carbonate–silicate microbialites from Hawaiian basaltic sea caves. *Chem Geol.* 2000; 169: 339–55.
[168] Souza-Egipsy V, Wierzchos J, Ascaso C, Nealson KH. Mg-silica precipitation in fossilization mechanisms of sand tufa endolithic microbial community, Mono Lake (California). *Chem Geol.* 2005; 217: 77–87.
[169] Souza-Egipsy V, Aguilera A, Mateo-Martí E, Martín-Gago JA, Amils R. Fossilization of acidophilic microorganisms. *Geomicrobiol J.* 2010; 27: 692–706.
[170] Cámara B, Souza-Egipsy V, Ascaso C, Artieda O, de los Ríos A, Wierzchos J. Biosignatures and microbial fossils in endolithic microbial communities colonizing Ca-sulfate crusts in the Atacama Desert. *Chem Geol.* 2016; 443: 22–31.
[171] Handley KM, Turner SJ, Campbell KA, Mountain BW. Silicifying biofilm exopolymers on a hot-spring microstromatolite: templating nanometer-thick laminae. *Astrobiology*. 2008; 8: 747–70.
[172] Merz M. The biology of carbonate precipitation by cyanobacteria. *Facies*. 1992; 26: 81–101.

[173] Bissett A, Reimer A, de Beer D, Shiraishi F, Arp G. Metabolic microenvironmental control by photosynthetic biofilms under changing macroenvironmental temperature and pH conditions. *Appl Environ Microbiol.* 2008; 74: 6306–12.

[174] Shiraishi F, Bissett A, de Beer D, Reimer A, Arp G. Photosynthesis, respiration and exopolymer calcium-binding in biofilm calcification (Westerhfer and Deinschwanger creek, germany). *Geomicrobiol J.* 2008; 25: 83–94.

[175] Arp G, Thiel V, Reimer A, Michaelis W, Reitner J. Biofilm exopolymers control microbialite formation at thermal springs discharging into the alkaline Pyramid Lake, Nevada, USA. *Sediment Geol.* 1999; 126: 159–76.

[176] Dupraz C, Visscher PT. Microbial lithification in marine stromatolites and hypersaline mats. *Trends Microbiol.* 2005; 13: 429–38.

[177] Arp G, Helms G, Karlinska K, et al. Photosynthesis versus exopolymer degradation in the formation of microbialites on the atoll of Kiritimati, Republic of Kiribati, Central Pacific. *Geomicrobiol J.* 2012; 29: 29–65.

[178] Kawaguchi T, Decho AW. Isolation and biochemical characterization of extracellular polymeric secretions (EPS) from modern soft marine stromatolites (Bahamas) and its inhibitory effect on $CaCO_3$ precipitation. *Prep Biochem Biotechnol.* 2002; 32: 51–63.

[179] Vasconcelos CO, McKenzie JA, Bernasconi SM, Grajic D, Tien AJ. Microbial mediation as a possible mechanism for natural dolomite formation at low temperature. *Nature.* 1995; 377: 220–2.

[180] Warthmann R, Vasconcelos C, Sass H, McKenzie JA. *Desulfovibrio brasiliensis* sp. nov., a moderate halophilic sulfate-reducing bacterium from Lagoa Vermelha (Brazil) mediating dolomite formation. *Extremophiles.* 2005; 9: 255–61.

[181] Sanchez-Roman M, Vasconcelos C, Schmid T, et al. Aerobic microbial dolomite at the nanometer scale: implications for the geologic record. *Geology.* 2008; 36: 879–82.

[182] García del Cura MA, Sanz-Montero ME, de los Ríos, A, Ascaso, C. Microbial dolomite in fresh water carbonate deposits. *Sedimentology.* 2013; 61: 41–55.

[183] Visscher PT, Reid PR, Bebout BM. Microscale observations of sulfate reduction: correlation of microbial activity with lithified micritic laminae in modern marine stromatolites. *Geology.* 2000; 28: 919–22.

[184] Braissant O, Decho AW, Dupraz C, Glunk C, Przekop KM, Visscher PT. Exopolymeric substances of sulfate-reducing bacteria: interactions with calcium at alkaline pH and implication for formation of carbonate minerals. *Geobiology.* 2007; 5: 401–11.

[185] Rodriguez-Navarro C, Jimenez-Lopez C, Rodriguez-Navarro A, Gonzalez-Munoz MT, Rodriguez-Gallego M. Bacterially mediated mineralization of vaterite. *Geochim Cosmochim Acta.* 2007; 71: 1197–213.

[186] Bosak T, Souza-Egipsy V, Corsetti F, Newman DK. Micro-scale porosity: a biosignature in carbonate crusts. *Geology.* 2004; 32: 701–4.

[187] Bosak T, Souza-Egipsy V, Newman DK. A laboratory model of abiotic peloid formation. *Geobiology.* 2004; 2: 189–98.

[188] Sanchez-Roman M, Rivadeneyra MA, Vasconcelos C, McKenzie JA. Biomineralization of carbonate and phosphate by moderately halophilic bacteria. *FEMS Microbiol Ecol.* 2007; 61: 273–84.

[189] Rivadeneyra MA, Martin-Algarra A, Sanchez-Roman M, Sanchez-Navas A, Daniel Martin-Ramos J. Amorphous Ca-phosphate precursors for Ca-carbonate biominerals mediated by *Chromohalobacter marismortui. ISME J.* 2010; 4: 922–32.

[190] Konhauser KO, Fyfe WS, Schultze-Lam S, Ferris FG, Beveridge TJ. Iron phosphate precipitation by epilithic microbial biofilms in arctic Canada. *Can J Earth Sci.* 1994; 31: 1320–4.

[191] Souza-Egipsy V, del Cura MA, Ascaso C, de los Ríos A, Wierzchos J, González Martín J. Interaction between calcite and phosphorus in biomineralization processes in tufa carbonates. *Int Rev Hydrobiol.* 2006; 91: 222–41.

[192] Lundberg J, McFarlane DA. Subaerial freshwater phosphatic stromatolites in Deer Cave, Sarawak—a unique geobiological cave formation. *Geomorphology*. 2011; 128: 57–72.
[193] Sanchez-Navas A, Martin-Algarra A, Nieto F. Bacterially-mediated authigenesis of clays in phosphate stromatolites. *Sedimentology*. 1998; 45: 519–33.
[194] Lefevre CT, Bazylinski DA. Ecology, diversity, and evolution of magnetotactic bacteria. *Microbiol Mol Biol Rev*. 2013; 77: 497–526.
[195] Beveridge TJ, Meloche JD, Fyfe WS, Murray RGE. Diagenesis of metals chemically complexed to bacteria—laboratory formation of metal phosphates, sulfides, and organic condensates in artificial sediments. *App Environ Microbiol*. 1983; 45: 1094–108.
[196] Fortin D, Ferris FG, Beveridge TJ. Surface mediated mineral development by bacteria. In: Banfield JF, Nealson KH, editors. Reviews in Mineralogy. Geomicrobiology: Interactions between Microbes and Minerals. Washington (DC): Mineralogical Society of America. 1997; 35. p. 161–80.
[197] Fortin D, Langley S. Formation and occurrence of biogenic iron-rich minerals. *Earth Sci Rev*. 2005; 72: 1–19.
[198] Miot J, Benzerara K, Morin G, Kappler A, Obst M, Brown G. Iron biomineralization by neutrophilic nitrate-reducing iron-oxidizing bacteria. *Geochem Cosmochem Acta*. 2009; 73: 696–711.
[199] Tazaki K, Okuno M, Furumoto M, Watanabe H. Biomineralization of pisoliths in hot springs. Mater Sci Eng C Biomimet Supramol Syst. 2006; 26: 617–23.
[200] Chan CS, Fakra SC, Emerson D, Fleming EJ, Edwards KJ. Lithotrophic iron-oxidizing bacteria produce organic stalks to control mineral growth: implications for biosignature formation. *ISME J*. 2011; 5: 717–27.
[201] Labrenz M, Druschel GK, Thomsen-Ebert T, et al. Formation of sphalerite (ZnS) deposits in natural biofilms of sulfate-reducing bacteria. *Science*. 2000; 290: 1744–7.
[202] Druschel GK, Labrenz M, Thomsen-Ebert T, Fowle DA, Banfield JF. Geochemical Modeling of ZnS in biofilms: an example of ore depositional processes. *Econ Geol*. 2002; 97: 1319–29.
[203] Picard A, Gartman A, Clarke DR, Girguis PR. Sulfate-reducing bacteria influence the nucleation and growth of mackinawite and greigite. *Geochim Cosmochim Acta*. 2018; 220: 367–84.
[204] Moreau J, Weber P, Martin M, et al. Submicron-scale isotopic variations within biogenic ZnS record the mechanism and kinetics of extracellular metal–sulfide biomineralization. *Geochim Cosmochim Acta*. 2006; 70, A428–A.
[205] Jones B, Renaut RW. Selective mineralization of microbes in Fe-rich precipitates (jarosite, hydrous ferric oxides) from acid hot springs in the Waiotapu geothermal area, North Island, New Zealand. *Sediment Geol*. 2007; 194: 77–98.
[206] Adamo P, Violante P. Weathering of rocks and neogenesis of minerals associated with lichen activity. *Appl Clay Sci*. 2000; 16: 229–56.
[207] Dorn RI, Oberlander TM. Microbial origin of desert varnish. *Science*. 1981; 213: 1245–7.
[208] Elbert W, Weber B, Burrows S, et al. Contribution of cryptogamic covers to the global cycles of carbon and nitrogen. *Nat Geosci*. 2012; 5: 459–62.
[209] De los Ríos A, Ascaso C. Contributions of *in situ* microscopy to current understanding of stone biodeterioration processes. *Int Microbiol*. 2005; 8: 181–8.
[210] Chen J, Blume HP, Beyer, L. Weathering of rocks induced by lichen colonization—a review. *Catena*. 2000; 39: 121–46.
[211] Adeyemi AO, Gadd GM. Fungal degradation of calcium-, lead- and silicon-bearing minerals. *Biometals*. 2005; 18: 269–81.
[212] Weed SB, Davey CB, Cook MG. Weathering of mica by fungi. *Soil Sci Soc Am Proc*. 1969; 33: 702–6.
[213] Ascaso C, Galván, J. Studies on the pedogenetic action of lichen acids. *Pedobiologia*. 1976; 16: 321–31.

[214] Ascaso C, Sancho LG, Rodriguez-Pascual C. The weathering action of saxicolous lichens in maritime Antarctica. *Polar Biol.* 1990; 11: 33–9.
[215] Barker WW, Welch SA, Chu S, Banfield, JF. Experimental observations of the effects of bacteria on aluminosilicate weathering. *Am Miner.* 1998; 83: 1551–63.
[216] Vandevivere P, Welch SA, Ullman WJ, Kirchman DL. Enhanced dissolution of silicate minerals by bacteria at near-neutral Ph. *Microb Ecol.* 1994; 27: 241–51.
[217] Purvis OW, Halls C. A review of lichens in metal-enriched environments. *Lichenologist.* 1996; 28: 571–601.
[218] Sterflinger, K. Fungi as geologic agents. *Geomicrobiol J.* 2000; 17: 97–124.
[219] Gadd GM. Fungal production of citric and oxalic acid: importance in metal speciation, physiology and biogeochemical processes. *Adv Microb Physiol.* 1999; 41: 47–92.
[220] Gadd GM. Fungi, rocks, and minerals. *Elements.* 2017,13: 171–6.
[221] Jones DL. Organic acids in the rhizosphere—a critical review. *Plant Soil.* 1998; 205: 25–44.
[222] Barker WW, Welch SA, Banfield JF. Biochemical weathering of silicate minerals. In: Banfield JF, Nealson KH, editors. Reviews in Mineralogy. Geomicrobiology: Interactions between Microbes and Minerals. Washington (DC): Mineralogical Society of America. 1997; 35. p. 391–428.
[223] Wierzchos J, Ascaso C. Mineralogical transformation of bioweathered granitic biotite, studied by HRTEM: evidence for a new pathway in lichen activity. *Clay Clay Min.* 1998; 46: 446–52.
[224] Bundeleva IA, Menez B, Auge T, Bodenan F, Recham N, Guyot F. Effect of cyanobacteria *Synechococcus* PCC 7942 on carbonation kinetics of olivine at 20 degrees C. *Miner Eng.* 2014; 59: 2–11.
[225] Zavarzina DG, Chistyakova NI, Shapldn AV, et al. Oxidative biotransformation of biotite and glauconite by alkaliphilic anaerobes: the effect of Fe oxidation on the weathering of phyllosilicates. *Chem Geol.* 2016; 439: 98–109.
[226] Lamerand C, Shirokova LS, Benezeth P, Rols J-L, Pokrovsky OS. Olivine dissolution and hydrous Mg carbonate and silicate precipitation in the presence of microbial consortium of photo-autotrophic and heterotrophic bacteria. *Geochim Cosmochim Acta.* 2020; 268: 123–41.
[227] Bennett PC, Rogers JR, Choi WJ. Silicates, silicate weathering, and microbial ecology. *Geomicrobiol J.* 2001; 18: 3–19.
[228] Ascaso C, Wierzchos J, Castello R. Study of the biogenic weathering of calcareous litharenite stones caused by lichen and endolithic microorganisms. *Int Biodeterior Biodegrad.* 1998; 42: 29–38.
[229] Ascaso C, Wierzchos J, Delgado J. Aires-Barros L, Henriquez FMA, Charola AE. Endolithic microorganisms in the biodeterioration of the Belem Tower. *Internationale Zeitschrift für Bauinstandsetzen.* 1998; 4: 627–40.
[230] Shorabi M, Favero-Longo SE, Pérez-Ortega S, et al. Lichen colonization and associated deterioration processes in Pasargadae, UNESCO world heritage site, Iran. *Int Biodeterior Biodegrad.* 2017; 117: 171–82.
[231] García-Pichel F, Ramirez-Reinat E, Gao Q. Microbial excavation of solid carbonates powered by P-type ATPase-mediated transcellular Ca^{2+} transport. PNAS 2010; 107: 21749–54.
[232] Weber B, Scherr C, Bicker F, Friedl T, Buedel B. Respiration-induced weathering patterns of two endolithically growing lichens. *Geobiology.* 2011; 9: 34–43.
[233] Krumbein WE. Patina and cultural heritage a geomicrobiologist's perspective. In: Biodeterioration and Its Control—Biotechnologies in Cultural Heritage Protection and Conservation. Proceedings 5th EC Conference, Polska Akademia Nauk, Krakow. 2002. p. 39–47.
[234] Gadd GM, Dyer TD. Bioprotection of the built environment and cultural heritage. *Microb Biotechnol.* 2017; 10: 1152–6.
[235] Nordstrom DK. Hydrogeochemical processes governing the origin, transport and fate of major and trace elements from mine wastes and mineralized rock to surface waters. *Appl Geochem.* 2011; 26: 1777–91.

[236] Rohwerder T, Gehrke T, Kinzler K, Sand W. Bioleaching review part A: progress in bioleaching: fundamentals and mechanisms of bacterial metal sulfide oxidation. *Appl Microbiol Biotechnol.* 2003; 63: 239–48.

[237] Welch SA, Vandevivere P. Effect of microbial and other naturally-occurring polymers on mineral dissolution. *Geomicrobiol J.* 1994; 12: 227–38.

[238] Fomina MA, Alexander IJ, Colpaert JV, Gadd GM. Solubilization of toxic metal minerals and metal tolerance of mycorrhizal fungi. *Soil Biol Biochem.* 2005; 37: 851–66.

[239] Ahmed E, Holmstrom SJM. Siderophores in environmental research: roles and applications. *Microb Biotechnol.* 2014; 7: 196–208.

[240] Lovley DR, Holmes DE, Nevin KP. Dissimilatory Fe(III) and Mn(IV) reduction. *Adv Microb Physiol* 2004; 49: 219–86.

[241] Reguera G, McCarthy KD, Mehta T, Nicoll JS, Tuominen MT, Lovley DR. Extracellular electron transfer via microbial nanowires. *Nature.* 2005; 435: 1098–101.

Pedro H. Lebre, Don A. Cowan, and Thulani P. Makhalanyane

1.2 The hypolithic habitat: microbial communities under translucent rocks

Hypoliths, cryptic assemblages found below translucent rocks, are widely distributed across both hot and cold deserts. In these deserts, a combination of environmental stressors such as katabatic wind episodes, high ultraviolet radiation, and substantial temperature fluctuations are expected to result in simple food webs. These food webs are thought to be disproportionally driven by microbial assemblages, which are remarkably phylogenetically diverse despite these restrictive environmental conditions. Previous studies on hypoliths have demonstrated that these niches harbor a wide array of genes linked to the sequestration of carbon, nitrogen, and phosphorus. These studies also suggest that, by exerting substantial metabolic control, hypolithons play important roles at local, regional, and global scales. In this chapter, we discuss key aspects regarding the ecology and distribution of hypolithic communities across hot and cold deserts. We describe both biotic and abiotic drivers of phylogenetic diversity and how these factors may affect hypolithic evolutionary processes. This review also evaluates recent advances in our understanding of functional processes in hypolithons, highlighting the regulatory roles played by bacteriophages. We conclude by synthesizing current insights on microbial community interactions and explore several key questions regarding the ecology of hypolithons.

1.2.1 Climate change, desertification, and refuge niches

Climate change poses a real threat to marine and terrestrial ecosystems [1]. To limit these ecosystem threats, substantial reductions in anthropogenic carbon emissions are urgently required [2–4]. The Paris Climate Agreement has set a series of ambitious targets, including aspirations to hold the increase in the global average temperature to below 2°C above preindustrial levels and reducing increases to 1.5°C above preindustrial levels [5].

The precise effects of changing climatic regimes on biodiversity and ecosystem services are still being quantified. However, studies suggest that increasing temperatures are likely to result in substantial losses in biodiversity and diminished capacity to provide crucial ecosystem services [6–8]. In the oceans, increased acidification and calcification have been directly attributed to rising anthropogenic CO_2 levels [9, 10]. In terrestrial ecosystems, a direct effect of climate change is the gradual transition toward greater aridity (desertification) [11–15].

Approximately 33% of the planets biomes may be classified as deserts [16, 17] (Fig. 1.2.1). In areas classified as hyperarid, where the precipitation to potential

https://doi.org/10.1515/9783110646467-002

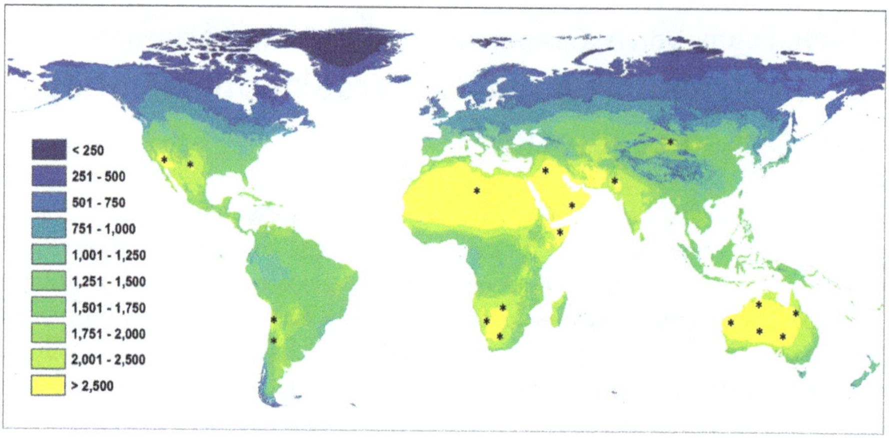

Fig. 1.2.1: Map showing the aridity gradient globally. Four key zones of aridity are further defined: subhumid (P/PET = 0.5–0.65), semiarid (0.2–0.5), arid (0.05–0.2), and hyperarid (<0.05) (Barrow 1992). A star denotes the location of a major desert system (excluding the Antarctic Dry Valleys). The map was sourced from Antonio and Robert [26].

evapotranspiration (P/PET) ratio is below 0.05, environmental conditions are often inhospitable. These conditions, including the prevalence of strong winds (which destabilize surface soils) and severe temperature fluctuations (which pose a severe challenge to thermal controls and adaptation), substantially limit the diversity of higher plants and animals [18, 19]. Consequently, desert life is often found in lithic-associated microhabitats (also called soil rock-surface communities), which are widespread in hot and cold deserts. The rocks provide a "shelter" for biological organisms, buffering the ecological stressors found in deserts. This biology is often microbial and dominated by bacteria, fungi, archaea, and viruses. These microorganisms possess the metabolic versatility required to rapidly adapt to challenging environmental stressors [17, 20–25].

We have previously reviewed the different lithic substrates and niches found in both hot and cold desert environments [27]. In this chapter, we focus exclusively on hypolithic communities. We provide an overview of hypoliths and their biogeography and discuss their phylogenetic diversity. We also detail their functional processes and nutrient sequestration capacity. We highlight the roles played by bacteriophages in the hypolithic niche and conclude by discussing possible areas for future research.

1.2.2 Hypoliths and their quartz substrate

Hyperarid regions represent an extreme challenge to life on earth [28–32]. These biomes are characterized by a combination of stressors that push life to the fringes of survival and subsistence [33, 34]. As such, these inhospitable landscapes are dominated by microbial communities that have adapted to a lifestyle in cryptic,

Fig. 1.2.2: The depauperate desert landscape in a cold and hot desert. (A) The Taylor Valley in the McMurdo Dry Valleys of Antarctica. (B) The Namib Desert landscape in Southern Africa. (C) Sample collection from the McMurdo Dry Valleys showing the distinctive desert pavement structure. (D) A quartz rock with hypolithic colonization in the Namib Desert.

protected niches. One such niche is the ventral surface of semitranslucent rocks, and the microbial communities that subsist in it are termed hypoliths [35] (Fig. 1.2.2). Semitranslucent and translucent stones (such as quartz) are a common feature in desert pavements [16], and consequently hypoliths represent a significant proportion of life in these extreme environments [17, 36]. For instance, hypoliths were found to be widespread in the Canadian high Artic [37, 38] as well as in the Antarctic continent [23, 39], whereas in the Namib Desert, hypoliths colonized 98% of quartz stones in quartz-dominated sites [40].

Quartz and other translucent stones provide an environment where microbial communities are physically shielded from various environmental stressors [27]. Studies have shown that translucent quartz harboring hypoliths significantly reduce the sunlight transmitted through the rock, between 60% and 99.9%, depending on the thickness and translucence of the rock [40, 41]. In addition, colonized semitranslucent rocks were shown to filter out most of light transmission at the UV-B/C range of the light spectrum, thus shielding underlying communities from UV-driven DNA damage [42]. The stone substrates also provide some degree of temperature isolation and moisture retention. In the coastal Antarctic Miers Valley, hypolithic communities have been shown to undergo smaller temperature fluctuations compared with open soils, while exhibiting higher relative humidity (RH) levels [43]. Similarly, in the Mojave Desert, hypolith communities were found to be cooler during the day and warmer during the night compared with open soil, whereas the mean RH in the hypolithic community was 30% higher than in the open soils [44]. Although most of these studies found a wide variety of stone sizes in which colonization was possible, one study in particular [40] described a depth threshold (around 5 cm) below which colonization was limited because of the lack of light penetration, highlighting the importance of photoautotrophy for the establishment of hypolith communities.

1.2.3 Phylogenetic diversity under rocks

Hypolith communities have been classified into two dominant types [43, 45], depending on distinct morphological characteristics. Type I communities were characterized by dominant cyanobacterial growth, whereas the relatively rare type II communities were dominated by mosses. Most studies on the biodiversity of hypoliths have been conducted on type I communities, which contain a plethora of photoautotrophic Cyanobacteria, with two genera in particular, *Chroococcidiopsis* and *Phormidium*, dominating most structures [46]. It has been hypothesized that the global distribution of different *Chroococcidiopsis* is a reflection of ancient lineages that predate contemporary climatic selective pressures [47]. Other diazotrophic Cyanobacteria showing variable abundances across hypolith communities include *Leptolyngbya* [48], Nostocales [49], *Oscillatoria*, and *Synechococcus* [50].

Four other bacterial phyla, Proteobacteria, Bacteroidetes, Chloroflexi, and Actinobacteria, have been shown to consistently present in hypolith communities across different biomes, although not as highly abundant as Cyanobacteria [46, 48, 49, 51–54]. Interestingly, most of the studies to date on the biodiversity of hypoliths are based on the use of terminal restriction fragment length polymorphisms (TRFLPs), and recent studies using next-generation sequencing (NGS) have since contested the dominance of cyanobacterial taxa. For instance, a comparative metagenomics of desert hypoliths from Namib and Antarctica deserts revealed that Actinobacteria and Proteobacteria were more dominant than the cyanobacterial population [54]. Another study focusing on the active fraction of the hypolith communities in the Namib Desert [55, 56] showed that Firmicutes, rather the Cyanobacteria, were the dominant population, and Actinobacteria and Proteobacteria also represented a significant portion of the hypolithic diversity. Results from the same study also suggested that Alphaproteobacteria, particularly Shingomonadales and Rhizobiales, might play a disproportionally important role as drivers of the community structure. Although NGS and TRFLP methods have been shown to be comparable in capturing the composition and abundances of microbial communities [57], the former achieves a higher resolution in terms of taxonomy and community diversity [58–60]. Therefore, it is likely that some of the earlier studies on hypoliths might have not captured the full diversity of hypolith communities. Lacap-Bugler et al. [46] showed that the global distribution of cyanobacteria in hypolithons and its ratio to heterotrophic bacteria (the producer/ consumer ratio) varied across a precipitation gradient, so that wetter areas such as the Taklimakan Desert, China, were dominated by heterotrophic bacteria, whereas in hyperarid Antarctic soils, Cyanobacteria accounted for 70% of the community. Very few studies have so far documented the eukaryotic diversity in hypoliths, partially due to their poor representation in such microenvironments [61]. Studies conducted on the hypolithons from Miers Valley, Antarctica, revealed that most hypoliths were dominated by the fungal phylum Ascomycota [62, 63], particularly *Acremonium*, whereas protists Amoebozoa and Cercozoa were present in both cyanobacteria-dominated and moss-dominated hypolithons [63].

1.2.4 Drivers of hypolith development

As suggested by Lacap-Bugler et al. [46], water is a major limiting factor to hypolith colonization. Mapping hypolith communities across the water gradient in the Atacama Desert, Warren-Rhodes et al. [64] demonstrated that water availability, rather than the chemical characteristics of the surrounding soils, was strongly correlated with the community diversity and functional capacity of quartz hypoliths. In coastal desert systems such as the Namib Desert, water is available in the form of dew, fog, and rain [65]. Dew has been shown to play a negligible role in hypolith colonization [66], whereas fog supports the hypolith communities only in specific coastal deserts (i.e., the Namib and Atacama deserts [40, 67]). A survey of hypoliths in the McMurdo Dry Valleys [43] showed that the diversity and type of hypolith communities varied with altitude, which was interpreted as a proxy for moisture.

In addition to water availability and light penetration, other factors have been shown to play a role in shaping hypolith development. For instance, hypolithic communities in the Namib Desert were found to be dependent on selective recruitment from the local soil microbiome, suggesting that the development of these communities is partly shaped by the heterogeneous nature of the underlying soil community and by dispersal limitation [53]. Studies have also highlighted environmental stochasticity as playing a significant role in shaping the microbial composition of hypoliths at both global and local scales [68, 69]. Makhalanyane et al. [52] suggested that this stochasticity becomes less important as hypolith communities develop over time, from cyanobacterial dominated to moss dominated. There is still much to be understood regarding the abiotic and biotic drivers of hypolith community development, as most studies to date have focused on only a small number of variables. The microenvironmental variability between hypoliths and the temporal biotic relationships within communities are still not well documented, and these will be crucial for an understanding of hypolith evolution and development.

1.2.5 Functional activity in environmentally challenging habitats

The combined effects of pronounced seasonal variability and extreme environmental stressors do not appear to limit the diversity (species richness) in Antarctic soil microbiomes. By providing access to the large majority of microbial taxa, which are difficult to culture under standard laboratory conditions, modern culture-independent approaches have revealed unprecedented species diversity in a range of Antarctic habitats (Fig. 1.2.3). These habitats include aquatic [70–73], terrestrial [74–76], and rock-associated niches [22, 23, 27, 48, 62, 77, 78]. Irrespective of the habitat, these studies appear to indicate similar levels of phylogenetic diversity at phylum and class levels. However, the implications of this broad phylogenetic composition on functional activity are less clearly understood.

The expectation is that Antarctic hypolithic microbial communities may play analogous functional roles as those from less extreme environments. By harboring phyla that have previously been shown to mediate carbon and nitrogen cycling, these microbiomes are expected to provide similar ecosystem services. Providing definitive evidence to support this proposal is difficult because of logistical restrictions and the inability to set up long-term *in situ* experiments in Antarctica. Without such

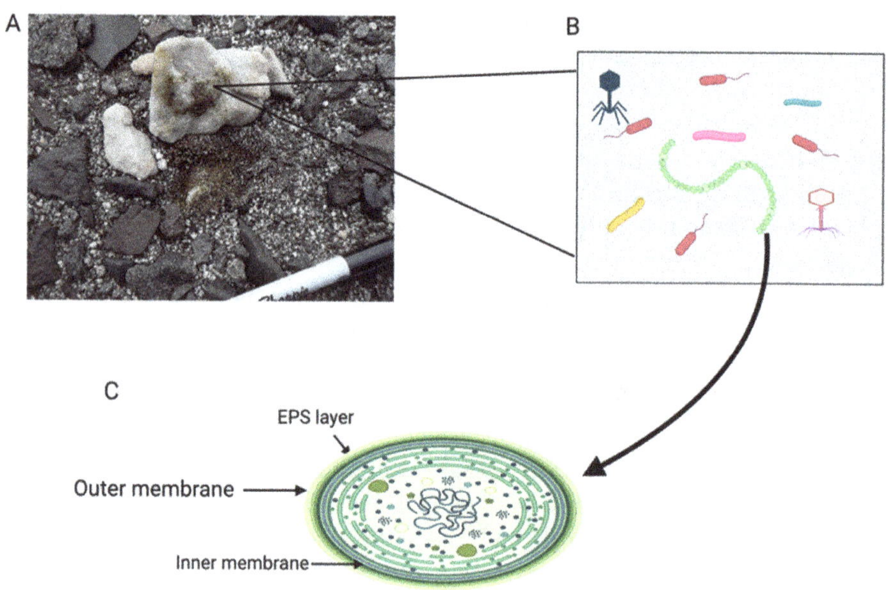

Fig. 1.2.3: (A) Hypolithons are a mixture of phylogenetically diverse groups found below quartz rocks. (B) These groups include morphologically distinct Actinobacteria, Bacteroidetes, Firmicutes, Proteobacteria, and Cyanobacteria. Recent studies have shown that several tailed phages are found in hypoliths and that these likely control metabolic interactions in hypoliths. (C) The cyanobacteria found in hypoliths have capacity to exhibit a wide range of metabolic processes. These functional roles include nitrogen assimilation, carbon sequestration, and other photoautotrophic processes.

experiments, it is impossible to gain a true mechanistic understanding of the relationship between microbial phylogeny and functional traits.

Some functional gene evidence does appear to, at least in part, confirm a capacity to provide these ecosystem services [43, 54, 78, 79]. Previous studies have shown the presence of genes linked to major primary metabolic pathways [78]. These pathways include mechanisms for autotrophic, heterotrophic, and diazotrophic metabolism in Antarctic soils, endoliths, and hypoliths [78].

Chan et al. [78] showed that Antarctic terrestrial niches, including hypoliths, harbor a wide array of enzymes linked to carbon fixation pathways, including ribulose-1,5-bisphosphate carboxylase/oxygenase, in the Calvin-Benson-Bassham cycle; propionyl-CoA/acetyl-CoA carboxylase (*pcc*) in the 3-hydroxypropionate/malyl-CoA cycle, the ATP citrate lyase (*acl*B); and carbon monoxide dehydrogenase in the reductive acetyl-CoA pathway. These functional genes were correlated to the several phyla, including Actinobacteria, Chloroflexi, Gammaproteobacteria, and Spirochaetes [78].

There is both quantitative and qualitative evidence for nitrogen cycling in Antarctic hypolithons [23, 43, 54, 80, 81]. In addition to sequence data (amplicon-based approaches and shotgun metagenomics), a previous study on hypoliths also provided quantitative nitrogen fixation estimates [43]. Using *nif*H gene clone libraries, the study showed a diverse assemblage of taxa with capacity to fix nitrogen. Cyanobacteria with nifH sequences similar to those reported from *Nostoc*, *Tolypothrix*, *Scytonema* spp., and several uncultured nitrogen-fixing cyanobacteria were dominant [43]. In addition, *nif*H genes similar to those described in several *Proteobacteria* including *Azotobacter vinelandii* were also identified. The use of acetylene reduction assays in hypolithic samples provided the first evidence showing direct ecosystem services to underlying soil communities. Estimates suggest that hypolithic communities contribute approximately 14,200 mmol N (0.38 kg N) to the three valley systems investigated [43]. These data suggest that hypolithic communities are substantial contributors to the total nitrogen budgets in oligotrophic Antarctic soils. This finding is surprising because of the limited nitrogen in Antarctic soils [76] but highlights the importance of cryptic niches to the local and regional ecology.

A study by Le et al. [54] used metagenomes from Antarctic hypoliths and other biomes to demonstrate that these communities harbor a range of stress response genes. These genes include those implicated in replication, recombination, and DNA repair in both hot and cold desert hypoliths [54]. Several genes linked to oxidative stress and oxygen limitation were also found in Antarctic hypolithons, supporting the proposed diversity of functional traits.

In addition to well-known genes linked to carbon cycling, nitrogen acquisition, and stress response, a study by Guerrero et al. [79] provided evidence of remarkable functional versatility in phototrophic proteins in hypolithons. Previous studies have shown that a limited number of edaphic [82, 83] and marine [84, 85] microbiomes harbor a genetic capacity to synthesize rhodopsins. This capacity appears to be largely restricted to oligotrophic environments, which are known to select and shape microbial communities. In such habitats, microbial genomes often undergo genome miniaturization [83, 86]. Microbial rhodopsins translocate ions through cell membranes using solar energy to generate a proton motive force [84, 87]. Guerrero et al. [79] showed the presence of functional rhodopsin systems in hypolithic communities. These include several novel putative H+, Na+, and Cl+ pumping rhodopsins, associated with Proteobacteria, Bacteroidetes, Actinobacteria, and Cyanobacteria. In the first application of genome-resolved metagenomics in Antarctic soils, Guerrero et al. [79] showed that these genes could be linked to high-quality Proteobacteria and Bacteroidetes draft genomes [54]. The correlations between functional genes show that these abundant taxa also appear to drive key functional processes in hypolithic communities.

The considerable body of evidence appears to show that numerically dominant taxa drive key ecosystem services in hypoliths [17, 25, 55, 88, 89]. This is in contrast to evidence from other systems, which suggest that ecologically rare taxa may play key functional roles [90–94]. Although there are no direct report of co-occurrence analyses from Antarctic hypoliths, insights on Namib Desert hypoliths appear to confirm this assertion [56, 95].

1.2.6 Viruses as drivers of diversity in hypoliths

Viruses have been hypothesized to play an important role in hypolith community structure by actively challenging hosts in a predator-prey association [96]. In the only study to date documenting the viral diversity of hypoliths, Zablocki et al. [97] described a wide range of viral families targeting the main bacterial phyla described above, as well as dsDNA eukaryotic viruses such as the giant Mimiviridae, which target amoebae. These studies also showed a high proportion of tailed phages from common soil families, including Siphoviridae, Myoviridae, and Podoviridae. Although it is difficult to gain a mechanistic understanding of interactions between bacteria and viruses in environmental systems, a recent study has provided some insights. A study on hypolithic communities by Bezuidt et al. [96] has recently revealed that these communities harbor widespread antiphage innate immunity systems. This study provides the first (albeit indirect) evidence of a dynamic and continuous interaction between bacterial hosts and phages in hypoliths. The immune systems include the DISARM and the BREX mechanisms, which have not been reported previously in cold desert soils.

Several key areas for future research include studies aimed at testing hypotheses linked to metabolic reprogramming by viruses in hypolithic communities [98]. Studies in other systems have shown that bacteriophages access functional capacities from their hosts through the acquisition of auxiliary metabolic genes [99, 100]. The accumulation of metagenomic sequence data makes such studies more feasible, allows us to directly link bacteriophages to their hosts, and tests the evolutionary trajectory of hypolithic host-encoded alleles.

1.2.7 Concluding remarks

The past three decades of research have revealed remarkable insights into the diversity and functional gene repertoire of hot and cold desert hypolithic communities [17, 55, 56, 95]. These studies have shown, in contrast to expectations, that the harsh environmental conditions do not severely limit phylogenetic diversity in these niches [80]. In fact, recent 16S rRNA gene and ITS-region amplicon-based analyses have shown that the diversity in hypoliths mirrors that of more mesic edaphic systems [80]. These communities also appear to harbor a wide range of mechanisms for sequestering carbon and nitrogen and supporting other important biogeochemical cycles [101]. For instance, the demonstration of a suit of genes linked to the synthesis of microbial rhodopsins has added another dimension to the capacity of phototrophic processes in these niche communities [79].

The combined insights into the structure and functional processes in hypolithons have provided some clarity on the evolutionary ecology of microbial communities in extreme environments. Until recently, the diversity of viruses and bacteriophages in Antarctic hypoliths was unknown. However, recent metavirome studies have shown that these communities harbor phylogenetically diverse viruses and bacteriophages [97]. These studies suggest that bacteriophages may play an important role in controlling and regulating bacterial communities in Antarctic hypolithic environments, although this suggestion is yet to be experimentally verified.

Although these recent studies have clarified microbial community interactions and functional capacity in cold desert hypolithons, there are still many unanswered questions. First, the potential for microbial gene flow from the rest of the world into Antarctic soils remains underexplored. Recent aeromicrobiology studies have provided some hints regarding such gene flow processes [102]. Second, although detailed assessments of phylogeny are important for understanding the ecology of hypolithic systems, we lack a sufficiently comprehensive baseline for testing the effects of global change on Antarctic microbiomes, including hypolithic systems. It will be important to understand how community structures, and the interactions between taxa within microbiomes, are likely to change under augmented water availability, increased temperatures, and altered nutrient regimes [103]. Third, despite the accumulation of amplicon sequence data from surveys of hypolithic communities, we yet have little understanding of the functional roles of many taxa across the tree of life. The metagenomic data acquired so far have not been sufficient for large-scale binning of nonbacterial genomes. For example, we lack a detailed understanding of archaeal phylogenetic and functional diversity. Recent Antarctic soil microbiome studies have shown that amplicon-based approaches may substantially underestimate archaeal diversity [89, 104, 105]. Metagenome assembled genomes (MAGs) of functionally important taxa, such as Thaumarchaeota lineages, will facilitate phylogenomic analyses and may provide insights into critical biogeochemical cycling processes in hypoliths. Finally, the extent of metabolic functionality in hypolithic microbiomes is unclear. Although a combination of functional gene approaches [78, 79] and acetylene reduction assays [43] have provided some insights into carbon and nitrogen cycling in hypoliths, the true extent of these metabolic processes, and how they respond to changing local environmental conditions, is unknown. In Antarctic desert hypolithons, it has been suggested that metabolic activity is restricted to the austral summer period [80]. However, there is little evidence to support such predictions, and detailed quantitative data on the temperature- and water activity-dependent rates of carbon and nitrogen fixation are essential for an accurate estimation of the annual productivity in these communities.

Recent studies showing a wide range of carbon and energy (trace gas) scavenging mechanisms in Antarctic soils, combined with evidence of microbial community activity at subzero temperatures, suggest that some metabolic activities may continue during the austral winter [106, 107]. Assessing the extent to which atmospheric trace gases support primary production in hypolithic communities may extend our understanding of metabolic activity and food webs in these niches.

Acknowledgments

The authors are grateful to the National Research Foundation (NRF) (grant no. 118981), the South African National Antarctic Program (SANAP; grant nos. 110717 and 129225), and the University of Pretoria for funding. Their computational work on hypolithic communities would not have been possible without the Centre for High Performance Computing (Cape Town, South Africa) and the University of Pretoria's Centre for Bioinformatics and Computational Biology. They wish to thank colleagues and friends with whom they have collaborated on hypolith studies, including Craig S. Cary, Ian Hogg, Angel Valverde, Evelien Adriaenssens, Surendra Vikram, Oliver Bezuidt, Leandro Guerrero, Jean-Baptiste Ramond, Phuong Le, Rian Pierneef, Donna Lacap-Bugler, and Yves Van de Peer.

References

[1] Cavicchioli R, Ripple WJ, Timmis KN, Azam F, Bakken LR, Baylis M, Behrenfeld MJ, Boetius A, Boyd PW, Classen AT. Scientists' warning to humanity: microorganisms and climate change. *Nat Rev Microbiol.* 2019; 17: 569–86.

[2] Joos F, Plattner G-K, Stocker TF, Marchal O, Schmittner A. Global warming and marine carbon cycle feedbacks on future atmospheric CO_2. *Science.* 1999; 284: 464–7.

[3] Kindermann G, Obersteiner M, Sohngen B, Sathaye J, Andrasko K, Rametsteiner E, Schlamadinger B, Wunder S, Beach R. Global cost estimates of reducing carbon emissions through avoided deforestation. *Proc Natl Acad Sci USA.* 2008; 105: 10302–7.

[4] Van der Werf GR, Morton DC, DeFries RS, Olivier JG, Kasibhatla PS, Jackson RB, Collatz GJ, Randerson JT. CO_2 emissions from forest loss. *Nat Geosci.* 2009; 2: 737–8.

[5] Rogelj J, Den Elzen M, Höhne N, Fransen T, Fekete H, Winkler H, Schaeffer R, Sha F, Riahi K, Meinshausen M. Paris Agreement climate proposals need a boost to keep warming well below 2°C. *Nature*. 2016; 534: 631–9.

[6] Duffy JE. Biodiversity loss, trophic skew and ecosystem functioning. *Ecol Lett*. 2003; 6: 680–7.

[7] Barlow J, Lennox GD, Ferreira J, Berenguer E, Lees AC, Mac Nally R, Thomson JR, de Barros Ferraz SF, Louzada J, Oliveira VHF. Anthropogenic disturbance in tropical forests can double biodiversity loss from deforestation. *Nature*. 2016; 535: 144–7.

[8] Chase JM, Blowes SA, Knight TM, Gerstner K, May F. Ecosystem decay exacerbates biodiversity loss with habitat loss. *Nature*. 2020; 584: 238–43.

[9] Riebesell U, Zondervan I, Rost B, Tortell PD, Zeebe RE, Morel FM. Reduced calcification of marine plankton in response to increased atmospheric CO_2. *Nature*. 2000; 407: 364–7.

[10] Riebesell U, Gattuso J-P. Lessons learned from ocean acidification research. *Nat Clim Chang*. 2015; 5: 12–14.

[11] Reynolds JF, Smith DMS, Lambin EF, Turner B, Mortimore M, Batterbury SP, Downing TE, Dowlatabadi H, Fernández RJ, Herrick JE. Global desertification: building a science for dryland development. *Science*. 2007; 316: 847–51.

[12] Maestre FT, Eldridge DJ, Soliveres S, Kéfi S, Delgado-Baquerizo M, Bowker MA, García-Palacios P, Gaitán J, Gallardo A, Lázaro R. Structure and functioning of dryland ecosystems in a changing world. *Ann Rev Ecol Evol Syst*. 2016; 47.

[13] Schlaepfer DR, Bradford JB, Lauenroth WK, Munson SM, Tietjen B, Hall SA, Wilson SD, Duniway MC, Jia G, Pyke DA. Climate change reduces extent of temperate drylands and intensifies drought in deep soils. *Nat Commun*. 2017; 8: 1–9.

[14] Borrelli P, Robinson DA, Panagos P, Lugato E, Yang JE, Alewell C, Wuepper D, Montanarella L, Ballabio C. Land use and climate change impacts on global soil erosion by water (2015–2070). *PNAS*. 2020; 117: 21994–2001.

[15] Jansson JK, Hofmockel KS. Soil microbiomes and climate change. *Nat Rev Microbiol*. 2020; 18: 35–46.

[16] Laity JJ. *Deserts and desert environments*. Chichester (UK): Wiley-Blackwell; 2009.

[17] Makhalanyane TP, Valverde A, Gunnigle E, Frossard A, Ramond J-B, Cowan DA. Microbial ecology of hot desert edaphic systems. *FEMS Microbiol*. 2015; 39: 203–21.

[18] Lester ED, Satomi M, Ponce A. Microflora of extreme arid Atacama Desert soils. *Soil Biol Bioch*. 2007; 39: 704–8.

[19] Chamizo S, Cantón Y, Lázaro R, Solé-Benet A, Domingo F. Crust composition and disturbance drive infiltration through biological soil crusts in semiarid ecosystems. *Ecosystems*. 2011; 15: 148–61.

[20] Pointing SB, Warren-Rhodes KA, Lacap DC, Rhodes KL, McKay CP. Hypolithic community shifts occur as a result of liquid water availability along environmental gradients in China's hot and cold hyperarid deserts. *Environ Microbiol*. 2007; 9: 414–24.

[21] Warren-Rhodes KA, Rhodes KL, Boyle LN, Pointing SB, Chen Y, Liu S, Zhuo P, McKay CP. Cyanobacterial ecology across environmental gradients and spatial scales in China's hot and cold deserts. *FEMS Microbiol Ecol*. 2007; 61: 470–82.

[22] Lacap DC, Warren-Rhodes KA, McKay CP, Pointing SB. Cyanobacteria and chloroflexi-dominated hypolithic colonization of quartz at the hyper-arid core of the Atacama Desert, Chile. *Extremophiles*. 2011; 15: 31–8.

[23] Chan Y, Lacap DC, Lau MC, Ha KY, Warren-Rhodes KA, Cockell CS, Cowan DA, McKay CP, Pointing SB. Hypolithic microbial communities: between a rock and a hard place. *Environ Microbiol*. 2012; 14: 2272–82.

[24] Pointing SB, Belnap J. Disturbance to desert soil ecosystems contributes to dust-mediated impacts at regional scales. *Biodivers Conserv*. 2014: 1–9.

[25] Makhalanyane TP, Van Goethem MW, Cowan DA. Microbial diversity and functional capacity in polar soils. *Curr Opin Biotechnol*. 2016; 38: 159–66.

[26] Antonio T, Robert Z. Global Aridity Index and Potential Evapotranspiration (ET0) Climate Database v2; 2019.
[27] Makhalanyane TP, Pointing SB, Cowan DA. Lithobionts: cryptic and refuge niches. In: Cowan DA, editor. Antarctic Terrestrial Microbiology: Physical and Biological Properties of Antarctic Soils. Springer; 2014. p. 163–79.
[28] Bull AT, Asenjo JA. Microbiology of hyper-arid environments: recent insights from the Atacama Desert, Chile. *Antonie Van Leeuwenhoek*. 2013; 103: 1173–9.
[29] Bull AT, Idris H, Sanderson R, Asenjo J, Andrews B, Goodfellow M. High altitude, hyper-arid soils of the Central-Andes harbor mega-diverse communities of actinobacteria. *Extremophiles*. 2018; 22: 47–57.
[30] Schulze-Makuch D, Wagner D, Kounaves SP, Mangelsdorf K, Devine KG, de Vera J-P, Schmitt-Kopplin P, Grossart H-P, Parro V, Kaupenjohann M. Transitory microbial habitat in the hyperarid Atacama Desert. *Proc Natl Acad Sci USA*. 2018; 115: 2670–5.
[31] Sayed AM, Hassan MH, Alhadrami HA, Hassan HM, Goodfellow M, Rateb ME. Extreme environments: microbiology leading to specialized metabolites. *J Appl Microbiol*. 2020; 128: 630–57.
[32] Wierzchos J, Artieda O, Ascaso C, García FN, Vítek P, Azua-Bustos A, Fairén AG. Crystalline water in gypsum is unavailable for cyanobacteria in laboratory experiments and in natural desert endolithic habitats. *Proc Natl Acad Sci USA*. 2020; 117: 27786–7.
[33] Berner T, Evenari M. The influence of temperature and light penetration on the abundance of the hypolithic algae in the Negev Desert of Israel. *Oecologia*. 1978; 33: 255–60.
[34] Büdel B, Schultz M. A way to cope with high irradiance and drought: inverted morphology of a new cyanobacterial lichen, *Peltula inversa* sp. nov., from the Nama Karoo, Namibia. *Bibl Lichenol*. 2003; 86: 225–32.
[35] Golubic S, Friedmann I, Schneider J. The lithobiontic ecological niche, with special reference to microorganisms. *J Sediment Res*. 1981; 51: 475–8.
[36] Pointing SB, Belnap J. Microbial colonization and controls in dryland systems. *Nat Rev Microbiol*. 2012; 10: 551–62.
[37] Cockell CS, Stokes MD. Widespread colonization by polar hypoliths. *Nature*. 2004; 431: 414.
[38] Cockell CS, Stokes MD. Hypolithic colonization of opaque rocks in the Arctic and Antarctic polar desert. *Arc Antarct Alp Res*. 2006; 38: 335–42.
[39] Cowan DA. Cryptic microbial communities in Antarctic deserts. *Proc Natl Acad Sci USA*. 2009; 106: 19749–50.
[40] Warren-Rhodes KA, McKay CP, Boyle LN, Wing MR, Kiekebusch EM, Cowan DA, Stomeo F, Pointing SB, Kaseke KF, Eckardt F. Physical ecology of hypolithic communities in the central Namib Desert: the role of fog, rain, rock habitat, and light. *J Geophys Res Biogeosci*. 2013; 118: 1451–60.
[41] Jolitz RD, McKay CP. Quantitative 3D model of light transmittance through translucent rocks applied to the hypolithic microbial community. *Microb Ecol*. 2013; 66: 112–9.
[42] Smith HD, Baqué M, Duncan AG, Lloyd CR, McKay CP, Billi D. Comparative analysis of cyanobacteria inhabiting rocks with different light transmittance in the Mojave Desert: a Mars terrestrial analogue. *Int J Astrobiology*. 2014; 13: 271.
[43] Cowan DA, Sohm JA, Makhalanyane TP, Capone DG, Green TGA, Cary SC, Tuffin IM. Hypolithic communities: important nitrogen sources in Antarctic desert soils. *Environ Microbiol Rep*. 2011; 3: 581–6.
[44] Ekwealor JT, Fisher KM. Life under quartz: hypolithic mosses in the Mojave Desert. *PLOS One*. 2020; 15: e0235928.
[45] Cowan DA, Khan N, Pointing SB, Cary SC. Diverse hypolithic refuge communities in the McMurdo Dry Valleys. *Antarct Sci*. 2010; 22: 714–20.
[46] Lacap-Bugler DC, Lee KK, Archer S, Gillman LN, Lau MC, Leuzinger S, Lee CK, Maki T, McKay CP, Perrott JK. Global diversity of desert hypolithic cyanobacteria. *Front Microbiol*. 2017; 8: 867.

[47] Bahl J, Lau MC, Smith GJ, et al. Ancient origins determine global biogeography of hot and cold desert cyanobacteria. *Nat Commun*. 2011; 2: 163.

[48] Pointing SB, Chan Y, Lacap DC, Lau MC, Jurgens JA, Farrell RL. Highly specialized microbial diversity in hyper-arid polar desert. *Proc Natl Acad Sci USA*. 2009; 106: 19964–9.

[49] Wong FK, Lacap DC, Lau MC, Aitchison JC, Cowan DA, Pointing SB. Hypolithic microbial community of quartz pavement in the high-altitude tundra of central Tibet. *Microb Ecol*. 2010; 60: 730–9.

[50] Wei ST, Lacap-Bugler DC, Lau MC, Caruso T, Rao S, de los Rios A, Archer SK, Chiu JM, Higgins C, Van Nostrand JD. Taxonomic and functional diversity of soil and hypolithic microbial communities in Miers Valley, McMurdo Dry Valleys, Antarctica. *Front Microbiol*. 2016; 7: 1642.

[51] Caruso T, Taormina M, Migliorini M. Relative role of deterministic and stochastic determinants of soil animal community: a spatially explicit analysis of oribatid mites. *J Anim Ecol*. 2012; 81: 214–21.

[52] Makhalanyane TP, Valverde A, Birkeland N-K, Cary SC, Marla Tuffin I, Cowan DA. Evidence for successional development in Antarctic hypolithic bacterial communities. *ISME J*. 2013; 7: 2080–90.

[53] Makhalanyane TP, Valverde A, Lacap DC, Pointing SB, Tuffin MI, Cowan DA. Evidence of species recruitment and development of hot desert hypolithic communities. *Environ Microbiol Rep*. 2013; 5: 219–24.

[54] Le PT, Makhalanyane TP, Guerrero LD, Vikram S, Van de Peer Y, Cowan DA. Comparative metagenomic analysis reveals mechanisms for stress response in hypoliths from extreme hyperarid deserts. *Genome Biol Evol*. 2016; 8: 2737–47.

[55] Van Goethem MW, Makhalanyane TP, Valverde A, Cary SC, Cowan DA. Characterization of bacterial communities in lithobionts and soil niches from Victoria Valley, Antarctica. *Fems Microbiol Ecol*. 2016: 92.

[56] Van Goethem MW, Makhalanyane TP, Cowan DA, Valverde A. Cyanobacteria and Alphaproteobacteria may facilitate cooperative interactions in niche communities. *Front Microbiol*. 2017: 8.

[57] van Dorst J, Bissett A, Palmer AS, Brown M, Snape I, Stark JS, Raymond B, McKinlay J, Ji M, Winsley T. Community fingerprinting in a sequencing world. *FEMS Microbiol Ecol*. 2014; 89: 316–30.

[58] Pilloni G, Granitsiotis MS, Engel M, Lueders T. Testing the limits of 454 pyrotag sequencing: reproducibility, quantitative assessment and comparison to T-RFLP fingerprinting of aquifer microbes. *PLOS One*. 2012; 7: e40467.

[59] Kasai C, Sugimoto K, Moritani I, Tanaka J, Oya Y, Inoue H, Tameda M, Shiraki K, Ito M, Takei Y. Comparison of human gut microbiota in control subjects and patients with colorectal carcinoma in adenoma: terminal restriction fragment length polymorphism and next-generation sequencing analyses. *Oncol Rep*. 2016; 35: 325–33.

[60] De Vrieze J, Ijaz UZ, Saunders AM, Theuerl S. Terminal restriction fragment length polymorphism is an "old school" reliable technique for swift microbial community screening in anaerobic digestion. *Sci Rep*. 2018; 8: 1–12.

[61] Vikram S, Guerrero LD, Makhalanyane TP, Le PT, Seely M, Cowan DA. Metagenomic analysis provides insights into functional capacity in a hyperarid desert soil niche community. *Environ Microbiol*. 2016; 18: 1875–88.

[62] Khan N, Tuffin M, Stafford W, Cary C, Lacap DC, Pointing SB, Cowan D. Hypolithic microbial communities of quartz rocks from Miers Valley, McMurdo Dry Valleys, Antarctica. *Polar Biol*. 2011; 34: 1657–68.

[63] Gokul J, Valverde A, Tuffin M, Cary S, Cowan D. Micro-eukaryotic diversity in hypolithons from Miers Valley, Antarctica. *Biology*. 2013; 2: 331–40.

[64] Warren-Rhodes KA, Rhodes KL, Pointing SB, Ewing SA, Lacap DC, Gomez-Silva B, Amundson R, Friedmann EI, McKay CP. Hypolithic cyanobacteria, dry limit of photosynthesis, and microbial ecology in the hyperarid Atacama Desert. *Microb Ecol*. 2006; 52: 389–98.

[65] Eckardt FD, Soderberg K, Coop LJ, Muller AA, Vickery KJ, Grandin RD, Jack C, Kapalanga TS, Henschel J. The nature of moisture at Gobabeb, in the central Namib Desert. *J Arid Environ*. 2013; 93: 7–19.

[66] McKay CP. Water sources for cyanobacteria below desert rocks in the Negev Desert determined by conductivity. *Glob Ecol Conserv*. 2016; 6: 145–51.

[67] Azua-Bustos A, Gonzalez-Silva C, Mancilla RA, Salas L, Gomez-Silva B, McKay CP, Vicuna R. Hypolithic cyanobacteria supported mainly by fog in the coastal range of the Atacama Desert. *Microb Ecol*. 2011; 61: 568–81.

[68] Caruso T, Chan Y, Lacap DC, Lau MC, McKay CP, Pointing SB. Stochastic and deterministic processes interact in the assembly of desert microbial communities on a global scale. *ISME J*. 2011; 5: 1406–13.

[69] Lebre PH, Bottos E, Makhalanyane TP, Hogg I, Cowan DA. Islands in the sand: are all hypolithic microbial communities the same? *FEMS Microbiol Ecol*. 2020. https://doi.org/10.1093/femsec/fiaa216.

[70] Mikucki JA, Priscu JC. Bacterial diversity associated with Blood Falls, a subglacial outflow from the Taylor Glacier, Antarctica. *Appl Environ Microbiol*. 2007; 73: 4029–39.

[71] Williams TJ, Long E, Evans F, Demaere MZ, Lauro FM, Raftery MJ, Ducklow H, Grzymski JJ, Murray AE, Cavicchioli R. A metaproteomic assessment of winter and summer bacterioplankton from Antarctic Peninsula coastal surface waters. *ISME J*. 2012; 6: 1883–900.

[72] Wilkins D, Yau S, Williams TJ, Allen MA, Brown MV, DeMaere MZ, Lauro FM, Cavicchioli R. Key microbial drivers in Antarctic aquatic environments. *FEMS Microbiol Rev*. 2013; 37: 303–35.

[73] Yau S, Lauro FM, Williams TJ, Demaere MZ, Brown MV, Rich J, Gibson JA, Cavicchioli R. Metagenomic insights into strategies of carbon conservation and unusual sulfur biogeochemistry in a hypersaline Antarctic lake. *ISME J*. 2013; 7: 1944–61.

[74] Yergeau E, Kang S, He Z, Zhou J, Kowalchuk GA. Functional microarray analysis of nitrogen and carbon cycling genes across an Antarctic latitudinal transect. *ISME J*. 2007; 1: 163–79.

[75] Yergeau E, Kowalchuk GA. Responses of Antarctic soil microbial communities and associated functions to temperature and freeze-thaw cycle frequency. *Environ Microbiol*. 2008; 10: 2223–35.

[76] Lee CK, Barbier BA, Bottos EM, McDonald IR, Cary SC. The Inter-Valley Soil Comparative Survey: the ecology of Dry Valley edaphic microbial communities. *ISME J*. 2012; 6: 1046–57.

[77] Rao S, Chan Y, Lacap D, Hyde K, Pointing S, Farrell R. Low-diversity fungal assemblage in an Antarctic Dry Valleys soil. *Polar Biol*. 2012; 35: 567–74.

[78] Chan Y, Van Nostrand JD, Zhou J, Pointing SB, Farrell RL. Functional ecology of an Antarctic Dry Valley. *Proc Natl Acad Sci USA*. 2013, doi:10.1073/pnas.1300643110.

[79] Guerrero LD, Vikram S, Makhalanyane TP, Cowan DA. Evidence of microbial rhodopsins in Antarctic Dry Valley edaphic systems. *Environ Microbiol*. 2017; 19: 3755–67.

[80] Cowan DA, Makhalanyane TP, Dennis PG Hopkins DW. Microbial ecology and biogeochemistry of continental Antarctic soils. *Front Microbiol*. 2014; 5: 154.

[81] Cowan DA, Ramond J-B, Makhalanyane TP, De Maayer P. Metagenomics of extreme environments. *Curr Opin Microbiol*. 2015; 25: 97–102.

[82] Nakamura Y, Kaneko T, Sato S, Mimuro M, Miyashita H, Tsuchiya T, Sasamoto S, Watanabe A, Kawashima K, Kishida Y. Complete genome structure of *Gloeobacter violaceus* PCC 7421, a cyanobacterium that lacks thylakoids. *DNA Res*. 2003; 10: 137–45.

[83] Gushchin I, Chervakov P, Kuzmichev P, Popov AN, Round E, Borshchevskiy V, Ishchenko A, Petrovskaya L, Chupin V, Dolgikh DA. Structural insights into the proton pumping by unusual proteorhodopsin from nonmarine bacteria. *Proc Natl Acad Sci USA*. 2013; 110: 12631–6.

[84] Béja O, Aravind L, Koonin EV, Suzuki MT, Hadd A, Nguyen LP, Jovanovich SB, Gates CM, Feldman RA, Spudich JL. Bacterial rhodopsin: evidence for a new type of phototrophy in the sea. *Science*. 2000; 289: 1902–6.

[85] Finkel OM, Béja O, Belkin S. Global abundance of microbial rhodopsins. *ISME J.* 2013; 7: 448–51.
[86] Holtzendorff J, Partensky F, Mella D, Lennon J-F, Hess WR, Garczarek L. Genome streamlining results in loss of robustness of the circadian clock in the marine cyanobacterium Prochlorococcus marinus PCC 9511. *J Biol Rhythms.* 2008; 23: 187–99.
[87] Sharma AK, Spudich JL, Doolittle WF. Microbial rhodopsins: functional versatility and genetic mobility. *Trends Microbiol.* 2006; 14: 463–9.
[88] Adriaenssens EM, Kramer R, Van Goethem MW, Makhalanyane TP, Hogg I, Cowan DA. Environmental drivers of viral community composition in Antarctic soils identified by viromics. *Microbiome.* 2017: 5.
[89] Barnard S, Van Goethem MW, de Scally SZ, Cowan DA, van Rensburg PJ, Claassens S, Makhalanyane TP. Increased temperatures alter viable microbial biomass, ammonia oxidizing bacteria and extracellular enzymatic activities in Antarctic soils. *FEMS Microbiol Ecol.* 2020; 96: fiaa065.
[90] Lynch MD, Neufeld JD. Ecology and exploration of the rare biosphere. *Nat Rev Microbiol.* 2015; 13: 217–9.
[91] Thompson LR, Sanders JG, McDonald D, Amir A, Ladau J, Locey KJ, Prill RJ, Tripathi A, Gibbons SM, Ackermann G. A communal catalogue reveals Earth's multiscale microbial diversity. *Nature.* 2017; 551: 457–3.
[92] Calatayud J, Andivia E, Escudero A, Melián CJ, Bernardo-Madrid R, Stoffel M, Aponte C, Medina NG, Molina-Venegas R, Arnan X. Positive associations among rare species and their persistence in ecological assemblages. *Nat Ecol Evol.* 2020; 4: 40–5.
[93] Jiao S, Lu Y. Soil pH and temperature regulate assembly processes of abundant and rare bacterial communities in agricultural ecosystems. *Environ Microbiol.* 2020; 22: 1052–65.
[94] Phoma BS, Makhalanyane TP. Depth-dependent variables shape community structure and functionality in the Prince Edward Islands. *Microb Ecol.* 2020: 1–14.
[95] Valverde A, Makhalanyane TP, Seely M, Cowan DA. Cyanobacteria drive community composition and functionality in rock-soil interface communities. *Mol Ecol.* 2015; 24: 812–21.
[96] Bezuidt OKI, Lebre PH, Pierneef R, León-Sobrino C, Adriaenssens EM, Cowan DA, Van de Peer Y, Makhalanyane TP. Phages actively challenge niche communities in the Antarctic soils. *bioRxiv* 2020; 2001: 2017.911081.
[97] Zablocki O, van Zyl L, Adriaenssens EM, Rubagotti E, Tuffin M, Cary SC, Cowan D. High-level diversity of tailed phages, eukaryote-associated viruses, and virophage-like elements in the metaviromes of antarctic soils. *Appl Environ Microbiol.* 2014; 80: 6888–97.
[98] Hurwitz BL, U'Ren JM. Viral metabolic reprogramming in marine ecosystems. *Curr Opin Microbiol.* 2016; 31: 161–8.
[99] Crummett LT, Puxty RJ, Weihe C, Marston MF, Martiny JB. The genomic content and context of auxiliary metabolic genes in marine cyanomyoviruses. *Virology.* 2016; 499: 219–29.
[100] Hurwitz BL, Hallam SJ, Sullivan MB.) Metabolic reprogramming by viruses in the sunlit and dark ocean. *Genome Biol.* 2013; 14: R123.
[101] Boison G, Mergel A, Jolkver H, Bothe H. Bacterial life and dinitrogen fixation at a gypsum rock. *Appl Environ Microbiol.* 2004; 70: 7070–7.
[102] Archer SD, Lee KC, Caruso T, Maki T, Lee CK, Cary SC, Cowan DA, Maestre FT, Pointing SB. Airborne microbial transport limitation to isolated Antarctic soil habitats. *Nat Microbiol.* 2019; 4: 925–32.
[103] Convey P, Peck LS. Antarctic environmental change and biological responses. *Sci Adv.* 2019; 5: eaaz0888.
[104] Ortiz M, Bosch J, Coclet C, Johnson J, Lebre P, Salawu-Rotimi A, Vikram S, Makhalanyane T, Cowan D. Microbial nitrogen cycling in Antarctic soils. *Microorganisms.* 2020; 8: 1442.

[105] Zhang E, Thibaut LM, Terauds A, Raven M, Tanaka MM, van Dorst J, Wong SY, Crane S, Ferrari BC. Lifting the veil on arid-to-hyperarid Antarctic soil microbiomes: a tale of two oases. *Microbiome*. 2020; 8: 1–12.
[106] Cowan DA, Makhalanyane TP. Energy from thin air. *Nature*. 2017; 552: 1–2.
[107] Ji M, Greening C, Vanwonterghem I, Carere CR, Bay SK, Steen JA, Montgomery K, Lines T, Beardall J, Van Dorst J. Atmospheric trace gases support primary production in Antarctic desert surface soil. *Nature*. 2017; 552: 400–3.

2 Organisms

Claudia Coleine and Laura Selbmann

2.1 Black fungi inhabiting rock surfaces

Fungi represent one of the most diverse and ancient branches of the tree of life and have a worldwide distribution. They successfully colonize every biome, including extreme environments such as polar and desert regions, playing a pivotal role in global biogeochemical processes. In the fungal kingdom, "black fungi" (or microcolonial fungi or meristematic fungi) are a slow-growing group that reproduces mostly asexually; it includes ascomycetous taxa that are phylogenetically quite heterogeneous in the classes Dothideomycetes, Eurotiomycetes, and Arthoniomycetes.

In contrast to the majority of fungi, which live more or less comfortably inside their hosts or substrates, few black fungal lineages have evolved and adapted to thrive on bare naked rock surfaces (e.g., in the Mediterranean basin or in hot and cold dry deserts) and are therefore referred to rock-inhabiting fungi (RIF). For their ability to thrive in harshest conditions such as drought, temperature and pH extremes, excessive radiation, and oligotrophy, RIF are accounted as the most resistant eukaryotic organisms known to date. In this chapter, we will provide an overview of the diversity, taxonomy, occurrence, and ecology of RIF colonizing both worldwide natural rocks and stone monuments, also exploring the future perspectives, including their biotechnological, bioremediation, and radioprotection potential.

2.1.1 Introduction

The fungal kingdom includes millions of species [1], ranging from yeast to mushrooms, and represents one of the most diverse and ancient branches of the tree of life, being critical in both terrestrial and aquatic ecosystems. Based on current knowledge, nineteen phylum-level clades can be defined, as reported by Wijayawardene et al. [2]. Fungi have a worldwide distribution because of their small size and their cryptic lifestyle in soil, and decomposing matter, as symbionts with algae, bryophyte, plants, and animals [3, 4]. They are found in every biome such as polar, temperate, and tropical environments, as well in the bottom of the Dead Sea, where they play an essential role in global geological processes, namely, "geomycology," including organic and inorganic transformations and element cycling, bioweathering, and mycogenic mineral formation [5]. Fungi are highly resilient and capable of successfully occupying extreme environments.

In contrast to the majority of fungi, which live more or less comfortably inside of hosts or substrates, black fungal lineages have evolved and adapted to thrive on bare naked rock surfaces in nutrient-deprived conditions and excessive radiation, among other challenging stressors. The phenotypes of these commonly called "black fungi" have evolved in different highly adapted and specialized lineages of ascomycetes;

https://doi.org/10.1515/9783110646467-003

they are phylogenetically more diverse than previously thought and primarily belong to the classes Dothideomycetes, Eurotiomycetes, and Arthoniomycetes.

This chapter will first provide an overview of the current understanding or RIF. We will then describe the diversity, taxonomy, occurrence, and ecology of black fungi colonizing both worldwide natural rocks and stone monuments, with a particular focus on insights into the whole-genome sequencing approach. We will finally explore the future perspectives, including the biotechnological, bioremediation, and radioprotection potential.

2.1.2 Rocks as substratum for fungi

Rock represents the earliest terrestrial niche for life since the time when Earth was inhabited exclusively by microbes. Subaerial biofilms (SABs) are microbial communities that develop on mineral surfaces exposed to the atmosphere and are mainly composed of algae, bacteria, cyanobacteria, and fungi [6]. Lithic-associated microhabitats and their communities' inhabitants are termed "lithobionts" [7]; this lifestyle allows microbes to withstand environmental stressors such as oligotrophic conditions, high or low and changing temperatures, UV radiation, and osmotic stress. Thus, rock-inhabiting microbes are a major focus of many investigations of life in harsh environments or studies with astrobiological implications [8]. Although rocks are habitable places for microbes in extreme environments, microbial diversity on the lithic substratum is still underestimated. Lithic microbial colonization has been recorded in various sedimentary rocks such as sandstone [9], limestone [10], gypsum [11], dolomite [12], carbonate [13], and halite [14].

Rock-inhabiting communities can be epilithic, hypolithic, or endolithic [15]. Epiliths occur on the surface of rocks and building stone; hypoliths are found under and attached to pebbles, particularly in hot and cold deserts; and endoliths inhabit the rock subsurface, sometimes forming distinct masses or brightly colored layers. Endolithic microorganisms can occur as (i) chasmoendoliths, growing in preexisting cracks and fissures within rocks; (ii) cryptoendoliths, growing inside cavities and among crystal grains and cannot be observed from the rock surface; and (iii) euendoliths, which can actively penetrate into rocks. The irregular system of pores and fissures provides microbes with an efficient protective network, promoting, to some extent, water retention [16, 17].

Physical properties (e.g., porosity) and elemental composition of the host rock (e.g., carbon, phosphorus, potassium, sulfur, and metal content) as well as environmental factors such as climatic exposure, nutrient sources, and water availability may govern initial establishment, growth, and survival of microbial communities [18]. The bioreceptivity is the ability of a material to be colonized by living organisms, and it is directly dependent on petrographic parameters [19].

Many rock-dwelling microbes have evolved symbiotic strategies to thrive in extreme environments with lichens, a symbiotic relationship between a fungus and a photobiont (either algae or cyanobacteria) being the best-known example. However, free-living heterotrophic bacteria, fungi, and non-lichenized algae and cyanobacteria are also frequently found in rock surfaces where they can survive as biofilms [5]. Fungi-colonizing rocks can be split into two groups that are ecologically and taxonomically different: (i) hyphomycetes of soil and epiphytic origin [20] and (ii) black (highly melanized) slow-growing, typically with meristematic development, fungi (Ascomycetes, mainly orders Chaetothyriales, Dothideales, and Capnodiales) that form peculiar, compact microcolonies [21–25]. The former dominates under favorable conditions and proliferates during milder seasons, and when in higher humidity, abundant nutrients are present. They might be considered occasional colonizers that may be secondarily adapted to stone surfaces when conditions become permissive due, for instance, to the proximity to the soil. Rock hyphomycetes, as *Aureobasidium pullulans* or *Hormonema*-like fungi reported on marble artworks in Sicily, have been ascribed as derived from surrounding vegetation (leaves, barks, or soils) [20]. *Phoma* species have been frequently isolated from rock surfaces in humid as well as in semiarid areas [26] and also reported together with *Epicoccum* species from monument surfaces in Vienna and historical quarry in a rural area [25, 27]. *Phoma* species form highly resistant chlamydospores and dictyochlamydospores, which resemble the growth of microcolonial fungi (MCF) in form and in function that may enable this species to endure, by chance, on rock surfaces even if they do not represent their favorite niches. There is a presence of monument alteration because the melanin excreted causes evident discoloration on surfaces [23]. Conversely, black meristematic fungi prevail under harsh and hostile environments where they are not outcompeted by fast-growing competitive fungi [21, 28–31]. They have been variously named as black yeasts, meristematic, MCF, or RIF to evocate perfectly adapted organisms to life on and in the rocks and are being described in detail in the paragraph below.

2.1.3 Black meristematic RIF

RIF belong to the morphoecological group of "black meristematic fungi," "black yeasts," or simply "black fungi," characterized by basic traits such as thick cell wall, yeast-like polar budding, ability to reproduce by unicellular growth, deep melanization, meristematic growth, at least for a part of their life-cycle, and usually a very slow growth rate [32, 33].

Black fungi in general are exceptionally skilled to exploit practically all kinds of extremes, including saltpans, hydrocarbon-contaminated sites, glaciers, deserts, high mountains, solar panels, building roofs, and exposed rocks, even in Polar regions (e.g., 34–41). Some black fungi can colonize human environments like dishwashers and steam bath or sauna facilities, whereas others have been isolated from a silicone seal and in tap water [42–45]; some of them may switch to opportunism, which makes them important model organisms also with respect to clinical mycology. They are also able to survive chemical and physical stresses such as extreme pH, extreme temperature, UV and ionizing radiation, alpha particles, and even space conditions [46–50]. An excellent capacity to resurrect from dry conditions has been observed [e.g., 51]; dried fungal colonies are metabolically inactive and can survive up to 120°C, whereas temperatures between 35°C and 75°C are lethal for hydrated colonies [52]. In some particularly resistant species, vitality may also be unaffected after 1-h exposition to 90°C [53].

Their exceptional skill to cope with stresses sits on their ability to switch to meristematic development optimizing the volume/surface ratio [6] (Fig. 2.1.1). Many species of black fungi also encompass the ability to switch between filamentous and yeast growth forms depending on the environmental conditions, namely, phenotypic plasticity [54]. Moreover, in addition to melanins as protective pigments, these fungi tend to accumulate other protective compounds such as mycosporines, trehalose, polyalcohols, betaine, and carotenoids that confer them additional ability to endure stresses [55].

Despite morphological and ecological homogeneity, they are phylogenetically quite diverse and comprise lineages mainly in two classes of Ascomycota:

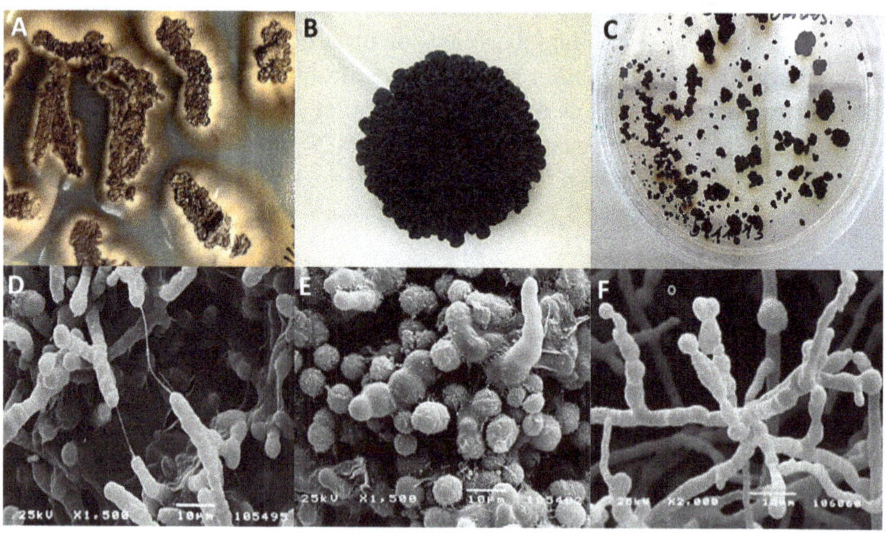

Fig. 2.1.1: Cultures grown on Petri dishes. (A) *Exophiala mesophila* CCFEE 6314; (B) *Knufia petricola* A95 [56]; (C) *Cryomyces antarcticus* [57]. Microscopic images. (D) *Cryomyces minteri* CCFEE 5187 [31]; (E) *Friedmanniomyces endolithicus* [31]; (F) *Recurvomyces mirabilis* CCFEE 5264 [58].

Dothideomycetes and Eurotiomycetes; they have also been retrieved in the class Arthoniomycetes from Antarctica and the Mediterranean but never formally described [59].

Rock is a recurrent substratum for black fungi, and the highest diversity has been detected right in rocky environments. It has been hypothesized that RIF may represent the ancestors of black fungal lineages, which later evolved into other lifestyles [60], including lichens. Within Dothideomycetes, the order Capnodiales is particularly rich in extremotolerant black fungal species and is exceptionally recurrent as rock colonizers, from Antarctic rocky communities [31, 61], high-altitude mountains [38], and hot deserts [62], whereas representative species are more barely found in Pleosporales, Myriangiales, and Dothideales. Black fungi in Eurotiomycetes are mainly found in the order Chaetothyriales [60]. Although they are mostly known as saprophytic and pathogenic fungi [63], they also comprise a relevant number of species able to spread also on rocks as, for instance, the genera *Knufia*, *Bradymyces*, *Cladophialophora*, and *Capronia* [25, 65–67]. It has been postulated that black fungi in Eurotiomycetes are ancestral to present lineages of pathogenic and lichenized fungi in Chaetothyriales and Verrucariales [68] because they appear particularly diverse in early diverging lineages of these orders.

RIF must be accounted as the most persistent inhabitants of perennially exposed subaerial rock surfaces [69–71]. They are surprisingly common but have long been overlooked because all blackish material on stone was considered to be fly ash particles or dirt [24]. Their interactions with other rock inhabitant settlers as well as with the mineral substrate are complex and drive the development of these communities. For example, even if RIF occurring directly on exposed rocks do not necessarily form associations with algae, they may be able to do so in culture [72, 73]. The genus *Lichenothelia*, for instance, may inhabit rocks or lichens or grow loosely associated with algae as borderline lichens [62]; for this reason, it has been hypothesized as a link between rock-inhabiting and lichenized fungi [74, 75]. To thrive on the rocks, RIF evolved a very fast rehydration and upregulation of the metabolic activity [76] as this extreme environment is particularly characterized by extreme changes from humidity to long periods of desiccation and extreme temperature fluctuations. Any atmospheric change influences life on the rocks, which in turn affects atmospheric composition through its metabolic activity and biologically induced weathering. In Zacharova et al. [77], the ability to survive after desiccation and the speed of rehydration as well as the changes of the whole cell protein pattern have been proved; unlike the mesophilic strains, the extreme-tolerant Antarctic RIF *Cryomyces antarcticus* does not show any response to desiccation but seems just to downregulate its metabolism.

RIF form black, clumplike colonies consisting of isodiametrically dividing cells in cracks, pores, and fissures of the rock and in micropits, created by their own deteriorative activity. RIF are also defined as poikilotrophes, i.e., able to deal with varying microclimatic conditions [70].

On bare rock surfaces with a limited and discontinuous presence of liquid water, black fungi must be able to exploit a wide range of carbon sources deposited by dust, water, or in the form of volatile organic compounds [78]. For instance, *Knufia petricola* (Chaetothyriales, Eurotiomycetes) has been proved to tolerate and grow on media containing monoaromatic compounds, confirming its capacity to use recalcitrant carbon sources eventually spurned by other microorganisms [79]. It was also observed, using ^{14}C-labeling, that black fungi isolated from Antarctic desert may actually uptake CO_2 [80]. The authors supposed that Antarctic fungi may actually incorporate CO_2 by carboxylation of pyruvate [81]. A similar process was reported earlier for a number of fungi; such mechanisms might be irrelevant for fungi living in nutrient-rich habitats but could represent a life-sustaining option for slow-growing oligotrophic RIF in the far extreme oligotrophic environments [82, 83].

Rock fungi from cold environments usually also produce a high amount of extracellular polymeric substances (EPS) to increase their resistance against freeze-thaw damage and to retain water over long periods, characteristics that are critical to the maintenance of life under desiccation [39]. The interactions of MCF with the rock substrate can lead to a variety of surface coatings or varnishes through the accumulation of metals and minerals in cell walls and extracellular materials [5, 6, 84]. Indeed, MCF can cause "micropitting" (the formation of lesions in a size range of up to 2 cm in diameter and depth on stone) in rocks, which leads to cavities that can contain the fungal colonies [24].

The protective role of melanin in black fungi

Melanins are an ancient group of diverse pigments of high molecular mass, which are usually dark brown or black and found in all kingdoms of life. Melanins are widely distributed, from microorganisms to plants and animals, and in the fungal kingdom, melanization is observed across all phyla. Some fungal species are constitutively melanized (i.e., melanotic fungi) whereas others only under specific developmental phases (i.e., conidia, yeast filamentous growth), in response to environmental stresses (i.e., *Cryptococcus neoformans*) [85]. Fungal melanins, regardless of their precursor, may share similar functional groups and comparable physicochemical properties [86] while the structure remains scarcely understood. Melanin production in fungi is a result of three pathways known as the polyketide, 3,4-dihydroxyphenylalanine (L-DOPA), and L-tyrosine degradation synthetic pathways, which produce two chemically different compounds with similar properties [87]. The main unifying characteristic of black fungi is the accumulation of melanins, representing an important virulence factor, especially in Eurotiomycetes (e.g., *Exophiala dermatitidis*) [88]. Beyond pathogenicity, melanin is a major contributing factor in the survival of black fungi under various physical and chemical stresses, making them polyextremophiles. The best-known function of melanins in black fungi is to protect them against UV radiation, which is of particular importance for the RIF thriving on bare rock surfaces. Melanins also provide radiation resistance [89], and protection against enzymatic lysis [90], extreme temperatures [91], solar radiation [92], oxidizing agents, desiccation [93], osmotic stress, and extremely high NaCl concentrations that are typical of hypersaline environments (e.g., *Hortaea werneckii* isolated from salterns) [94]. Similar to other light-harvesting biological pigments (i.e., chlorophylls and carotenoids), microbial melanins can absorb radiation energy and transduce it onto energy harvesting [46]. In 2007, Dadachova et al. [95] provided experimental evidence that fungal melanin mediates an energy transduction process from harmful ionizing radiation into useful metabolic energy.

Melanotic fungi inhabit some of the most extreme environments known, and it is generally believed that the presence of melanin contributes to survival through a variety of mechanisms in the harshest environments. For instance, the reactor at Chernobyl and the surrounding soils that host a large population of melanotic fungal species are perhaps the best-known example of fungi living in a high-radiation environment. Even the Antarctic rocks, where many black rock fungi thrive exposed on the surface and interior of rocks, are also a high-radiation environment [39], especially during the long austral summer and considering the recent historical weakening of the atmospheric protection in the southern hemisphere through ozone depletion.

2.1.4 Natural rocks as main substrate for black fungi

Freshly exposed rock surfaces are easily colonized by microbial biofilms, which comprise algal, fungal, bacterial, and cyanobacterial associations, which are normally succeeded by macroscopic vegetation [6]; yet, under permanent extreme conditions, microbes remain the only settlers of this peculiar niche. Rocks, as the most abundant natural substrate of oligotrophic black fungi, are colonized in all climatic zones, including the harshest environments on Earth such as the McMurdo Dry Valleys in Antarctica, the Atacama Desert, and the high Alpine mountains (Fig. 2.1.2).

The RIF biodiversity is enormous and still largely neglected: to date, only a relatively small number of species have been described. Nevertheless, several hundred strains have been documented, and numerous novel genera have been discovered, particularly from the Antarctic desert, although our knowledge from worldwide rocky communities remains patchy and scattered [96].

Black fungi on natural rocks of the Antarctic desert

The extremely low temperature, low water availability, frequent freeze-thaw cycles, low annual precipitation, strong winds, high sublimation, and evaporation together constitute limiting factors for most life forms in Antarctica. In addition, life on exposed rocks in Antarctic ice-free areas must to cope with strong solar radiation and UV radiation; during the austral summer, the exposition is more intense than under hot conditions because of the ozone hole persisting at the South Pole. Abundance and diversity of organisms decrease from the maritime to the continental Antarctic zone

Fig. 2.1.2: (A, B) The diagrams indicate the possible lithobiontic habitats of microorganisms: epilithic (rock surface), hypolithic (rock underside in contact with the soil), and endolithic (further divided into cryptoendolithic, chasmoendolithic, and hypoendolithic) [15]. (C) Endolithic communities at Linnaeus Terrace, McMurdo Dry Valleys, Southern Victoria Land, and Continental Antarctica. (D, E) Natural rocks in the Karst landform area in Guizhou Province, China [97]. (F) Linnaeus Terrace, McMurdo Dry Valleys, Southern Victoria Land, and Continental Antarctica. Photo credit for panels C and F: Italian National Program for Antarctic Researches (PNRA).

because of harsher conditions. There, epilithic growth becomes rare and represented by lichens only until almost complete disappearance, whereas endolithic colonization becomes predominant [98] and microbial life finds more buffered and stable environmental conditions inside rocks [99–101].

In particular, the McMurdo Dry Valleys (Southern Victoria Land) are among the most extreme ice-free environments known on Earth and accounted as the closest terrestrial Martian analogue and were thought to be lifeless until the discovery of these cryptic life forms [102]. Both eukaryotic and prokaryotic endolithic communities have been reported [103], but the most widely distributed in the ice-free areas of continental Antarctica is the cryptoendolithic "lichen-dominated community" found in sandstone [34]. This community appears under the rock crust as a conspicuous zone up to 10 mm deep, formed by parallel and differently colored bands, composed by a black zone under the crust followed by a white, a green, and sometimes a blue-green zone. In these endolithic ecosystems, black fungi play a primary role in the protection of the whole community forming a black "sunscreen" barrier just above the other microbial members [104].

The biodiversity, the co-occurrence, and the ecology of black fungi in Antarctic endolithic communities have been recently reviewed by Coleine et al. [96], highlighting that RIF in Dothideomycetes, together with lichenized fungi, are the most abundant whereas Eurotiomycetes are found less frequently [105–107]. Several endemic taxa of black meristematic fungi such as *Cryomyces antarcticus*, *C. minteri*, *Rachicladosporium antarcticum*, *R. murdoi*, *Extremus antarcticus*, *Meristemomyces frigidus*, *Vermiconia antarctica*, and *Oleoguttula mirabilis* have been described so far [31, 58]. Other species such as *Recurvomyces mirabilis* and *Elasticomyces elasticus* show a wider distribution [38, 61].

The endemic *Friedmanniomyces endolithicus* is the most widespread and abundant species in these ecosystems, suggesting a high degree of specialization and adaptation.

Because of their exceptional stress resistance and adaptations that enable them to establish in and explore hostile environments, Antarctic RIF were selected as eukaryotic models for astrobiological studies that contributed to shed light into lithopanspermia theory, planetary protection, and habitability [108, 109]. Several experiments, in the frame of European Space Agency–funded programs, have been performed to test survival and tolerance of *C. antarcticus* studies to extreme conditions ranging from high and low temperatures, desiccation, and lack of nutrients to lethal doses of ionizing and UV radiation (Martian-simulated and space conditions [110–112]). This species was also identified as "core" taxa of these communities, suggesting that they may have a wider distribution in the Continental Antarctica than previously hypothesized [113–115].

RIF belonging to *Exophiala* spp. were also found in the milder environmental conditions of the Argentine Islands in Antarctica Peninsula [116], whereas *Lichenothelia antarctica* was isolated from rocks collected in the Signy and Lynch Islands [117].

Together with black fungi, lichenized fungi dominate these communities with the orders Lecanorales, Verrucariales, and Acarosporales [107]. The role of primary producers of algae living in obligate association with lichenized fungi and the protective role played by black fungi create the appropriate conditions for the settlement of other microbial species, including fungi, that are not, as a rule, associated with rocks. In particular basidiomycetous and ascomycetous yeasts are recurrent rock inhabitants in these communities, such as species in the genera *Cryptococcus*, *Naganisha*, and *Taphrina* [118].

Conversely, investigations on functionality of Antarctic RIF are still in its infancy. A recent work based on untargeted metabolomics compared the adaptation strategies of two Antarctic endolithic communities collected in the same locality of the McMurdo Dry Valleys (accounted as the closest the Martian analogue on Earth) but subjected to very different environmental pressure because of different sun exposure. Given the significant differences in stress responses, the communities expressed metabolites strictly related to protection from the solar irradiation of photosystems (allantoin) in the north exposed surface, whereas protectants to multiple stresses (melanins) are in the south exposed surface [119].

Black fungi on natural rocks worldwide

RIF remained for a long time just a subject for a few specialists, at first mainly focused on the biodeterioration of artworks. Pioneering studies were therefore devoted in isolating and identifying black fungi from stone monuments and building surfaces [120, 121]. With time, they have become an ever-expanding field of study, offering research opportunities in many basic fields, such as microbial ecophysiology, evolution, and adaptation to extremes, as well as applied research as human pathogenicity, bioremediation, and astrobiology, and the research interest also moved therefore on natural rocks as reservoirs of unexplored genetic resources of potential biotechnological interest. The first broad surveys on natural rocks beyond Antarctica were carried out by Ruibal and coworkers, who have performed an extensive sampling of rock formations in Mallorca and central mountain system in Spain [64, 65]. These works clearly highlighted the huge presence of black fungi on rocks with an unexpected and still undescribed biodiversity. Some of these taxa were later formally described as new genera and species (i.e., *Constantinomyces*, *Lapidomyces*, *Hyphoconis*, *Petrophila*, and *Recurvomyces*) [40, 58, 61].

First reports of microcolonial rock fungi from inhospitable environments such as hot dry deserts of Arizona and Negev were published by Nash and colleagues [122] and, a few years later, by Staley et al. [21]. Black fungi commonly colonize rocks of highest mountain peaks worldwide and have been reported from the Alps (where the new genus *Saxomyces* was described) [38], Andes, Indian Himalayan, and Agoncagua [61]. Bare rocks in arid and semiarid climates in China harbor a bewildering biodiversity of RIF that has been overlooked until recently. Huge surveys, where thousands of RIF isolates were obtained, revealed a conspicuous presence of undescribed biodiversity unrelated with what was described at the moment. Two new dothideomycetous genera *Spissiomyces* and *Rupestriomyces* (with the species *S. aggregatus*, *S. ramosus*, *R. sinensis*, *R. ampulliformis*, and *R. torulosus*) and one new genus *Anthracina* in Chaetothyriales were described [97, 123]. Notably, some RIF species show a worldwide distribution while some others seem to occur in specific restricted areas. For instance, the genera *Elasticomyces elasticus* and *Recurvomyces mirabilis* have been reported from Antarctica, Andes, and Himalaya and from Antarctica and Alps, respectively; however, most of the species show a quite restricted distribution and occur in specific areas: *Bradymyces alpinus*, for instance, has been recorded from Alpine rocks at high altitude only, and *B. yuannensis* has been described from rocks in China. *Lithohypha guttulata* occurs in the Mediterranean area and *Lithohypha catenulata* from Tibet; the species *Knufia petricola*, *K. marmoricola*, *K. karalitana*, *K. vaticanii*, and *K. perforans* occur in the Mediterranean area, whereas *K. separata* and *K. calcareola* are from limestone in Beijing and from sandstone in Yunnan province (China), respectively. The species *C. antarcticus* and *C. minteri* were found in Antarctic desert only, whereas the species *C. funiculosus* and *C. montanus* are from rocks above 3000 m asl in the Alps only. In some cases, the limited distribution encompass the whole genus, such as *Oleoguttula* and *Friedmanniomyces* (Antarctica), *Monticola* (Alps), *Perusta* (Spain), *Saxomyces* (Alps), *Spissiomyces*, *Rupestriomyces*, and *Anthracina* (China).

The reason for such a different ability to spread is still to be elucidated. None of these fungi may produce airborne, easily dispersed conidia; rather, they disseminate through active, mechanical weathering of rocks, and colonized powdered stones act as a means of dispersal. Therefore, dispersal appears to be managed through the same equally efficient mechanism in different areas.

Insights on black RIF whole genomes

In the last decades, the sequencing of fungal genomes has become routine and straightforward, enabling these data as the starting point for an increasing number and types of research [4]. Sequencing of large numbers of fungal genomes, in particular of extremophilic fungi, will allow us to understand the diversity of genes encoding enzymes and pathways that produce several novel compounds [63]. With the rapid accumulation of sequenced fungal genomes, the observed diversity of genes, their pathways known to be critical against different types of stress preserving cells, for instance from both UV and desiccation damages [124], has increased exponentially. The 1000 Fungal Genomes Project by Spatafora et al. had considerably implemented the number of fungal genomes throughout the Fungal Tree of Life, and to date, more than 1500 reference genomes are available [125] and several lineages remain neglected, including the extremophiles and extreme-tolerant fungi inhabiting rocks. In particular, sequencing the whole genomes of these fungi will shed light into the understanding of processes that govern their success in the extremes. Although we are only beginning to interpret the details of the biology of RIF with a genomics approach, several efforts have been performed to untangle the genetic traits of this intriguing group of fungi.

To date, the genomes of eight RIF from Antarctic endolithic communities have been sequenced: *Rachicladosporium antarcticum* CCFEE 5527, *Rachicladosporium* sp. CCFEE 5018 [126], *Hortaea thailandica* CCFEE 6315, *Friedmanniomyces endolithicus* CCFEE 5311, *F. simplex* CCFEE 5184 [127], *C. antarcticus* CCFEE 534 [57], and *C. minteri* CCFEE 5187 belong to the class Dothideomycetes, whereas *Exophiala mesophila* CCFEE 6314 belongs to the class Eurotiomycetes [128].

In Coleine et al. [127], *F. endolithicus* genome was sequenced and compared with a few representatives black fungi (*F. simplex*, *Acidomyces acidophilus*, *Baudoinia panamericana*, *H. thailandica*, and *H. werneckii*) isolated from different extreme environments. The 45- to 50-Mb genomes of *Friedmanniomyces* spp. were larger than those of the currently available black fungi genomes sequences [63, 129]. Moreover, the similar genome size between *H. werneckii* (49.9 Mbp) and the two *Friedmanniomyces* spp. (45–50 Mbp) strains might suggest evolutionary advantages because of a large-scale genome duplication in the Antarctic species' genome to adapt and survive to the hostile conditions of the Antarctic desert. A few genomic features were unique for *Friedmanniomyces* spp. strains only, such as responses to X-rays radiation, DNA damage, and salt tolerance stress, whereas genetic traits associated with meristematic growth and cold adaptation were unique for *F. endolithicus* that may allow this species to be particularly adapted to such harshest conditions and the most widespread in the endolithic communities along the Victoria Land [105, 107].

Conversely, the genome of the hyper-adapted Antarctic endolithic RIF *C. antarcticus* showed a relatively small size (24 Mbp), and the comparative analysis did not reveal any significant deviations of this genome from comparative species and mesophilic hyphomycetes [57]. These findings may result from a poor-quality assembly; whole-genome sequencing based both on Illumina (short-reads) and PacBio (long-reads) is planned for further comparative analysis.

A few years ago, the genome of *Knufia petricola* (*synonym*, *Knufia chersonesos*, order Chaetothyriales) was obtained from red sandstone collected in the Arctic [130]. The genus *Knufia* is a relatively small clade prevalently accommodating extremotolerant fungi inhabiting bare rock surfaces. In particular, *K. petricola*, originally isolated from marble in Crimea [131] and frequently isolated in a wide range of extreme environments from the Mediterranean region to the Polar regions], has been previously proposed as a model organism to elucidate its rock lifestyle as well as stress survival. Furthermore, the ability to degrade aliphatic-aromatic copolyesters [132] makes *K. petricola* a suitable candidate for bioremediation and biotechnology. In the case of the *K. petricola* MA5789 (i.e., the melanized strain), a 27.78-Mbp genome was generated, whereas the size of the nonmelanized strain (MA5789) genome was 27.73 Mbp. As *K. petricola* is the only species with a nonmelanized pink mutant that spontaneously originated under laboratory conditions, further studies based on genome and transcriptome sequence can help to shed light on melanin function.

Ametrano et al. [133] used a phylogenomic approach to resolve relationships in Dothideomycetes, focusing on two genera of melanized, extremotolerant RIF belonging to the widespread worldwide dothidealean genera *Lichenothelia* (*L. convexa* L1844 isolated from shale and *L. intermixta* L2282 isolated from limestone in Mojave Desert) and *Saxomyces* (*S. alpinus* CCFEE 5470 isolated from Italian Alps and *S. americanus* L1853 isolated from Mojave desert), which have been suggested to be early diverging lineages.

Further studies on black fungi are needed to elucidate the evolution, adaptation, and processes that govern their outstanding success in the extremes. The project "Shedding light in the Dark Lineages of the Fungal Tree of Life" (http://www.stresblackfungi.org/) funded by the Department of Energy–Joint Genome Institute (DOE JGI) aims to sequence up to 92 species of black fungal reference genomes, mostly from unsampled lineages, including RIF, for a comprehensive study of evolutionary and adaptation processes. Moreover, as it is still not possible to directly infer the involvement of genes in particular functions by using DNA sequences only, information obtained from genome sequencing will be complemented with transcriptomics and metabolomics experiments.

2.1.5 Monuments and role of black fungi in deterioration

Rock and mineral substrates used in buildings, monuments, statues, gravestones, and other constructions of cultural heritage are subject to fungal colonization and transformation termed "geomycology" [5]. Biological greyish-black patina is one of the most common SABs widespread on the surfaces of calcareous substrates and responsible for the serious phenomena of biodeterioration of outdoor exposed artworks. Microbial attack not only causes aesthetic changes but also produces mechanical and chemical damages affecting both surfaces and inner zones. The most frequent stone-colonizing agents are algae, cyanobacteria, bacteria, fungi, and lichens. In particular, epilithic and endolithic fungi play a major role in the weathering of monuments made of rock, ranging from discoloration and staining to biodeterioration and the formation of new biogenic minerals and rock coatings. Indeed, RIF as well as free-living filamentous species and lichens may significantly change the rock structure and appearance. A high bioreceptivity together with favorable environmental conditions can induce a rapid biological colonization. When living on monuments, some RIF are prone to cope with sun exposition, prolonged periods of drought, and nutrient deficiency, whereas others have a competitive advantage to thrive in polluted environments as monuments are often located in urban environments, developing the ability to assimilate toxic hydrocarbons [78]. Most of the blackening and brown patinas on marbles and limestones in monuments are primarily caused by air pollution, fly ash, and the oxidation of manganese and iron [134] as well by the presence and activity of melanotic fungi that have the capacity to settle on the surface of rocks, to attach firmly to the surface, and to penetrate deeper into the rock [135, 136], producing biopitting (the formation of lesions in a size range of up to 2 cm in diameter and depth on stone). Besides, the presence of melanin in the cell wall has been hypothesized to enhance the fungal turgor pressure, conferring penetration potential to these microorganisms. Moreover, EPS produced by fungi facilitates fungal biofilm formation and the attachment to the rock and increases mechanical pressure, giving rise to shrinking and swelling [137].

RIF represent a challenge for restorers because of the high resistance to many types of chemical treatment [138]; moreover, using chemical formulates with biocidal action may induce risks for the substrate, operator, and environment.

Synthetic materials used for the treatment of stone during conservation practices can favor the growth of fungi, representing an increase in nutrient sources. However, the application of biocide in combination with these compounds can reduce the risk of fungal colonization [139].

Sterflinger et al. [140] and Burford et al. [137] reported some lists of common RIF species isolated from building stone and rocks (sandstone, marble, granite, limestone, soapstone, quartzite, andesite, and basalt) in various geographical and climatic zones (Figs. 2.1.3 and 2.1.4).

In particular, the Mediterranean Basin represents an elective site for these investigations due to both the richness in cultural heritage and the conditions of temperatures and sun exposition that are particularly favorable for RIF development. RIF have been found in historical sites in Greece [140], more recently from Ukraine and Turkey [136], and, to a lesser extent, Italy [141–144]. These fungi have also been barely isolated from tropical climates [145, 146] and the Qinghai-Tibet plateaus [97]. *Hortaea*, *Exophiala*, *Coniosporium*, *Sarcinomyces*, *Capnobotryella*, and *Pseudotaeniolina* have been isolated from small black colonies visible on and inside the monuments, often occurring in close association with lichens [147].

A wide sampling of MCF has been from Italian monuments in historical sites characterized by air pollution and salt concentrations due to the proximity of the sea, including the Cortile della Pigna, the Vatican Museum and St. Peter's colonnade (Vatican City State), the monumental cemetery of Bonaria, and other monuments in the city of Cagliari. Based on multilocus phylogeny, the new genus and species *Lithohypha guttulata* and five new species *Knufia marmoricola*, *K. vaticanii*,

Fig. 2.1.3: Rock-inhabiting fungi (RIF) colonizing natural rock and historical stone monuments. (A–C) Temple of Hephaestus in Athens, Greece. (D–F) Monuments in Beijing in Guizhou Province, China [97].

K. karalitana, *K. mediterranea*, and *Exophiala bonariae* have been described in Chaetothyriales, whereas a new genus and species *Saxophila tyrrhenica* and two new species *Vermiconia calcicola* and *Devriesia sardiniae* have been described in Capnodiales [148].

In particular, the genus *Knufia* includes ecologically diverse species ranging from lichenicolous taxa as *K. peltigerae*, human opportunists as *K. epidermidis* [149], insect-associated species as *K. aspidioti* [150], and plant pathogens as *K. cryptophialidica* and *K. endospora* [151], whereas *K. petricola* was described as the remains one of the most recurrent species on stone monuments in the Mediterranean area [152]. *Exophiala* genus belongs to the family Herpotrichiellaceae that includes many human opportunists [153]. The ability of fungi in this family to metabolize hydrocarbons, particularly alkylbenzenes, may explain their success in colonizing anthropogenic polluted habitats [154]. *Vermiconia* genus exclusively comprises saxicolous species and has a worldwide distribution on high altitude in Spain, at the Alps, and in the Antarctic desert [61]. In similar environments, MCF in Eurotiomycetes were isolated from two very valuable statues exposed in Florence, the "Ratto delle Sabine" and the "Copia del David" [143], from the Vallerano cave [144], and from a marble statue exposed in the Gardens of the Quirinal Palace in Rome [142]. Other famous monuments covered and deteriorated by black fungi are the Acropolis of Athens, the marble monuments of the Crimea, and the antique temples of Delos [23]. Recently, members of both Capnodiales and Chaetothyriales have been recorded from 149 tombstones across three continents using high-throughput sequencing [155].

Although much data were recently accumulated, a lot remains to be done in terms of knowledge of the actual wideness of RIF biodiversity. Physical control methods, such as ionizing radiation, gamma radiation, and UV radiation, were applied on several artworks [156], as an alternative for the conservation of stone colonized by biodeteriogens. An innovative new physical controlling method has been proposed by Cuzman et al. [157] that used microwave radiation (nonionizing radiation), for the first time on *Sarcinomyces* sp., *Pithomyces* sp., and *Scolecobasidium* sp., revealing a loss of vitality heating for 3 min at 65°C. A multidisciplinary approach based on Dimethyl sulfoxide gel has been demonstrated to be efficient with no undesired side effects on the substrate and exhibited a number of advantages in the field as reported by Toreno et al. [158]. One important reason why fungi are a great problem for conservation of cultural heritage is a lack of information and training for restorers, curators, and other museum personnel. Nowadays, multidisciplinary approaches, based on applied mycology and biotechnology, represent invaluable tools to understand fungi thriving in stone monuments for present and future preservation and restoration of cultural heritage. Nevertheless, the use of modern -omics techniques for the detection of fungi on and in materials of cultural heritage will provide a deeper insight into and understanding of fungal community structures and their consequences for the material [159].

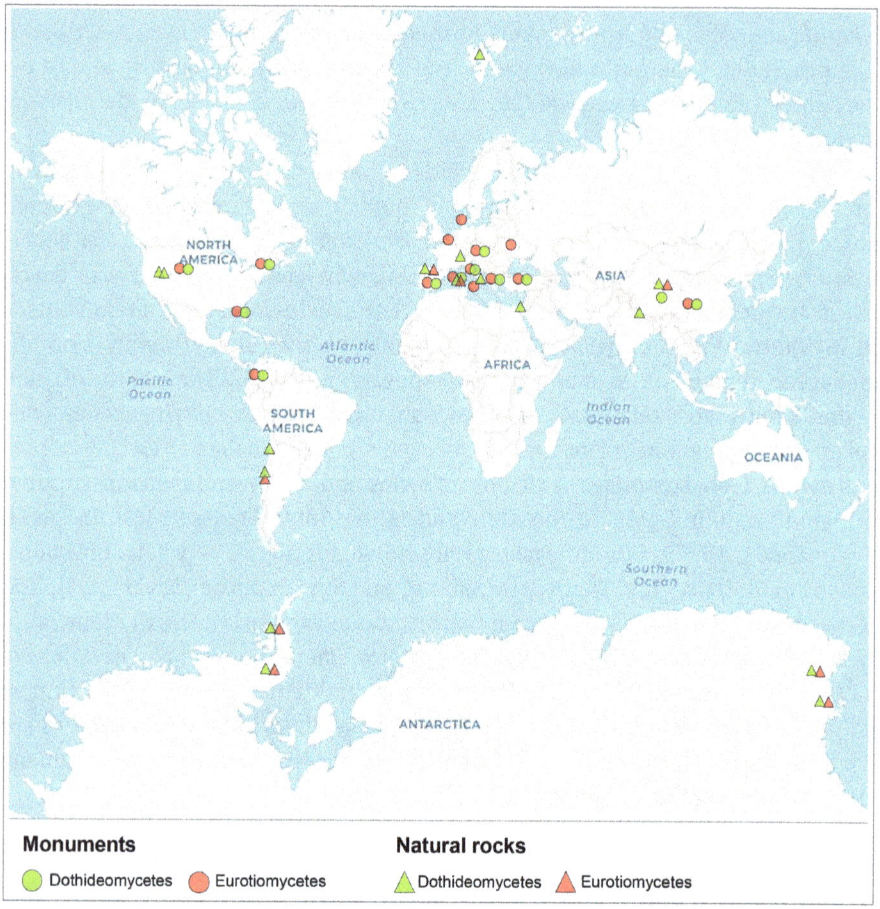

Fig. 2.1.4: Map of worldwide distribution of rock-inhabiting fungi (RIF). Map has been created using Quantum GIS v3.4.15 (https://qgis.org/en/site/forusers/download.html).

2.1.6 Potentialities

RIF have developed the unique ability to not only survive in the extremes environment (i.e., the Polar Regions, the damaged nuclear reactor in Chernobyl and oil-contaminated soil) but also actually thrive in such severe conditions. Indeed, melanization is associated with protection and adaptation to multiple chemical and mechanical stressors such as temperature, radiation, humidity, and toxicity by different pollutants, making melanin-producing microorganisms particularly useful to provide new tools for achieving a sustainable future and protecting humans from pollutants and radiation damage [160, 161]. Research on environmental melanotic fungi has far-reaching implications on numerous topics, including survival in extreme conditions, response to nuclear disasters, space travel, and global ecology, and it may have a huge application in different fields, including radioprotection, bioremediation, and biomedical applications.

Astrobiology, radioprotection, and biomedical application

Although the radiation-shielding performance of fungal melanins remains to be further untangled, our current understanding suggests that they present several advantages for protection against various forms of ionizing radiation and reducing toxicity. Electromagnetic radiation is both necessary (e.g., photosynthesis) and potentially harmful for life, depending on the frequency and time of exposure. Many deleterious effects are associated with exposure to high-energy electromagnetic wavelengths or ionizing radiation, including gamma, X-rays, and ultraviolet frequencies. Depending on the type and length of exposure, radiation can result cytotoxic reactive oxygen species (ROS) that damage intracellular molecules (i.e., DNA, proteins), likely leading to serious short- and long-term health problems to astronauts, including cancer, cataracts, acute radiation sickness, and neurological damage [162]. The role of melanins as photoprotective agents is widely known in the human skin, being capable of not only absorbing light but also dissipating the energy within the structure [163], making them unique in terms of energy harvesting [95]. Indeed, similar to the tanning of human skin, fungal melanogenesis is also stimulated by exposure to ionizing radiation and resulting in melanized strains, which are more resistant to radiotoxicity than their albino counterparts [164, 165]. Black fungi, indeed, can survive ionizing radiation levels that are lethal to the most life forms [108, 109, 166] with regard to the limits of life.

Melanin mediates radioprotection by both (i) absorbing radiation energy and dissipating it in the form of heat while limiting the generation of ROS and/or (ii) trapping and neutralizing the free radicals or ROS generated by the ionization of molecules. The first experimental evidence demonstrating a direct role of fungal melanin in radioprotection was provided by Dadachova et al. [46] in a study comparing survival rates of melanized and nonmelanized yeasts after exposure to very high doses of gamma radiation (1,600,000 times higher than the average lethal dose to humans). However, the photoprotection capacity may differ among melanin types, radiation frequencies, and types of irradiation exposure. It is also possible that photoprotection can turn into photodamage after a certain threshold of irradiation where melanin itself can produce cytotoxic radical species.

The discovery of melanized fungi in extreme environments has prompted researchers to consider whether melanotic fungi can survive in space and to test this in simulations as well as on actual space missions. The capacity of melanins to absorb or screen multiple types of ionizing radiation makes those biomolecules strong candidates for biomaterials for incorporation into irradiation protection in future manned space missions [167]. Melanized fungi have been found growing on the Mir Space Station and the International Space Station. These studies informed us on implications for space travel and colonization, extraterrestrial life, and habitability [168]. Zakharova et al. [77] demonstrated that black fungi, including both the Antarctic RIF *C. antarcticus* and *K. perforans* and *Exophiala jeanselmei*, were able to not only survive in Martian-simulated conditions but also undergo essential metabolism under such conditions. A recent study compared melanized and nonmelanized forms of *Cryptococcus neoformans* and *C. antarcticus* after irradiation with deuteron 2H particles and X-rays; the results show that melanized forms of both species are more resistant to 1.5 kGy deuteron dose (300,000 times higher than the human lethal dose) [169]. Understanding the molecular composition of fungal melanin could help to understand this peculiar capability. The melanin extracted from *C. antarcticus* has been recently characterized by various chemical, spectrophotometric, and spectroscopic techniques. Results demonstrated that, despite having a specific type of melanin as the majority of fungi, this black fungus possesses the ability to produce both DHN and L-DOPA melanins, opening interesting scenarios for the protection role against radiation [170].

Future space missions, including round trips to Mars, will require effective materials for protection against various forms of ionizing radiation. Existing solutions to these problems are often expensive, unscalable, and/or use materials that deplete natural resources. As melanin is a natural product, the abovementioned studies have largely encouraged the use of this pigment in applications in radioprotection of humans [171].

Finally, broadening our understanding of how melanin mediates its effects on biological systems is important for various reasons. By exploiting the shielding properties of melanin, this pigment could be used as cost-effective, sensitive biological detectors of nuclear accidents or protect individuals undergoing radiation treatments or exposed to nuclear accidents. Besides, uncovering the capacity for melanin-related energy transduction in fungi will allow us to consider the role that the fungi could contribute as energy generators, which could be of great importance as our global environment changes in a global warming scenario.

Bioremediation

Aromatic hydrocarbons like benzene, toluene, ethylbenzene, and the xylene isomers (collectively known as BTEX) from industrial processes present a major hazard, and their removal from the environment is difficult and costly [172]. Sustainable biological processes such as gas biofiltration present several advantages over conventional methods (i.e., physicochemical adsorption, condensation, and incineration), including the relatively lower costs of investment, performance, and maintenance. In recent decades, promising and cost-effective biological alternatives to physicochemical environmental remediation have emerged. For instance, the group of black fungi, particularly those species belonging to Chaetothyriales (Eurotiomycetes), have begun to be recognized for their bioremediation potential because these microorganisms can potentially bind radionuclides and many other toxic substances. Indeed, melanized fungi can tolerate acidic and dry conditions, making them attractive biocatalysts in air biofilters, as well as in the *in situ* bioremediation of polluted soils in extreme environments. Besides, the chelating properties of melanins toward free radicals and metals mediate protection and pose this group of fungi as powerful tools in bioremediation and other biotechnologies [173].

As mentioned above, the ability to metabolize aromatics seems to be restricted to the black yeasts belonging to the family Herpotrichiellaceae (order Chaetothyriales); the same group harbors numerous species able to cause infections in immunocompetent human hosts [78]. In these fungi, the assimilation of phenolic compounds and hydrocarbons may act as additional virulence factor because it may enable to infect the central nervous system, which has a high content of monoaromatic catecholamine neurotransmitters and aliphatic amino alcohols.

An ecological connection between aromatic hydrocarbon assimilation and human pathogenicity has been previously hypothesized [78]. Elucidation of ecological and evolutionary parameters is of utmost significance to prevent biohazard in engineered bioremediation applications that understand those factors of habitat choice that seem to enhance opportunism on the human host [174].

Several slow-growing black fungi related to rock-inhabiting species have been isolated from different highly polluted sources such as industrial spills, car gasoline tanks, railway sleepers, and air biofilters and can capture and degrade volatile aromatic hydrocarbons, including toluene, ethylbenzene, and styrene [175, 176]. These findings support the hypothesis that the presence of fungi inhabiting black rock in polluted sites is still underestimated. They may thrive in the polluted environments because of their polyextreme tolerance, scavenging carbon sources from the gas phase as they do when they colonize the rocky niches. Besides, a few species are able to use these compounds as their sole carbon source for growth [177]. The genera *Exophiala* and *Cladophialophora* have a high potential to grow in polluted environments and to metabolize hydrocarbons as the sole source of carbon and energy [175]. In particular, the genus *Exophiala* has frequently been isolated from sites polluted with hydrocarbons or with the aid of alkyl benzene enrichment [178] as well as from Antarctic endolithic ecosystems [128] and has been proven to be potential causative agents of human or animal disorders [179]. Therefore, it has been of paramount importance to develop and optimize isolation and screening methods for black fungi to control and prevent the spread of dangerous biohazard species. For instance, Isola and coworkers [154] developed a new isolation method and a screening technique based on a 2,6-dichlorophenolindophenol assay for black fungi, with potential use in bioremediation. Fifty-six black fungal strains (e.g., *Coniosporium uncinatum*, *Exophiala equina*, *Exophiala mesophila*, *Exophiala oligosperma*, *Exophiala sideris*, *Exophiala xenobiotica*, *K. epidermidis*, and *Ochroconis humicola*) were isolated from hydrocarbon-polluted sites and indoor environments (gasoline car tanks and washing machine soap dispensers). Some of the strains tested were able to use toluene as a carbon source. Among the studied strains, the authors proposed *E. xenobiotica* as the best model for such studies because of its phylogenetic position, its competence to degrade xenobiotics, and its ability to colonize humans, human-dominated environments, as well as rocks. More recently, Blasi et al. [176] developed a fast and efficient prescreening method with the aim of finding new fungal species with pollutant degradation abilities. The authors found that >150 strains of black yeastlike fungi, chosen as representatives of relevant environmental pollutants, were screened for the ability to grow on hexadecane, toluene, and polychlorinated biphenyl 126 as the sole carbon and energy source. In particular, *E. mesophila* (CBS 120910) and *Cladophialophora immunda* (CBS 110551), isolated from a patient with chronic sinusitis and a polluted soil sample, showed a high potential to grow in polluted environments and to metabolize hydrocarbons as the sole source of carbon and energy [180]. Zhao et al. [178] applied an enrichment technique based on solid state–like incubations in a controlled atmosphere containing monoaromatic volatile hydrocarbons, usually toluene, as sole carbon source, whereas a high-throughput methodology based on the solid state–like enrichment on volatile monoaromatic hydrocarbons for the isolation of black fungi has been proposed recently by Quan et al. [181].

The biological processes responsible for the xenobiotics-degrading ability of these fungi have not yet been studied with next-generation sequencing approaches. To shed light into the mechanisms of toluene tolerance and degradation, Blasi et al. [176] proposed the first genomic and transcriptomic analysis of *C. immunda* upon growth with toluene as sole carbon and energy source. The genes involved in toluene degradation and several stress response mechanisms, which allowed the fungus to survive the toluene exposure, were identified, and their expression levels during the toluene exposure were assessed. The comparison of the fungal and bacterial toluene degradation pathways revealed that a few genes in *C. immunda* involved toluene degradation; moreover, proteins involved in complex carbohydrates degradation and sugar transport are overrepresented, suggesting that this species evolved in an oligotrophic and polluted environment. Toluene further triggered the expression of antioxidants as well as genes involved in cell detoxification (e.g., glyoxylase II, glutathione S-transferase, and ascorbate peroxidase).

The rock-associated *Knufia chersonesos* has the ability to resort to recalcitrant carbon sources, making it an interesting candidate for degradation purposes. A secretome screening toward polyesterases was carried out, for the first time, for this fungus and its nonmelanized counterpart, grown in the presence of the synthetic polyester polybutylene adipate terephthalate (PBAT) as additional or sole carbon source, and it resulted in the identification of novel esterolytic and lipolytic enzymes across the established cultivation conditions.

Further studies, based on -omics high-throughput approaches can provide new insights into the ecophysiology of polymer degrading fungi and ultimately aid in the identification of novel enzymes with potential application in polymer processing, recycling, and degradation.

2.1.7 Conclusions and future research directions

Research on black RIF has implications for multiple fields such as microbiology, astrobiology, ecology, medicine, biotechnology, and bioremediation.

To date, the diversity, origin, and ecology of this fascinating group of fungi are beginning to be understood comprehensively by a combination of -omics approaches. In particular, further transcriptomic and metabolic analyses coupled with genomic and comparative genomic data may untangle their differential gene expressions and metabolic activity variations under stress (e.g., how rapidly the transcriptional responses occur under fluctuating hydration conditions).

Continued studies on RIF ecophysiology, resistance, and adaptation strategies will provide valuable insights into the origins and limits of Earth's life and shed light on the evolution of extremophiles.

Acknowledgments

Financial support was granted by the Italian National Program for Antarctic Research (PNRA) and the Italian Antarctic National Museum (MNA).

References

[1] Konopka JB, Casadevall A, Taylor JW, Heitman J, Cowen L. Onehealth: fungal pathogens of humans, animals, and plants. Colloquium report. Washington (DC): American Society for Microbiology; 2019.

[2] Wijayawardene NN, Hyde KD, Al-Ani LKT, Tedersoo L, Haelewaters D, Rajeshkumar KC, et al. Outline of fungi and fungus-like taxa. *Mycosphere*. 2020;11(1):1060–456.

[3] Das M, Royer TV, Leff LG. Diversity of fungi, bacteria, and actinomycetes on leaves decomposing in a stream. *Appl Environ Microbiol*. 2007;73:756–67.

[4] Spatafora JW, Aime MC, Grigoriev IV, Martin F, Stajich JE, Blackwell M. The fungal tree of life: from molecular systematics to genome-scale phylogenies. *Microbiol Spectr*. 2017;5(5).

[5] Gadd GM. Fungi, rocks, and minerals. *Elements*. 2017;13(3):171–6.

[6] Gorbushina AA. Life on the rocks. *Environ Microbiol*. 2007;9(7):1613–31.

[7] Golubic S, Friedmann EI, Schneider, J. The lithobiontic ecological niche, with special reference to microorganisms. *J Sediment Res*. 1981;51(2):475–8.

[8] Wierzchos J, Davila AF, Artieda O, Cámara-Gallego B, de los Ríos A, Nealson KH, et al. Ignimbrite as a substrate for endolithic life in the hyper-arid Atacama Desert: implications for the search for life on Mars. *Icarus*. 2013;224:334–46.

[9] Omelon CR, Pollard WH, Ferris FG. Chemical and ultrastructural characterization of high arctic cryptoendolithic habitats. *Geomicrobiol J*. 2006;23:189–200.

[10] Wong FK, Lau MC, Lacap DC, Aitchison JC, Cowan DA, Pointing SB. Endolithic microbial colonization of limestone in a high-altitude arid environment. *Microb Ecol*. 2010;59:689–99.

[11] Dong H, Rech JA, Jiang H, Sun H, Buck BJ. Endolithic cyanobacteria in soil gypsum: occurrences in Atacama (Chile), Mojave (United States), and Al-Jafr Basin (Jordan) deserts. *J Geophys Res Biogeosci*. 2007;112:G02030.

[12] Sigler WV, Bachofen R, Zeyer J. Molecular characterization of endolithic cyanobacteria inhabiting exposed dolomite in central Switzerland. *Environ Microbiol.* 2003;5:618–27.
[13] Hoppert M, Flies C, Pohl W, Günzl B, Schneider J. Colonization strategies of lithobiontic microorganisms on carbonate rocks. *Environ Geol.* 2004;46:421–8.
[14] Wierzchos J, Ascaso C, McKay CP. Endolithic cyanobacteria in halite rocks from the hyperarid core of the Atacama Desert. *Astrobiology.* 2006;6:415–22.
[15] Wierzchos J, de los Ríos A, Ascaso C. Microorganisms in desert rocks: the edge of life on Earth. *Int Microbiol.* 2012;15:171–81.
[16] Warren-Rhodes KA, Rhodes KL, Pointing SB, Ewing SA, Lacap DC, Gomez-Silva B, et al. Hypolithic cyanobacteria, dry limit of photosynthesis, and microbial ecology in the hyperarid Atacama Desert. *Microb Ecol.* 2006;52(3):389–98.
[17] Wierzchos J, DiRuggiero J, Vítek P, Artieda O, Souza-Egipsy V, Skaloud P, Tisza M, Davila AF, Vílchez C, Garbayo I, Ascaso C. Adaptation strategies of endolithic chlorophototrophs to survive the hyperarid and extreme solar radiation environment of the Atacama Desert. *Front Microbiol.* 2015;6:934.
[18] Gleeson DB, Kennedy NM, Clipson N, Melville K, Gadd GM, McDermott FP. Mineralogical influences on bacterial community structure on a weathered pegmatitic granite. *Microb Ecol.* 2006;51:526–34.
[19] Guillitte O. Bioreceptivity: a new concept for building ecology studies. *Sci Total Environ.* 1995;167(1–3):215–20.
[20] De Leo F, Criseo G, Urzi C. Impact of surrounding vegetation and soil on the colonization of marble statues by dematiaceous fungi. In: Proceeding of the Eighth International Congress on Deterioration and Conservation of Stone; 1996. p. 625–30.
[21] Staley JT, Palmer F, Adams JB. Microcolonial fungi: common inhabitants on desert rocks? *Science.* 1982;215(4536):1093–5.
[22] Gorbushina AA, Krumbein WE, Hamman CH. Role of fungi in the color change and biodeterioration of antique marbles. *Geomicrobiol J.* 1995;11:205–11.
[23] Diakumaku E, Gorbushina AA, Krumbein WE, Panina L, Soukharjevski S. Black fungi in marble and limestones—an aesthetical, chemical and physical problem for the conservation of monuments. *Sci Total Environ.* 1995;167(1–3):295–304.
[24] Wollenzien U, de Hoog GS, Krumbein WE, Urzì C. On the isolation of microcolonial fungi occurring on and in marble and other calcareous rocks. *Sci Total Environ.* 1995;167:287–94.
[25] Sterflinger K, Krumbein WE, Rullkötter J. Patination of marble, sandstone and granite by microbial communities. *Zeitschrift der deutschen geologischen Gesellschaft.* 1999:299–11.
[26] Anagnostidis K, Gehrmann CK, Gross M, Krumbein WE, Lisi S, Pantazidou A, et al. Biodeterioration of marbles of the Parthenon and Propylaea, Acropolis, Athens-associated organisms, decay and treatment suggestions. In: La conservation des monuments dans le bassin méditerranéen: actes du 2ème symposium international; 1991. p. 305–25.
[27] Sterflinger K, Prillinger H. Molecular taxonomy and biodiversity of rock fungal communities in an urban environment (Vienna, Austria). *Antonie van Leeuwenhoek.* 2001;80(3–4):275–86.
[28] Grote G, Krumbein WE. Microbial precipitation of manganese by bacteria and fungi from desert rock and rock varnish. *Geomicrobiol J.* 1992;10(1):49–57.
[29] Hirsch PFEW, Eckhardt FEW, Palmer Jr RJ. Fungi active in weathering of rock and stone monuments. *Can J Bot.* 1995;*73*(S1):1384–90.
[30] Sterflinger K. Fungi as geological agents. *Geomicrobiol J.* 2000;17:97–24.
[31] Selbmann L, De Hoog GS, Mazzaglia A, Friedmann EI, Onofri S. Fungi at the edge of life: cryptoendolithic black fungi from Antarctic desert. *Stud Mycol.* 2005;51(1):1–32.
[32] de Hoog GS, Hermanides-Nijhof EJ. The black yeasts and allied Hyphomycetes. *Stud Mycol.* 1977: 15.
[33] Sterflinger K. Black yeasts and meristematic fungi: ecology, diversity and identification. In: Biodiversity and Ecophysiology of Yeasts. Berlin: Springer; 2006. p. 501–14.

[34] Friedmann EI. Endolithic microorganisms in the Antarctic cold desert. *Science*. 1982;215(4536):1045–53.

[35] Abdel-Hafez SII. Studies on soil mycoflora of desert soils in Saudi Arabia. *Mycopathologia*. 1994;80:3e8.

[36] Adams BJ, Bardgett RD, Ayres E, Wall DH, Aislabie J, Bamforth S, et al. Diversity and distribution of Victoria Land biota. *Soil Biol Biochem*. 2006;38:3003–18.

[37] Gunde-Cimerman N, Zalar P, de Hoog S, Plemenitaš A. Hypersaline waters in salterns–natural ecological niches for halophilic black yeasts. *FEMS Microbiol Ecol*. 2000;32(3):235–40.

[38] Selbmann L, Isola D, Egidi E, Zucconi L, Gueidan C, de Hoog GS, Onofri S. Mountain tips as reservoirs for new rock-fungal entities: *Saxomyces* gen. nov. and four new species from the Alps. *Fungal Divers*. 2014;65:167–82.

[39] Selbmann L, Zucconi L, Isola D, Onofri S. Rock black fungi: excellence in the extremes. From the Antarctic to space. *Curr Genet*. 2015;61:335–45.

[40] Ruibal C, Selbmann L, Avci S, Martin-Sanchez P, Gorbushina A. Roof-inhabiting cousins of rock-inhabiting fungi: novel melanized microcolonial fungal species from photocatalytically reactive subaerial surfaces. *Life*. 2018;8(3):30.

[41] Perini L, Gostinčar C, Anesio AM, Williamson C, Tranter M, Gunde-Cimerman N. Darkening of the Greenland ice sheet: fungal abundance and diversity are associated with algal bloom. *Front Microbiol*. 2019;10:557.

[42] Matos T, de Hoog GS, de Boer AG, De Crom I, Haase G. High prevalence of the neurotrope *Exophiala dermatitidis* and related oligotrophic black yeasts in sauna facilities. *Mycoses*. 2002;45:373–7.

[43] Seyedmousavi S, Badali H, Chlebicki A, Zhao J, Prenafeta-Boldu FX, de Hoog GS. *Exophiala sideris*, a novel black yeast isolated from environments polluted with toxic alkyl benzenes and arsenic. *Fungal Biol*. 2011;115(10):1030–7.

[44] Blasi B, Tafer H, Tesei D, Sterflinger K. From glacier to sauna: RNA-Seq of the human pathogen black fungus *Exophiala dermatitidis* under varying temperature conditions exhibits common and novel fungal response. *PLOS One*. 2015;10(6):e0127103.

[45] Gümral R, Özhak-Baysan B, Tümgör A, Saraçlı MA, Yıldıran ŞT, Ilkit M, et al. Dishwashers provide a selective extreme environment for human-opportunistic yeast-like fungi. *Fungal Divers*. 2016;76(1):1–9.

[46] Dadachova E, Casadevall A. Ionizing radiation: how fungi cope, adapt, and exploit with the help of melanin. *Curr Opin Microbiol*. 2008;11(6):525–31.

[47] Selbmann L, Isola D, Zucconi L, Onofri S. Resistance to UV-B induced DNA damage in extreme-tolerant cryptoendolithic Antarctic fungi: detection by PCR assays. *Fungal Biol*. 2011;115:937–44.

[48] Onofri S, de Vera JP Zucconi L, Selbmann L, Scalzi G, Venkateswaran KJ, Rabbow E, de la Torre R, Horneck G. Survival of Antarctic cryptoendolithic fungi in simulated Martian conditions on board the International Space Station. *Astrobiology*. 2015;15:1052–9.

[49] Onofri S, Selbmann L, Pacelli C, Zucconi L, Rabbow E, de Vera JP. Survival, DNA, and Ultrastructural integrity of a cryptoendolithic antarctic fungus in Mars and lunar rock analogs exposed outside the International Space Station. *Astrobiology*. 2019;19(2):170–82.

[50] Pacelli C, Bryan RA, Onofri S, Selbmann L, Zucconi L, Shuryak I, Dadachova E. Survival and redox activity of *Friedmanniomyces endolithicus*, an Antarctic endemic black meristematic fungus, after gamma rays exposure. *Fungal Biol*. 2018;122(12):1222–7.

[51] Gorbushina AA, Kotlova ER, Sherstneva OA. Cellular responses of microcolonial rock fungi to long-term desiccation and subsequent rehydration. *Stud Mycol*. 2008;61:91–7.

[52] Sterflinger K. Temperature and NaCl-tolerance of rock-inhabiting meristematic fungi. *Antonie van Leeuwenhoek* 1998;74(4):271–81.

[53] Onofri S, Barreca D, Selbmann L, Isola D, Rabbow E, Horneck G, *et al.* Resistance of Antarctic black fungi and cryptoendolithic communities to simulated space and Martian conditions. *Stud Mycol*. 2008;61:99–109.

[54] Slepecky RA, Starmer WT. Phenotypic plasticity in fungi: a review with observations on *Aureobasidium pullulans*. *Mycologia*. 2009;101(6):823–32.
[55] Moreno LF, Vicente VA, de Hoog S. Black yeasts in the omics era: achievements and challenges. *Med Mycol*. 2018;56:S32,S41.
[56] Flieger K, Knabe N, Toepel J. Development of an improved carotenoid extraction method to characterize the carotenoid composition under oxidative stress and cold temperature in the rock inhabiting fungus *Knufia petricola* A95. *J Fungi*. 2018;4(4):124.
[57] Sterflinger K, Lopandic K, Pandey RV, Blasi B, Kriegner A. Nothing special in the specialist? Draft genome sequence of *Cryomyces antarcticus*, the most extremophilic fungus from Antarctica. *PLOS One*. 2014;9(10):e109908.
[58] Selbmann L, De Hoog GS, Zucconi L, Isola D, Ruisi S, van den Ende AG, et al. Drought meets acid: three new genera in a dothidealean clade of extremotolerant fungi. *Stud Micol*. 2008;61:1–20.
[59] Ruibal C, Gueidan C, Selbmann L, Gorbushina AA, Crous PW, Groenewald JZ, et al. Phylogeny of rock-inhabiting fungi related to Dothideomycetes. *Stud Mycol*. 2009;64:123–3.
[60] Gueidan C, Ruibal C, de Hoog GS, Schneider H. Rock-inhabiting fungi originated during periods of dry climate in the late Devonian and middle Triassic. *Fungal Biol*. 2011;115:987–96.
[61] Egidi E, De Hoog GS, Isola D, Onofri S, Quaedvlieg W, De Vries M, Verkly GJM, Stielow JB, Zucconi L, Selbmann L. Phylogeny and taxonomy of meristematic rock-inhabiting black fungi in the Dothideomycetes based on multi-locus phylogenies. *Fungal Divers*. 2014;65(1):127–65.
[62] Muggia L, Kocourková J, Knudsen K. Disentangling the complex of *Lichenothelia* species from rock communities in the desert. *Mycologia*. 2015;107(6):1233–53.
[63] Teixeira MM, Moreno LF, Stielow BJ Muszewska A, Hainaut M, Gonzaga L, Abouelleil A, et al. Exploring the genomic diversity of black yeasts and relatives (Chaetothyriales, Ascomycota). *Stud Mycol*. 2017;86:1–28.
[64] Ruibal C, Platas G, Bills GF. Isolation and characterization of melanized fungi from limestone formations in Mallorca. *Mycol Prog*. 2005;4:23–38.
[65] Ruibal C, Platas G, Bills GF. High diversity and morphological convergence among melanised fungi from rock formations in the Central Mountain System of Spain. *Persoonia*. 2008;21:93–110.
[66] Hubka V, Réblová M, Řehulka J, Selbmann L, Isola D, de Hoog SG, Kolařík M. *Bradymyces* gen. nov. (Chaetothyriales, Trichomeriaceae), a new ascomycete genus accommodating poorly differentiated melanized fungi. *Antonie van Leeuwenhoek*. 2014;106(5):979–92.
[67] Réblová M, Hubka V, Thureborn O, Lundberg J, Sallstedt T, Wedin M, Ivarsson M. From the tunnels into the treetops: new lineages of black yeasts from biofilm in the Stockholm metro system and their relatives among ant-associated fungi in the Chaetothyriales. *PLOS One*. 2016;11:e0163396.
[68] Gueidan C, Villaseñor CR, De Hoog GS, Gorbushina, AA, Untereiner WA, Lutzoni F. A rock-inhabiting ancestor for mutualistic and pathogen-rich fungal lineages. *Stud Mycol*. 2008;61:111–9.
[69] Sterflinger K, Gorbushina AA. Morphological and molecular characterization of a rock inhabiting and rock decaying dematiaceous fungus isolated from antique monuments of Delos (Cyclades, Greece) and Chersonesus (Crimea, Ukraine). *Syst Appl Microbiol*. 1997;20:329–35.
[70] Gorbushina AA, Krumbein WE. Subaerial microbial mats and their effects on soil and rock. In: Riding R, Awramik S, editors. Microbial sediments. Berlin: Springer; 2000. p. 161–70.
[71] Gorbushina AA, Whitehead K, Dornieden T, Niesse A, Schulte A, Hedges JI. Black fungal colonies as units of survival: hyphal mycosporines synthesized by rock-dwelling microcolonial fungi. *Can J Bot*. 2003;81(2):131–8.
[72] Gorbushina AA, Broughton WJ. Microbiology of the atmosphere-rock interface: how biological interactions and physical stresses modulate a sophisticated microbial ecosystem. Ann Rev Microbiol 2009;63.
[73] Ametrano CG, Selbmann L, Muggia L. A standardized approach for co-culturing Dothidealean rock-inhabiting fungi and lichen photobionts in vitro. *Symbiosis*. 2017;73(1):35–44.

[74] Hawksworth DL. *Lichenothelia*, a new genus for the *Microthelia aterrima* group. *Lichenologist.* 1981;13(2):141–53.
[75] Muggia L, Gueidan C, Knudsen K, Perlmutter G, Grube M. The lichen connections of black fungi. *Mycopathologia.* 2013;175(5–6):523–35.
[76] Sterflinger K, Tesei D, Zakharova K. Fungi in hot and cold deserts with particular reference to microcolonial fungi. *Fungal Ecol.* 2012;5:453–62.
[77] Zakharova K, Tesei D, Marzban G, Dijksterhuis J, Wyatt T, Sterflinger K. Microcolonial fungi on rocks: a life in constant drought? *Mycopathologia.* 2013;175(5–6):537–47.
[78] Prenafeta-Boldú FX, Summerbell R, de Hoog SG. Fungi growing on aromatic hydrocarbons: biotechnology's unexpected encounter with biohazard? *FEMS Microbiol Rev.* 2006;30:109–30.
[79] Nai C, Wong HY, Pannenbecker A, Broughton WJ, Benoit I, de Vries RP, et al. Nutritional physiology of a rock-inhabiting, model microcolonial fungus from an ancestral lineage of the Chaetothyriales (Ascomycetes). *Fungal Genet Biol.* 2013;56:54–66.
[80] Palmer RJ, Friedmann EI. Incorporation of inorganic carbon by antarctic cryptoendolithic fungi. *Polarforschung.* 1988;58(2/3):189–91.
[81] Moses V, Holm-Hansen O, Calvin M. Non photosynthetic fixation of carbon dioxide by three microorganisms. *J Bacteriol.* 1959;77:70–8.
[82] Selbmann L, de Hoog GS, Zucconi L, Isola D, Onofri S. Black yeasts in cold habitats. In: Cold-Adapted Yeasts. Berlin: Springer; 2014. p. 173–89.
[83] Ametrano CG, Muggia L, Grube M. Extremotolerant black fungi from rocks and lichens. In: Fungi in Extreme Environments: Ecological Role and Biotechnological Significance. Springer; 2019. p. 119–43.
[84] Gadd GM. Fungi, rocks, and minerals. *Elements.* 2017;13(3):171–6.
[85] Cordero RJ, Casadevall A. Functions of fungal melanin beyond virulence. *Fungal Biol Rev.* 2017;31(2):99–112.
[86] Fogarty RV, Tobin JM. Fungal melanins and their interactions with metals. *Enzyme Microb Technol.* 1996;19:311e317.
[87] Langfelder K, Streibel M, Jahn B, Haase G, Brakhage AA. Biosynthesis of fungal melanins and their importance for human pathogenic fungi. *Fungal Genet Biol.* 2003;38:143–58.
[88] de Hoog GS, Vicente VA, Najafzadeh MJ, Harrak MJ, Badali H, Seyedmousavi S. Waterborne *Exophiala* species causing disease in cold-blooded animals. *Persoonia.* 2011;27:46.
[89] Zhdanova NN, Zakharchenko VA, Vember VV, Nakonechnaya LT. Fungi from Chernobyl: mycobiota of the inner regions of the containment structures of the damaged nuclear reactor. *Mycol Res.* 2000;104(12):1421–6.
[90] Butler MJ, Lachance MA. Inhibition of melanin synthesis in the black yeast *Phaeococcomyces* sp. by growth on low pH ascorbate medium: production of spheroplasts from "albinized" cells. *Can J Microbiol.* 1987;33(2):184–7.
[91] Rosas ÁL, Casadevall A. Melanization affects susceptibility of *Cryptococcus neoformans* to heat and cold. *FEMS Microbiol Lett.* 1997;153(2):265–72.
[92] Schumacher J, Gorbushina AA. Light sensing in plant-and rock-associated black fungi. *Fungal Biol.* 2020
[93] Zhdanova NN, Pokhodenko VD. Possible participation of melanin pigment in protecting the fungal cell from desiccation. *Mikrobiologiia.* 1973;42(5):848.
[94] Kejžar A, Gobec S, Plemenitaš A, Lenassi M. Melanin is crucial for growth of the black yeast *Hortaea werneckii* in its natural hypersaline environment. *Fungal Biol.* 2013;117(5):368–79.
[95] Dadachova E, Bryan RA, Huang X, Moadel T, Schweitzer AD, Aisen P, Casadevall A. Ionizing radiation changes the electronic properties of melanin and enhances the growth of melanized fungi. *PLOS One.* 2007;2(5):e457.
[96] Coleine C, Stajich JE, de los Ríos A, Selbmann L. Beyond the extremes: rocks as ultimate refuge for fungi in drylands. *Mycologia.* 2020. doi 10.1080/00275514.2020.1816761.

[97] Sun W, Su L, Yang S, Sun J, Liu B, Fu R, et al. Unveiling the hidden diversity of rock-inhabiting fungi: Chaetothyriales from China. *J Fungi*. 2020;6(4):187.
[98] Zucconi L, Onofri S, Cecchini C, Isola D, Ripa C, Fenice M, Madonna S, Reboleiro-Rivas P, Selbmann L. Mapping the lithic colonization at the boundaries of life in Northern Victoria Land, Antarctica. *Polar Biol*. 2016;39(1):91–102.
[99] Friedmann EI, Ocampo R. Endolithic blue-green algae in the dry valleys: primary producers in the Antarctic desert ecosystem. *Science*. 1976;193(4259):1247–9.
[100] Wierzchos J, Ascaso C. Life, decay and fossilisation of endolithic microorganisms from the Ross Desert, Antarctica. *Polar Biol*. 2001;24(11):863–8.
[101] Wierzchos J, Ascaso C. Microbial fossil record of rocks from the Ross Desert, Antarctica: implications in the search for past life on Mars. *Int J Astrobiology*. 2002;1(1):51–9.
[102] Friedmann EI. Endolithic microbial life in hot and cold deserts. In: Limits of Life. Dordrecht: Springer; 1980. p. 33–45.
[103] Friedmann EI, Hua M, Ocampo-Friedmann R. Cryptoendolithic lichen and cyanobacterial communities of the Ross Desert, Antarctica. *Polarforschung*. 1988;58(2/3):251–9.
[104] Selbmann L, Egidi E, Isola D, Onofri S, Zucconi L, de Hoog GS, et al. Biodiversity, evolution and adaptation of fungi in extreme environments. *Plant Biosyst*. 2013;147(1):237–46.
[105] Selbmann L, Onofri S, Zucconi L, Isola D, Rottigni M, Ghiglione C, et al. Distributional records of Antarctic fungi based on strains preserved in the Culture Collection of Fungi from Extreme Environments (CCFEE) Mycological Section associated with the Italian National Antarctic Museum (MNA). *MycoKeys*. 2015;10:57–71.
[106] Selbmann L, Onofri S, Coleine C, Buzzini P, Canini F, Zucconi L. Effect of environmental parameters on biodiversity of the fungal component in lithic Antarctic communities. *Extremophiles*. 2017;21(6):1069–80.
[107] Coleine C, Stajich JE, Zucconi L, Onofri S, Pombubpa N, Egidi E, Francks A, Buzzini P, Selbmann L. Antarctic cryptoendolithic fungal communities are highly adapted and dominated by Lecanoromycetes and Dothideomycetes. *Front Microbiol*. 2018;9:1392.
[108] Onofri S, de la Torre R, de Vera JP, Ott S, Zucconi L, Selbmann L, Scalzi G, Venkateswaran KJ, Rabbow E, Sánchez Iñigo FJ, Horneck G. Survival of rock colonizing organisms after 1.5 years in outer space. *Astrobiology*. 2012;12:508–16.
[109] Onofri S, Selbmann L, Pacelli C, De Vera JP, Horneck G, Hallsworth JE, Zucconi L. Integrity of the DNA and cellular ultrastructure of cryptoendolithic fungi in space or Mars conditions: a 1.5-year study at the International Space Station. *Life*. 2018;8(2):23.
[110] Onofri S, de Vera JP Zucconi L, Selbmann L, Scalzi G, Venkateswaran KJ, Rabbow E, de la Torre R, Horneck G. Survival of Antarctic cryptoendolithic fungi in simulated Martian conditions on board the International Space Station. *Astrobiology*. 2015;15:1052,1059.
[111] Selbmann L, Pacelli C, Zucconi L, Dadachova E, Moeller R, de Vera JP, Onofri S. Resistance of an Antarctic cryptoendolithic black fungus to radiation gives new insights of astrobiological relevance. *Fungal Biol*. 2018;122(6):546–54.
[112] Pacelli C, Selbmann L, Zucconi, L, Coleine, C, de Vera JP, Rabbow E, Böttger U, Dadachova E, Onofri S. Responses of the black fungus *Cryomyces antarcticus* to simulated Mars and space conditions on rock analogs. *Astrobiology*. 2019;19(2):209–20.
[113] Coleine C, Zucconi L, Onofri S, Pombubpa N, Stajich, JE, Selbmann L. Sun exposure shapes functional grouping of fungi in cryptoendolithic Antarctic communities. *Life*. 2018 8(2),19.
[114] Coleine C, Pombubpa N, Zucconi L, Onofri S, Turchetti B, Buzzini P, Stajich JE, Selbmann L. Uncovered microbial diversity in antarctic cryptoendolithic communities sampling three representative locations of the Victoria Land. *Microorganisms*. 2020;8(6):942.
[115] Coleine C, Pombubpa N, Zucconi L, Onofri S, Stajich JE, Selbmann L. Endolithic fungal species markers for harshest conditions in the McMurdo Dry Valleys, Antarctica. *Life*. 2020;10(2):13.

[116] Kondratiuk TO, Beregova TV, Parnikoza IY, Kondratyuk SY, Thell A. Microscopic fungi of lithobiont communities of Argentine Islands Region: data from the 22nd Ukrainian Antarctic Expedition. *Acta Bot Hung*. 2020;62(1–2):49–68.

[117] Øvstedal DO, Lewis Smith RI. Lichens of Antarctica and South Georgia: a guide to their identification and ecology. Cambridge: Cambridge University Press; 2001.

[118] Selbmann L, Zucconi L, Onofri S, Cecchini C, Isola D, Turchetti B, Buzzini P. Taxonomic and phenotypic characterization of yeasts isolated from worldwide cold rock-associated habitats. *Fungal Biol*. 2014;118(1):61–71.

[119] Coleine C, Gevi F, Fanelli G, Onofri S, Timperio AM, Selbmann L. Specific adaptations are selected in opposite sun exposed Antarctic cryptoendolithic communities as revealed by untargeted metabolomics. *PLOS One*. 2020;15(5):e0233805.

[120] Sterflinger K, Krumbein WE, Rullkötter J. Patination of marble, sandstone and granite by microbial communities. *Zeitschrift der deutschen geologischen Gesellschaft*. 1999:299–311.

[121] De Leo F, Urzì C. Fungal colonization on treated and untreated stone surfaces. In: Molecular Biology and Cultural Heritage. Lisse (The Netherlands): Swets & Zeitlinger BV; 2003. p. 213–8.

[122] Nash TH, White SL, Marsh JE. Lichen and most distribution and biomass in hot desert 1385 ecosystems. *Bryologist*. 1977;80:470–79.

[123] Su L, Guo L, Hao Y, Xiang M, Cai L, Liu X. *Rupestriomyces* and *Spissiomyces*, two new genera of rock-inhabiting fungi from China. *Mycologia*. 2015;107(4):831–44.

[124] Butler MJ, Lachance MA. Inhibition of melanin synthesis in the black yeast *Phaeococcomyces* sp. by growth on low pH ascorbate medium: production of spheroplasts from "albinized" cells. *Can J Microbiol*. 1987;33(2):184–7.

[125] Choi J, Kim SH. A genome tree of life for the fungi kingdom. *Proc Natl Acad Sci USA*. 2017;114(35):9391–6.

[126] Coleine C, Masonjones S, Selbmann L, Zucconi L, Onofri S, Pacelli C, Stajich JE. Draft genome sequences of the antarctic endolithic fungi *Rachicladosporium antarcticum* CCFEE 5527 and *Rachicladosporium* sp. CCFEE 5018. *Genome Announc*. 2017;5(27).

[127] Coleine C, Masonjones S, Sterflinger K, Onofri S, Selbmann L, Stajich, JE. Peculiar genomic traits in the stress-adapted cryptoendolithic Antarctic fungus *Friedmanniomyces endolithicus*. *Fungal Biol*. 2020;124(5):458–67.

[128] Coleine C, Selbmann L, Masonjones S, Onofri S, Zucconi L, Stajich JE. Draft genome sequence of an Antarctic isolate of the black yeast fungus *Exophiala mesophila*. *Microbiol Resour Announc* 2019;8(19).

[129] Moreno LF, Mayer V, Voglmayr H, Blatrix R, Stielow JB, Teixeira MM, et al. Genomic analysis of ant domatia-associated melanized fungi (Chaetothyriales, Ascomycota). *Mycol Progr*. 2019;18(4),541–52.

[130] Tesei D, Tafer H, Poyntner C, Piñar G, Lopandic K, Sterflinger K. Draft genome sequences of the black rock fungus *Knufia petricola* and its spontaneous non melanized mutant. *Genome Announc*. 2017;5(44).

[131] Tsuneda A, Hambleton S, Currah RS. The anamorph genus *Knufia* and its phylogenetically allied species in *Coniosporium*, *Sarcinomyces*, and *Phaeococcomyces*. *Botany*. 2011;89:523–36.

[132] Tesei D, Quartinello F, Guebitz GM, Ribitsch D, Nöbauer K, Razzazi-Fazeli E, Sterflinger K. Shotgun proteomics reveals putative polyesterases in the secretome of the rock-inhabiting fungus *Knufia chersonesos*. *Sci Rep*. 2020;10(1):1–15.

[133] Ametrano CG, Selbmann L, Muggia L. A standardized approach for co-culturing Dothidealean rock-inhabiting fungi and lichen photobionts in vitro. *Symbiosis*. 2017;73(1):35–44.

[134] Krumbeln WE, Petersen K, Schellnhuber HJ. On the geomicrobiology of yellow, orange, red, brown and black films and crusts developing on several different types of stone and objects of art. In: G. Alessandrini, editor. Le pellicole ad ossalato. Milan: CNR Centro Gino Bozza; 1989. p. 379–80.

[135] Urzì C, De Leo F, De Hoog GS, Sterflinger K. Of microbes and art: the role of microbial communities in the degradation and protection of cultural heritage. In: Ciferri O, Tiano P, Mastromei P, editors. New York: G. Plenum; 2000. p. 3–19.
[136] Sert HB, Sümbül H, Sterflinger K. Microcolonial fungi from antique marbles in Perge/Side/Termessos (Antalya/Turkey). *Antonie van Leeuwenhoek.* 2007;91:217–27.
[137] Burford EP, Kierans M, Gadd GM. Geomycology: fungal growth in mineral substrata. *Mycologist.* 2003;17: 98–107.
[138] Isola D, Selbmann L, Meloni P, Maracci E, Onofri S, Zucconi L. Detrimental rock black fungi and biocides: a study on the Monumental Cemetery of Cagliari. In: Rogerio-Candelera MA, Lazzari M, Cano E, editors. Science and Technology for the Conservation of Cultural Heritage. London: CRC Press; 2013. p. 83–6.
[139] Salvadori O, Charola AE. Methods to prevent biocolonization and recolonization: an overview of current research for architectural and archaeological heritage. In: Charola AE, McNamara C, Koestler RJ, editors. Biocolonization of Stone: Control and Preventive Methods. Washington (DC): Smithsonian Institution Scholarly Press; 2011. p. 37–50.
[140] Sterflinger K. Fungi as geological agents. *Geomicrobiol J.* 2000;17:97–24.
[141] De Leo F, Urzì C, de Hoog GS. Two *Coniosporium* species from rock surfaces. *Stud Mycol.* 1999;43:70–79.
[142] De Leo F, Antonelli F, Pietrini AM, Ricci S, Urzì C. Study of the euendolithic activity of black meristematic fungi isolated from a marble statue in the Quirinale Palace's Gardens in Rome, Italy. *Facies.* 2019;65(2):18.
[143] Marvasi M, Donnarumma F, Frandi A, Mastromei G, Sterflinger K, Tiano P, Perito B. Black microcolonial fungi as deteriogens of two famous marble statues in Florence, Italy. Int Biodeter Biodegradation 2012;68:36–44.
[144] Zucconi L, Gagliardi M, Isola D, Onofri S, Andaloro MC, Pelosi C, *et al.* Biodeterioration agents dwelling in or on the wall paintings of the Holy Saviour's cave (Vallerano, Italy). *Int Biodeter Biodegradation.* 2012;70:40–6.
[145] Kumar R, Kumar AV. Biodeterioration of Stone in Tropical Environments. An overview. The Getty Conservation Institute, 1999.
[146] Resende MA Rezende, GC, Viana EV, Becker TW, Warscheid T. Acid production by fungi isolated from historic monuments in the Brazilian state of Minas Gerais. In: Gaylarde CC, de SaA' EL, Gaylarde PM, editors. Biodegradation & Biodeterioration in Latin America. Porto Alegre: Mircen/UNEP/UNESCO/ICRO-FEPAGRO/UFRGS; 1996. p. 65–7.
[147] Sterflinger K. Fungi as geologic agents. *Geomicrobiol J.* 2010;17:97–124.
[148] Isola D, Zucconi L. Onofri S, Caneva G, de Hoog GS, Selbmann L. Extremotolerant rock inhabiting black fungi from Italian monumental sites. *Fungal Divers.* 2016;76(1):75–96.
[149] Li DM Chen XR. A new superficial fungal infection caused by *Coniosporium epidermidis. J Am Acad Dermatol.* 2010;63:725–7.
[150] He F, Lin B, Sun J, Liu X. Knufia aspidiotus sp. nov., a new black yeast from scale insects. *Phytotaxa* 2013;153:39–50.
[151] Tsuneda A, Currah RS. *Knufia endospora*, a new clematiaceous hyphomycete from trembling aspen. *Reports TMI.* 2004;42:1–9.
[152] Onofri S, Anastasi A, Del Frate G, Di Piazza S, Garnero N, Guglielminetti M, Isola D, et al. Biodiversity of rock, beach and water fungi. *Plant Biosyst.* 2011;145:978–87.
[153] Badali H, Carvalho VO, Vicente V, Attili-Angelis D, Kwiatkowski IB, van den Ende AHG, de Hoog GS. *Cladophialophora saturnica* sp. nov., a new opportunistic species of Chaetothyriales revealed using molecular data. *Med Mycol.* 2009;47:51–66.
[154] Isola D, Selbmann, L, Meloni P, Maracci E, Onofri S, Zucconi L. Detrimental rock black fungi and biocides: a study on the Monumental Cemetery of Cagliari. Science and Technology for the Conservation of Cultural Heritage. London: CRC Press; 2013. p. 83–6.

[155] Brewer TE, Fierer N. Tales from the tomb: the microbial ecology of exposed rock surfaces. *Environ Microbiol.* 2018;20(3):958–70.
[156] Guiomar Carneiro Tomazello MF. Wiendl F. The applicability of gamma radiation to the control of fungi in naturally contaminated paper. *Restaurator.* 2009;16:2,93–9.
[157] Cuzman OA, Olmi, R, Riminesi C, Tiano P. Preliminary study on controlling black fungi dwelling on stone monuments by using a microwave heating system. *Int J Conserv Sci.* 2013;4(2).
[158] Toreno G, Isola D, Meloni P, Carcangiu G, Selbmann L, Onofri S, et al. Biological colonization on stone monuments: a new low impact cleaning method. *J Cult Herit.* 2018;30:100–9.
[159] Adamiak J, Otlewska A, Tafer H, Lopandic K, Gutarowska B, Sterflinger K, Piñar G. First evaluation of the microbiome of built cultural heritage by using the Ion Torrent next generation sequencing platform. *Int Biodeter Biodegradation.* 2018;131:11–18.
[160] Revskaya E, Chu P, Howell RC, Schweitzer AD, Bryan RA, *et al.* Compton scattering by internal shields based on melanin-containing mushrooms provides protection of gastrointestinal tract from ionizing radiation. *Cancer Biother Radiopharm.* 2012;27;570–6.
[161] Cordero RJ, Vij R, Casadevall A. Microbial melanins for radioprotection and bioremediation. *Microb Biotechnol.* 2017;10(5):1186.
[162] Durante M, Cucinotta FA. Physical basis of radiation protection in space travel. *Rev Mod Phys.* 2011;83:1245–81.
[163] Wolbarsht ML, Walsh AW, George G. Melanin, a unique biological absorber. *Appl Opt.* 1981;20:2184–6.
[164] Wang Y, Casadevall A.. Decreased susceptibility of melanized *Cryptococcus neoformans* to UV light. *Appl Environ Microbiol.* 199460:3864e3866.
[165] Selbmann L, Isola D, Zucconi L, Onofri, S. Resistance to UV-B induced DNA damage in extreme-tolerant cryptoendolithic Antarctic fungi: detection by PCR assays. *Fungal Biol.* 2011;115:937e944.
[166] Dighton J, Tugay T, Zhdanova N. Fungi and ionizing radiation from radionuclides. *FEMS Microbiol Lett.* 2008 281:109–20.
[167] Cordero RJ. Melanin for space travel radioprotection. *Environ Microbiol.* 2017;19(7):2529–32.
[168] Gomoiu I, Chatzitheodoridis E, Vadrucci S, Walther I, Cojoc R. Fungal spores viability on the International Space Station. *Orig Life Evol Biosph.* 2016;46:403–18.
[169] Pacelli C, Bryan RA, Onofri S, Selbmann L, Shuryak I, *et al.* Melanin is effective in protecting fast and slow growing fungi from various types of ionizing radiation. *Environ Microbiol.* 2017;9:1612–24.
[170] Pacelli C, Cassaro A, Maturilli A, Timperio AM, Gevi F, Cavalazzi B, *et al.* Multidisciplinary characterization of melanin pigments from the black fungus Cryomyces antarcticus. *Appl Microbiol Biotechnol.* 202:01–11.
[171] Schweitzer AD, Revskaya E, Chu P, Pazo V, Friedman M, *et al.* Melanin-covered nanoparticles for protection of bone marrow during radiation therapy of cancer. *Int J Radiat Oncol Biol Phys.* 2010;78:1494–502.
[172] Irvine RL, Sikdar SK. Biodegradation Technology Developments: Principles and Practice. Vol 2. Florida: CRC Press; 1997.
[173] Apte M, Girme G., Bankar A, Ravikumar A, Zinjarde S. 3,4-Dihydroxy-L-phenylalanine-derived melanin from *Yarrowia lipolytica* mediates the synthesis of silver and gold nanostructures. *J Nanobiotechnol.* 2013;11:2.
[174] Vicente VA, Attili-Angelis D, Pie MR, Queiroz-Telles F, Cruz LM, Najafzadeh MJ, de Hoog GS, Zhao J, Pizzirani-Kleiner A. Environmental isolation of black yeast-like fungi involved in human infection. *Stud Mycol.* 2008;61:137–44.
[175] Blasi, B, Poyntner, C, Rudavsky T, Prenafeta-Boldú FX, De Hoog S, Tafer H, Sterflinger K. Pathogenic yet environmentally friendly? Black fungal candidates for bioremediation of pollutants. *Geomicrobiology J.* 2016;33(3–4):308–17.

[176] Blasi B, Tafer H, Kustor C, Poyntner C, Lopandic K, Sterflinger, K. Genomic and transcriptomic analysis of the toluene degrading black yeast *Cladophialophora immunda*. *Sci Rep*. 2017;7(1):1–13.
[177] Prenafeta-Boldú FX, Kuhn A, Luykx DM, Anke H, Van Groenestijn JW, De Bont JA. Isolation and characterisation of fungi growing on volatile aromatic hydrocarbons as their sole carbon and energy source. *Mycol Res*. 2001;105:477–84.
[178] Zhao J, Zeng J, de Hoog GS, Attili-Angelis D, Prenafeta-Boldú FX. Isolation of black yeasts by enrichment on atmospheres of monoaromatic hydrocarbons. *Microbiol Ecol*. 2010;60:149–56.
[179] De Hoog GS, Zeng JS, Harrak MJ, Sutton DA. *Exophiala xenobiotica* sp. nov., an opportunistic black yeast inhabiting environments rich in hydrocarbons. *Antonie Van Leeuwenhoek*. 2006;90:257–68.
[180] April TM, Abbott SP, Foght JM, Currah RS. Degradation of hydrocarbons in crude oil by the ascomycete *Pseudallescheria boydii* (Microascaceae). *Can J Microbiol*. 1998;44:270,278.
[181] Quan Y, van den Ende BG, Shi D, Prenafeta-Boldú FX, Liu Z, Al-Hatmi AM, et al. A comparison of isolation methods for black fungi degrading aromatic toxins. *Mycopathologia*. 2019;184(5):653–60.

Jan Kaštovský, Tomáš Hauer and Jiří Komárek

2.2 Cyanobacteria on rock surfaces

Despite their obvious presence on terrestrial rock surfaces worldwide, rock-inhabiting cyanobacteria are not very well known. In their habitats, these cyanobacteria represent important primary producers and often prepare the ground for further organisms. In this contribution, 53 cyanobacterial genera that include representatives of rock-inhabiting species are itemized in a determination key, and the genera are described on a global scale.

2.2.1 Short introduction

In comparison with the giant amount of studies on planktic types, the quantity of papers on rock-inhabiting cyanobacteria is rather low. It is understandable, for most people, that this kind of cyanobacteria represent less important entity. Nevertheless, in their habitats, these cyanobacteria represent important primary producers as their aquatic counterparts. In terms of aesthetics, however, aerial cyanobacteria surpass their water relatives: the variety of morphological shapes competes with the variety of photoprotective pigments (Figs. 2.3.1 and 2.3.2).

This chapter provides a guide for the determination of (sub)aerial cyanobacteria living on rocks and stones. Different aspects of their lives are either described in other chapters of this book or in a review [1]. Currently, it is highly important to accompany reviews like this with the following information: molecular-biological methods have brought many new impulses to the taxonomy and phylogenetics of cyanobacteria. They especially demonstrated the limits and incompleteness of our knowledge of the real diversity of this group. Various big genera fall into smaller ones (*Leptolyngbya*, *Phormidium*, and *Nostoc*), and various completely new taxa emerge (*Chalicogloea*, *Inacoccus*, *Lightfootiella*, and many others). So far, only a minority of cyanobacterial genera have undergone thorough modern evaluation; we are still very far away from the detail knowledge of the evolution of this group. It is even more true for the aerial taxa than for the planktic ones. We already know that the taxonomic importance of many morphological traits should not be overestimated; however, at the same time, the morphological traits are the only clue available for the routine floristic-ecological survey. The following determination compendium thus reflects the current state of knowledge on diversity in the dynamically evolving cyanobacterial world of the 21st century. Listed are the most typical aerial epilithic representatives.

The following text includes cyanobacteria from various "rocks" in full ecological meaning of this word—not only "common rocks" but also specific atmophytic (in reach of water vapor from thermal springs) or cave communities, as well as communities of marine splashing zones (supralittoral).

https://doi.org/10.1515/9783110646467-004

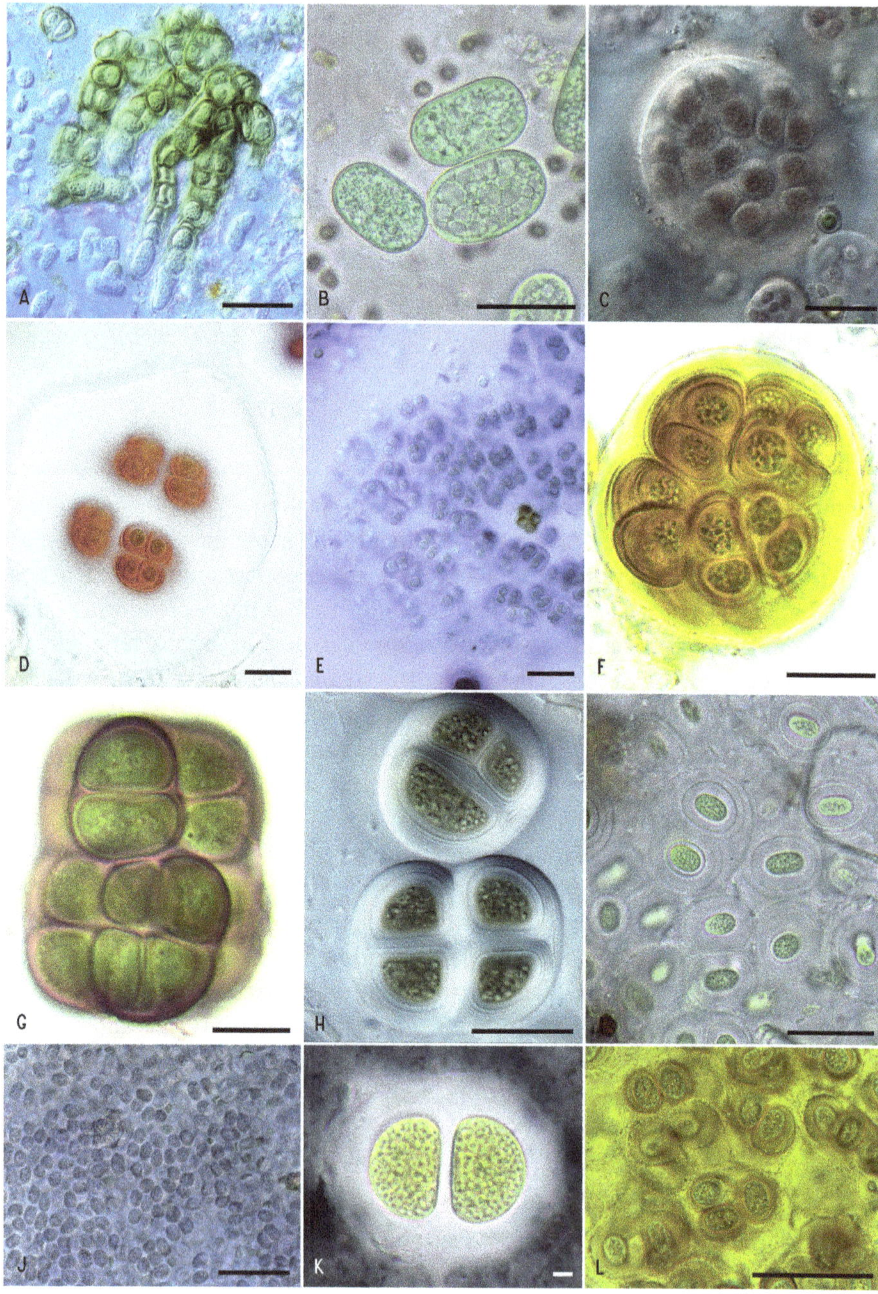

Fig. 2.2.1: (A) *Hyella balani* (courtesy of Alžběta Vondrášková). (B) *Cyanothece aeruginosa*. (C) *Asterocapsa* sp. (courtesy of Zuzana Hůlková). (D) *Gloeocapsa ralfsii*. (E) *Gloeocapsa atrata*. (F) *Gloeocapsa rupestris*. (G) *Gloeocapsopsis chroococcoides*. (H) *Chroococcus tenax*. (I) *Gloeothece tepidariorum*. (J) *Aphanothece* sp. (K) *Chroococcus spelaeus*. (L) *Gloeothece rupestris*. Scale bar length: 20 µm.

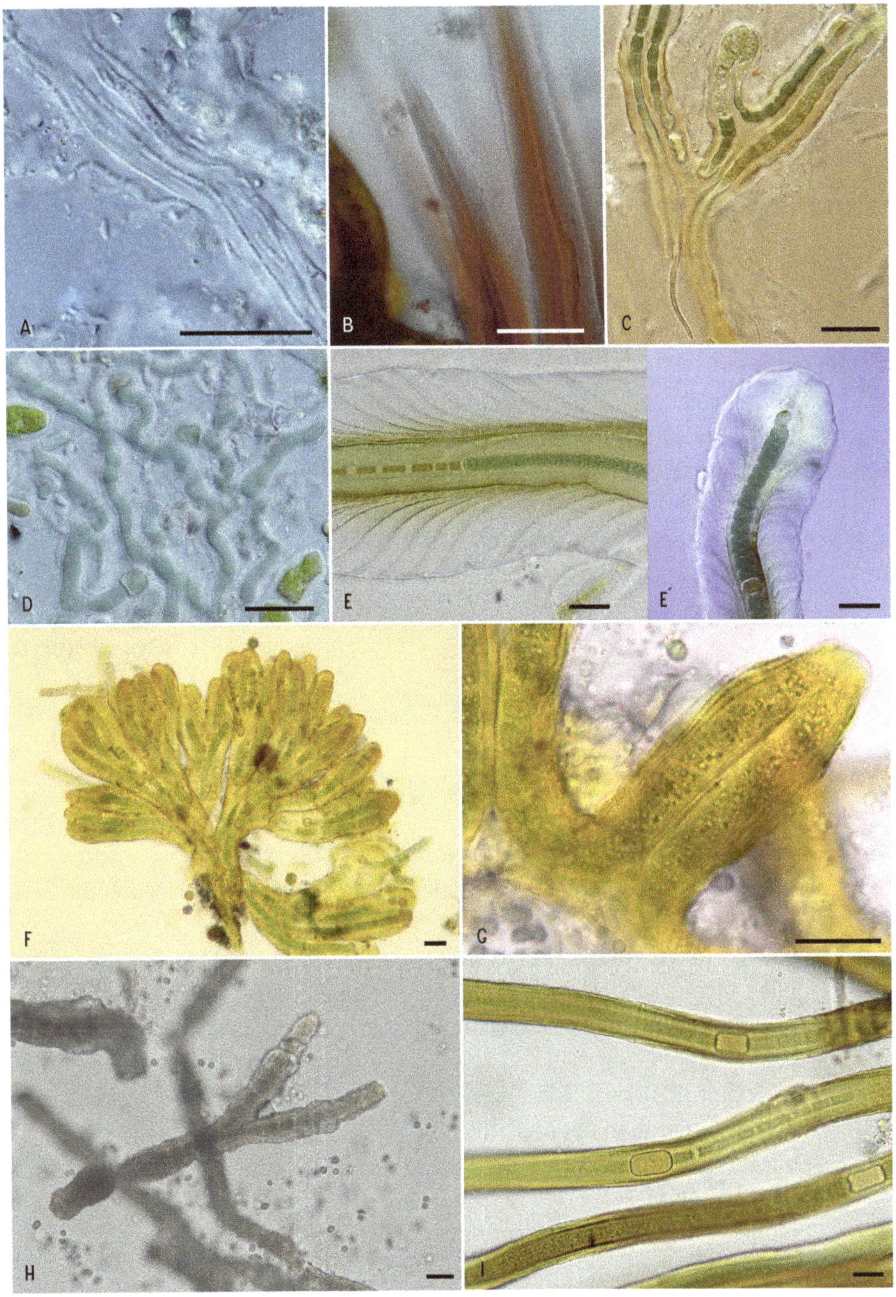

Fig. 2.2.2: (A) *Schizothrix* sp. (B) *Schizothrix latissimus*. (C) *Kyrtuthrix dalmatica* (courtesy of Alžběta Vondrášková). (D) *Mastigocoleus testarum* (courtesy of Alžběta Vondrášková). (E) *Petalonema alatum* (E′, terminal part of the filament). (F) *Tolypothrix elenkinnii*. (G) *Petalonema incrustans*. (H) *Geitleria calcarea*. (I) *Scytonema* sect. *Myochrotes* sp. Scale bar length: 20 µm.

The chapters are divided to sections of coccoid, simple trichal, and heterocytous taxa for practical reasons only. For the real phylogenetic system, see Komárek et al. [2].

2.2.2 Key to genera

In cases where there are only molecular differences, we note all possible genera.

Coccoid types

1a	Solitary cells with or without thin mucilaginous envelopes, never forming colonies or pseudofilaments	2
1b	Cells forming colonies or pseudofilaments, often with mucilaginous envelopes	5
2a	Cells spherical or hemispherical, dividing in two plains	*Synechocystis*
2b	Cells oval or cylindrical, dividing in one plain	3
3a	Cells cylindrical, at least twice as long as wide, sometimes bent	*Synechococcus/Thermosynechococcus*
3b	Cells not distinctly elongated	4
4a	Cells widely oval to short cylindrical, usually wider than 10 μm, often with net-like structured cell content	*Cyanothece*
4b	Cells cylindrical to widely oval with distinctly lengthwise-striated protoplast	*Cyanobacterium*
5a	Cells more or less elongated, heteropolar, usually with distinct sheath, attached to the substrate by their base, dividing in one plane, sometimes forming colonies	*Chamaesiphon*
5b	Cells not heteropolar	6
6a	Cells ± cylindrical, oval, ellipsoid, or rod shaped, arranged in spherical to amorphous but not polarized colonies, never forming pseudofilaments	7
6b	Cells (except for cell division) and colonies of other shapes	13
7a	Cells ellipsoid to oval, forming net-like colonies	*Lithococcus*
7b	Colonies other than net-like	8
8a	Cells elongated, cylindrical with not attenuated rounded ends, forming tube-like colonies	*Bacularia*
8b	Colonies are other shaped, cells are not arranged in one direction within the colony	9
9a	Cells solitary or in pairs on top of stratified mucilaginous stalks	*Hormothece*
9b	Cells in ± spherical or amorphous colonies	10
10a	Colonies as well as individual cell envelopes spherical with granular to spinylike surface, cells widely oval to irregular	*Asterocapsa*
10b	Colonies nor cell envelope's surface structured	11

11a Cells widely oval to rod shaped, usually with granules near both poles, mucilaginous envelopes, if present, always around particular cells, not common *Gloeobacter*

11b Cells never rod shaped, common mucilage present at least in part of the colony 12

12a All cells with distinct, often stratified, individual envelopes *Gloeothece*

12b All cells in common mucilage, only those in colony margin with ± distinct individual envelopes *Aphanothece*

13a Cells spherical, only after division hemispherical, mucilaginous envelopes around cells and subcolonies never stratified 14

13b Cells of different shapes or mucilaginous envelopes around cells or subcolonies stratified at least in part of colony 16

14a Flat, single-layered colonies with ± regularly arranged cells *Merismopedia*

14b Cells forming other than flat, single-layered colonies 15

15a ± cubic colonies with regularly arranged cells and subcolonies *Eucapsis*

15b Amorphous colonies *Aphanocapsa*

16a Cell always clearly elongated, spindle-like or at least attenuated at their ends, not forming tubelike colonies *Rhabdogloea*

16b Cells and colonies of different appearance 17

17a Cells spherical (hemispherical to oval during cell division), distant one from each other within the colony, with individual, usually stratified, mucilaginous envelopes, often colored 18

17b Cells, colonies, and envelopes of other appearance 19

18a Colonies spherical to amorphous, subcolonies ± spherical, gelatinous envelopes colored (yellow, red, blue, violet, black) in colonies/colony parts exposed to light, arthrospores are produced *Gloeocapsa*

18b Colonies spherical to amorphous, subcolonies ± spherical, gelatinous envelopes colored (orange to red), arthrospores are not produced *Lightfootiella*

19a Colonies ± hemispherical, subcolonies ± polygonal, densely attached to each other, gelatinous envelopes yellow to yellow-brown, cells spherical *Chondrocystis*

19b Cells and colonies of different shapes 20

20a Cells spherical, only after division hemispherical, dividing in two or more plains, on top of mucilaginous stalks *Cyanostylon*

20b Cells not placed on top of mucilaginous stalks 21

21a Individual cells spherical, in colonies polygonal, densely packed, colonies spherical, with colorless or slightly golden envelopes, no baeocytes production *Sinocapsa*

21b Cells and/or colonies of other shapes 22

22a Cells polygonal to almost spherical, colonies irregular, without common envelopes *Chalicogloea*

22b Cells and/or colonies of other shapes 23

23a Cells polygonal to hemispherical, colonies colored (red, yellow, violet) or colorless (then the cells are intensely blue) *Gloeocapsopsis*
23b Colonies not colored 24
24a Cell kidney shaped, colonies spherical with thick hyaline outer envelope *Nephrococcus*
24b Cells and/or colonies of other shapes 25
25a Cells spherical in basal, amorphous parts of colonies, cells in apical, pseudofilamentous parts of colonies ± oval to elongated, sheaths in apical parts open *Epilithia*
25b Cells and/or colonies of other shapes 26
26a Cells spherical to irregular with thin, delimited, unstructured sheaths, densely arranged in ± spherical colonies with thin firm envelopes *Chroococcidiopsis*
26b Cells and/or colonies of other shapes 27
27a Colonies ± spherical to lobate, usually hollow, ± spherical to irregular cells arranged near surface of the colonies *Placoma*
27b Cells and/or colonies of other shapes 28
28a Cells spherical, with individual envelopes, distant from each other, arranged in colonies in form of parallel rows *Lithocapsa*
28b Cells and/or colonies of other shapes 29
29a Cells spherical to oval, arranged in irregular rows in marginal parts of ± spherical to irregular colonies *Chlorogloea*
29b Cells and/or colonies of other shapes 30
30a Colonies polarized, tree shaped *Cyanoarbor*
30b Colonies of other shapes 31
31a Colonies slightly polarized, amorphous, when old composed of subcolonies arranged in irregular rows, cells spherical to irregular with individual envelopes *Entophysalis*
31b Cells and/or colonies of other shapes 32
32a Epilithic types, never forming tube-like or pseudofilamentous colonies 33
32b Epilithic and endolithic types forming tube-like, sac-like, or pseudofilamentous structures (morphology of epilithic and endolithic part of one thallus may differ!) 35
33a Packet-like colonies with thin, firm, delimited envelopes, cells spherical, hemispherical to irregular, adjacent tightly, no baeocytes production *Cyanosarcina*
33b Cells and/or colonies of other shapes 34
34a Cells ± hemispherical to polygonal, in colonies arranged radially *Pseudocapsa*
34b Young cells spherical, later hemispherical, mucilaginous envelopes following cell outlines, often stratified, cells in colonies not arranged radially *Chroococcus/Inacoccus*

35a Epilithic, amorphous colonies composed of drop- or pear-shaped cells in stratified mucilaginous envelopes *Ercegovicia*
35b Cells and/or colonies of other shapes 36
36a Epilithic crustose colonies with mono- or polyseriate pseudofilamentous, sometimes pseudodichotomously branched protuberances, cells irregular *Pleurocapsa*
36b Cells and/or colonies of other shapes 37
37a Endolithic, sac-like colonies of up to 12 oval to irregularly rounded cells *Cyanosaccus*
37b Colonies of different shape 38
38a Epilithic, ± oval to packet-like colonies composed of subcolonies, cells rounded polygonal to irregular *Myxosarcina*
38b Cells and/or colonies of other shapes 39
39a Epilithic, irregular colonies of ± spherical, drop- or pear-like cells in mucilaginous envelopes, arranged in irregular rows, marine types *Podocpasa*
39b Cells and/or colonies of other shapes 40
40a Pseudofilaments creeping on substrate or boring into it, clearly polarized; when old, forming amorphous part of the thallus *Hyella*
40b Thalli of different appearance 41
41a Endolithic part of thalli uniseriate, with cells considerably distant from each other enclosed in gelatinous stalks, epilithic part of thalli in form of calcified pseudofilaments *Solentia*
41b Thalli of different appearance 42
42a Endolithic part of thalli mostly multiseriate, epilithic part of thalli in form of amorphous gelatinous mass *Dalmatella*
42b Thalli of different appearance 43
43a Cells in endolithic part close to each other, epilithic parts reduced *Tryponema*
43b Cells in endolithic part distant from each other epilithic part well developed *Hormathonema*

Simple trichal types

1a Trichomes maximally 8–16 celled, constricted at cross walls *Borzia*
1b Trichomes longer if not fragmented and/or not constricted 2
2a Trichomes moniliform, without individual envelopes, enclosed in spherical mucilaginous colonies, cells mostly spherical to barrel shaped, terminal cells rounded *Yonedaella*
2b Organisms of different appearance 3
3a Trichomes heteropolar, tapering toward apical ends 4
3b Trichomes usually isopolar 6
4a Cells always discoid, only in apical parts (terminal hair) elongated *Homoeothrix*

4b	Cells ± isodiametric, shorter or longer than wide	5
5a	Parietal thylakoids, centro- and chromatoplasma often distinguishable	*Tapinothrix*
5b	Irregular thylakoids, cell content ± homogenous	*Phormidiochaete*
6a	Several trichomes enclosed in common sheath, with or without individual sheaths	7
6b	One trichome per sheath (if present), without common sheath	10
7a	Cells always shorter than wide, usually shorter than ½ of their width	*Blennothrix*
7b	Organisms of different appearance	8
8a	Trichomes motile, densely entangled in common sheath, sheath not tapering toward ends, cells ± isodiametric, terminal cells sometimes with calyptra	*Microcoleus* (part)
8b	Cells distinctly longer, terminal cells never with calyptra	9
9a	Trichomes with individual sheath, common sheath usually tapering toward end	*Schizothrix*
9b	Trichomes without individual sheath, densely entangled, common sheath not tapering toward end	*Trichocoleus*
10a	Cells always shorter than wide, usually shorter than ½ of their width (discoid)	11
10b	Cells in the trichome not always shorter than wide	12
11a	Trichomes without sheaths (in nature)	*Oscillatoria*
11b	trichomes in firm, later usually thick and lamellated, outside sometimes granular sheaths	*Lyngbya*
12a	Trichomes in thick, stratified and colored sheath	*Porphyrosiphon*
12b	Trichomes in other kind of sheath or without sheath	13
13a	Filaments entangled in dense, macroscopic, dark green erect fascicles	*Symploca*
13b	Organisms of different appearance	14
14a	Cells ± isodiametric or shorter than wide, with or without sheath, calyptra present in short-celled types only	*Phormidium*
14b	Organisms of different appearance	15
15a	Cells ± isodiametric, trichomes gradually narrowed toward ends, with calyptra on terminal cells	*Microcoleus* (part)
15b	Organisms of different appearance	15
16a	Cells ± isodiametric or shorter or longer than wide, terminal cells with thickened cell wall	*Pycnacronema*
16b	Organisms of different appearance	17
17a	Trichomes without sheath, but sometimes in diffluent slime	18
17b	Trichomes with sheath	20
18a	Trichomes moniliform	*Komvophoron*
18b	Trichomes not moniliform	19

19a Trichomes distinctly constricted on cross walls *Pseudanabaena*
19b Trichomes unconstricted *Jaaginema*
20a Filaments richly branched *Pseudophormidium*
20b Filaments unbranched or seldom branched 21
21a Trichomes unbranched, gradually narrowed and sometimes bent at ends, terminal cells rounded, without calyptra or thickened cell wall *Kamptonema*
21b Organisms of different appearance 22
22a Trichomes with yellow-brown sheath, cells longer than wide *Scytolyngbya*
22b Organisms of different appearance 23
23a Trichomes in thin sheath, constricted at cross walls, rarely branching, cells longer than wide, terminal cells with prominent rhodopsin granule at apical end *Oculatella*
23b Filaments thin, trichomes constricted or unconstricted, cells with variable length-width ratio *Leptolyngbya* and morphologically similar genera based on molecular differences: *Albertania, Cartusia, Kaiparowitsia, Komarkovaea, Kovacikia, Lusitaniella, Nodosilinea, Pegethrix, Tildeniella*, and *Timaviella*

Heterocytous types or types with akinetes

1a Isopolar unbranched filaments 2
1b Heteropolar and/or branched filaments 7
2a Cells of variable shapes, irregularly arranged in ensheathed filaments with transition to *Pseudanabaena*-like trichomes with conical cells, akinetes single or in series, thermal type *Cyanocohniella*
2b Organisms of different appearance 3
3a Filaments (if not broken) with terminal heterocytes on both ends, akinetes often with structured epispore exclusively attached to heterocytes *Cylindrospermum*
3b Organisms of different appearance 4
4a Filaments always without heterocytes, but producing akinetes *Isocystis*
4b Organisms of different appearance 5
5a Amorphous colonies up to 2 mm in diameter composed of tightly entangled filaments arranged in irregular spheres distant from each other *Compactonostoc*
5b Filaments not tightly entangled 6
6a Amorphous colonies with firm periderm on surface, filaments often with terminal heterocytes, less often with intercalary heterocytes *Nostoc*
6b Amorphous colonies without firm periderm on surface, filaments with common terminal and intercalary heterocytes *Desmonostoc*
7a Isopolar filaments with ± frequent false branching 8

7b	Heteropolar and/or true-branched filaments	12
8a	Trichomes distinctly tapering toward ends	*Scytonematopsis*
8b	Trichomes not tapering toward ends, but sometimes widening	9
9a	Filaments containing one or more trichomes, high number of hormogonia, cells capable of division in two planes	*Rexia*
9b	Filaments containing just one trichome, cell divide in one plane only	10
10a	Sheath very thick (several times thicker than cell width), funnel-like stratified, in apical part yellowish	*Petalonema*
10b	Sheath present, but not several times thicker than cell width, yellow-brown to brown	11
11a	Heterocytes distinctly "golden", vegetative cells ± violet	*Brasilonema*
11b	Heterocytes and vegetative cells ± olive green	*Scytonema*
12a	Heteropolar filaments with basal heterocytes and false branching	13
12b	True branching types	28
13a	Trichomes distinctly tapering toward apical end	14
13b	Trichomes not tapering or only slightly tapering toward apical end, base may be widened	22
14a	Trichomes never tapering to terminal hair	15
14b	Trichomes tapering into terminal hair (may be broken off)	17
15a	Cells always shorter than wide	*Toxopsis*
15b	Cells of variable length within one trichome	16
16a	Sheath thin, barely visible; growing on volcanic substrates	*Goleter*
16b	Sheath very thick, brown; growing on limestone	*Sacconema*
17a	Trichomes distinctly U-bent, arranged in one-layered colonies	*Kyrtuthrix*
17b	Trichomes not distinctly U-bent	18
18a	Trichomes branched, daughter trichomes usually remain attached to parent trichome	19
18b	Trichomes unbranched or seldom branched, branches shift off the parent trichome	20
19a	Bushlike colonies, never ± hemispherical, trichomes in thin, of the yellow-brownish individual sheaths	*Dichothrix*
19b	± hemispherical, distinctly layered colonies, often $CaCO_3$ incrusted	*Rivularia/Cyanomargarita*
20a	Trichomes with both single and double branching	*Nunduva*
20b	Trichomes with single branching only	21
21a	Basal heterocytes single only	*Calothrix*
21b	Basal heterocytes usually in pairs	*Macrochaete*
22a	Trichomes widening toward ends	23
22b	Trichomes not widening toward ends	24
23a	Cells in apical part up to 8 µm wide, at the base > 8 µm wide, on wet limestone rocks	*Croatella*

23b Cells in apical part up to 4 µm wide, at the base <1 µm wide, in limestone caves *Fortiea*

24a Filaments aggregated into dense fascicles, not enclosed in common sheath 25

24b Filaments not aggregated in fascicles 26

25a Branching very rare, all cells of ± same appearance *Streptostemon*

25b Branching common, cell morphology may substantially differ along the trichome *Seguenzaea*

26a Filaments distinctly crimped, growing from amorphous mass *Borzinema*

26b Filaments straight or arcuate, but not crimped 27

27a Filaments arcuate, branching very common, single-sided, cells always distinctly shorter than wide to discoid *Hassallia*

27b Filaments branched to both sides, cells not exclusively shorter than wide *Tolypothrix/Dactylothamnos*

28a Filaments uniseriate, completely enclosed in crust composed of latticelike limestone crystals, main and lateral branches not indistinguishable, no heterocytes known *Geitleria*

28b Filaments encrusted in different way or not encrusted 29

29a Filaments uniseriate, heterocytes located at ends of ± short branches 30

29b Heterocytes never located at ends of branches 32

30a Trichomes of same width along whole length, not constricted at cross walls; cave type *Baradlaia*

30b Trichomes at least in part of thalli constricted 31

31a Branches always tapering, marine type *Mastigocoleus*

31b Branches of variable morphology, usually macroscopic gelatinous colonies; subaerial non-cave type *Nostochopsis*

32a Both true and false branching present in one thallus 33

32b True branching only types 37

33a Young branches distinctly narrower than old filaments *Spelaeonaias*

33b Main filaments and branches of ± same width 34

34a Thallus composed mostly of erect filaments arranged in fascicles, hormocytes mostly solitary, cells of ± same shape in both main filaments and branches *Spelaeopogon*

34b Filaments not joined into fascicles 35

35a Thallus layered *Handeliella*

35b Thallus not layered 36

36a Filament incrusted with $CaCO_3$; cave type *Iphinoe*

36b Filament not incrusted; non-cave type *Schmidleinema*

37a Branches distinctly morphologically different (trichome shapes, cell shapes) form main filament 38

37b Branches morphologically ± similar to main filament 40

38a Both lateral and intercalary heterocytes present *Pelatocladus*

38b	Lateral heterocytes never present	39
39a	Trichomes uniseriate, sheaths not lamellated, reproduction by hormocytes often joint into groups, without arthrospores	*Leptopogon*
39b	Trichomes in main filaments often polyseriate, sheath thin, I older filaments widened and sometimes lamellated	*Fischerella*
40a	Heterocytes absent (in whole population)	41
40b	Heterocytes present	43
41a	Trichomes, not or slightly constricted at cross walls, enclosed in wide, diffluent mucilaginous envelopes	*Adrianema*
41b	Trichomes distinctly constricted at cross walls, enclosed in delimited sheath	42
42a	Trichomes T- and ± dichotomously branched	*Colteronema*
42b	Trichomes T-branched only	*Albrightia*
43a	Groups of trichomes enclosed in common, delimited, mucilaginous envelopes	*Brachytrichopsis*
43b	Groups of trichomes not enclosed in common, delimited enveloped	44
44a	Filaments arranged ± parallel into fascicles; thermal-atmophytic types	*Thalpophila*
44b	Filaments not arranged ± parallel into fascicles	45
45a	Trichomes cylindrical, uniseriate, not or slightly tapered	46
45b	Organisms of other appearance	50
46a	Gelatinous, spherical colonies, up to 1 mm in diameter, filaments arranged ± radially at periphery of colonies; on calcareous rocks	*Voukiella*
46b	Organisms of other appearance	47
47a	Most cells distinctly longer than wide, sheath thin, reproduction by terminal hormocytes	*Westiella*
47b	Organisms of other appearance	48
48a	Cells ± isodiametric, sheath thick, lamellate	*Loriella*
48b	Organisms of other appearance	49
49a	Sheaths always colorless, lamellate; cave type	*Loriellopsis*
49b	Sheaths yellow-brownish when older, not lamellate	*Symphyonema*
50a	Filaments uniseriate or polyseriate, trichomes never tapering, reproduction mainly by hormogonia, trichomes with T branching only	*Stigonema*
50b	Organisms of other appearance	51
51a	Terminal parts of trichomes strongly narrowed; marine types	*Brachytrichia*
51b	Sheath thick, delimited, lamellate, noncave inhabitant	*Herpyzonema*

2.2.3 Coccal taxa

***Aphanocapsa* Nägeli (Synechococcales, Merismopediaceae)**

Colonies micro- or macroscopic, many-celled, irregular, usually amorphic, gelatinous. Cells spherical, after division hemispherical, arranged irregularly, normally without own mucilaginous envelopes. Mucilage of colony colorless, yellowish, brownish, bluish, or red; fine; formless but delimited.

Aphanocapsa fusco-lutea with cells 1–1.5 µm; *Aphanocapsa muscicola* 2–3 (4.5) µm; *Aphanocapsa parietina* 4–7 µm, on limestone; *Aphanocapsa testacea* 7.7–9.5 µm, on granitic rocks [3]. Outside temperate zone was described *Aphanocapsa intertexta* 2.4—2.6 µm, different rocks, Puerto Rico [4].

***Aphanothece* Nägeli (Chroococcales, Aphanothecaceae)** (Fig. 2.2.1J)

Colonies micro- or macroscopic, many-celled, irregular, usually amorphic, gelatinous. Cells oval, cylindrical or rodlike, straight or slightly curved, arranged irregularly, cells facultatively in the margin of colonies with own envelopes. Mucilage of colony colorless or near the surface yellowish, yellow-brownish or reddish.

Aphanothece marina—wet limestone in splashing zone of Mediterranean sea; *Aphanothece bullosa* atmophytic with macroscopic colonies (to 15 cm) cosmopolitan; *Aphanothece saxicola* cells 1.5–4.5 µm long and 1–2.3 µm wide, colonies green or olive green, noncalcareous biotopes; *Aphanothece castagnei* cells 3.8–8 µm long and 2–4.8 µm wide, colonies brownish, cells with own envelopes at the margin of colony, on limestone; *Aphanothece pallida* 5.5–10 (16) µm long and 3.2–8 µm wide, cells on margin often with own yellow-brown envelopes. Outside temperate zones are known *Aphanothece cylindracea* 10–14 µm long and 6–7.5 µm wide from Puerto Rico [4] and *Aphanothece granulosa* 7–12(13) µm long 4.5–5 µm wide from Central America, limestone rocks (or metaphyton [5]).

***Asterocapsa* Chu (Chroococcales, Chroococcaceae)**

Colonies micro- or macroscopic, more or less spherical or irregular, common sheath of colony thick, in part of life cycle often with granular or spiny surface, colorless, yellow, pink, or brown. Cells spherical, subspherical, oval, or more or less irregular, with individual sheath, sometimes lamellate. Taxonomically unclear genus; several species possibly could be resting stage of *Gloeocapsa* or *Gloeocapsopsis*; by contrast, many of unknown taxa of *Asterocapsa* (Fig. 2.2.1C) are probably often determined as *Gloeocapsa* sp.

Definition of many existing species are unclear. There are three morphotypes:

a. small colonies (2–4 celled) or solitary cells with firm sheath, e.g., *Asterocapsa badia* from Himalaya [6].
b. colonies with many cells, sheath firm, more or less spherical, sometimes with spinelike structure on surface, nanocyte reproduction, e.g., *Asterocapsa divina*, south and Central America [6].
c. irregular-shaped cells, spiny sheaths, gelatinous mats, e.g., *Asterocapsa masayukiwatanabei*, Himalaya [6].

Bacularia Borzì (Synechococcales, Synechococcaceae)
Colonies microscopic, up to macroscopic, many-celled, tubular, usually tapering to one or both ends, sheaths colorless, usually homogeneous. Cells cylindrical, rod shaped, formed to irregular row in central part of colony, oriented more or less in the direction of a tubus of colony.

Bacularia caerulescens—marine splash zone, Mediterranean Sea [7].

Chalicogloea Roldán, Hernández-Mariné and Komárek (Chroococcales, Chroococcaceae)
Colonies microscopic, few celled, irregular, with densely aggregated cells, without common envelopes, sometimes only solitary cells. Cells polygonal, rounded or oval, with light centroplasma, individual envelopes colorless, homogenous or lamellated. Cell division in irregular planes, nanocyte or resting stages not observed. Thylakoid fasciculated.

Chalicogloea cavernicola—known from deep cave in Spain only [8].

Chamaesiphon Braun (Synechococcales, Chamaesiphonaceae)
Solitary, in group of cells or colonial. Cells spherical to cylindrical, heteropolar, basal part attached to the substrate, from apical part separate exocytes (daughter cells). Sheaths (pseudovagina) colorless, open at the apex.

Almost all species occur in water, as subaerophyte only sometimes *Chamaesiphon niger* (in small cell group only, blue, with violet or black sheath) or *Chamaesiphon oncobyrsoides* (colonial, slightly orange sheath), both known only from Europe; attached to stones or filamentous algae [3].

Chlorogloea Wille (Chroococcales, Entophysalidaceae)
Colonies micro- or macroscopic, irregular, gelatinous, in outline more or less spherical or amorphous. Mucilage colors or colored. Cells more or less spherical, distant from one another, surrounded often in individual envelopes, sometimes arranged in indistinct radial rows. Cells divide in three or more planes in successive generations, known also are planocyte and nanocyte formations.

Chlorogloea novacekii—only one typical aerophytic species of the genus, growing on limestone, on shady localities as caves, etc., known only from central Europe. *Chlorogloea rivularis* from Central European mountain waterfalls occur rare on the rock close to water, not on limestone [3].

Chondrocystis Lemmermann (Chroococcales, Chroococcaceae)
Colonies micro- or macroscopic; irregular, gelatinous, with numerous cubic or polygonal subcolonies. Cells spherical, with their own slightly layered envelopes. Taxonomically almost unknown genus.

Chondrocystis dermochroa—one aerophytic species of the genus, usually on limestone [3].

Chroococcus Nägeli (Chroococcales, Chroococcaceae)
Colonies usually microscopic, only few celled, more or less spherical, rarely many celled, macroscopic. Mucilage colorless or colored, homogeneous or lamellated, cells

often with individual envelopes. Cells spherical (young), later hemispherical or in a form of a sector of a sphere. Cell division in three directions perpendicular each to another in first successive generations.

Atmophytic localities: *Chroococcus membraninus*—3–8 µm, homogenous sheath; *Chroococcus globosus*—6–16 µm, homogenous, wide fine sheath; *Chroococcus thermalis*—slightly lamellated sheath, 5–10 (32) µm; *Chroococcus subsphaericus*—6–6.5 µm, purple-gray, Puerto Rico [4]

Common rocks: colorless sheath, nonlamellated: *Chroococcus varius*—cells 2–4 µm, colony few celled; *Chroococcus bituminosus*—small colonies, 2–3.5 µm; *Chroococcus cohaerens*—large colony, cells 2–5 (6) µm; *Chroococcus refractus* 4–8.5(12), cells subquadrate, triangular or multiangular, yellowish or brownish (known from the USA only [9]), *Chroococcus pallidus*—5–8 (13) µm; *Chroococcus turicensis*—13–15 µm, cells blue-green or yellowish, noncalcareous biotopes; *Chroococcus multicoloratus*—20 µm, cells yellowish-green, bluish-green, yellowish, brown, or blackish (known from the USA only[9]); *Chroococcus spelaeus*—15–30 µm, cells blue-green, olive green or slightly pinkish-violet, limestone (Fig. 2.2.1K). Sometimes is recorded from wet rocks *Chroococcus turgidus* too.

Colorless sheath, slightly lamellated: *Chroococcus helveticus* cells 4–9 µm; *Chroococcus westii*—13–27 µm, violet cells.

Colorless sheath, lamellated: *Chroococcus ercegovicii*—cells 4–11 (28) µm, limestone; *Chroococcus tenax*—15–21 µm (Fig. 2.2.1H).

Yellow or yellow-brown sheath of old colony: *Chroococcus lithophilus*—limestone; *Chroococcus montanus*—granite or other acidic rocks [3].

Purple—violet sheath: *Chroococcus muralis*—1.5–2 µm, Central America [4]

Granular sheath: *Chroococcus verrucosus*—Cameroon [10]

Chroococcidiopsis Geitler (Chroococcidiopsidales, Chroococcidiopsidaceae)

Unicellular or microscopic irregular colonies, common sheath, thin, and colorless. Cells spherical or irregularly spherical. Cell division in different planes in succeeding generations (daughter cells usually do not grow into the original shape and size before next division) or by baeocytes. Unclear genus, maybe polyphyletic.

Many of extremophilic cyanobacteria as hypo- and endolith of hot or cold deserts are determined as *Chroococcidiopsis* sp. Described were species *Chroococcidiopsis fissurarum*—chasmoendolith of marine splashing zone; *Chroococcidiopsis supralitoralis*—freshwater splashing zone, Lake Kineret [11]; *Chroococcidiopsis kashayi*—desert caves; *Chroococcidiopsis muralis*—caves from Hawaii [12]; *Chroococcidiopsis umbratilis*—shaded stone wall, Israel [11].

Cyanoarbor Wang (Chroococcales, Enthophysalidaceae)

Colonies macroscopic, polarized, erected from substrata, with clear pseudodichotomous or irregular ramification ("tree-like"). Cells in colonies arranged irregularly or indistinctly radially. Cells irregular, spherical to ellipsoidal. Sheaths narrow, distinct, colorless.

Only one species *Cyanoarbor rupestre* has been described from wet rocks among mosses in China [13], types similar to aquatic species *Cyanoarbor himalayensis* was found on concrete wall Brazil [14].

***Cyanobacterium* Rippka and Cohen-Bazire (Chroococcales, Cyanobacteriaceae)**
Solitary or in irregular groups, oval or shortly cylindrical, without or with very fine sheaths. Visible lengthwise position of thylakoids—cells are with lengthwise striation. Only one aerophytic species, *Cyanobacterium cedrorum* from bark of tree and wet rocks, has this striation almost invisible (it is more "central granule"); maybe this species does not belong to genus *Cyanobacterium* [3].

***Cyanobium* Rippka et Cohen-Bazire (Synechococcales, Synechococcaceae)**
Solitary cells (or in pairs after division), oval, ellipsoid to shortly rod shaped, 1–2 (4) µm long and 1 (–3) µm wide, without visible sheath, usually with well-visible chromatoplasma (parietal thylakoids). Division in one plain. *Cyanobium parvum* in only one species, which occur among other thing, aerophytically [15].

***Cyanosaccus* Lukas and Golubić (Pleurocapsales, Hyellaceae)**
Solitary, in pairs or in clusters with up to eight cells. Mostly endolithic, in intertidal marine splashing zone. Cells more or less pear shaped or club shaped, on both ends rounded, enclosed within ends of mucilaginous fingerlike branches of thallus, oriented by their narrowed ends to the surface of the substrate.

Cyanosaccus aegeus—thallus divaricated without formation of basal cells, cells 14–27 µm; *Cyanosaccus piriformis*—basal cells occasionally, 32–52 µm; *Cyanosaccus atticus*—basal cells present, 12–16 [3].

***Dalmatella* Ercegović (Pleurocapsales, Hydrococcaceae)**
Young thallus epilithic, forming amorphous mass with irregular small cells, in groups or forming irregular rows of cells or pseudofilaments. Older, endolithic pseudofilaments grow parallelly to one another and perpendicularly into the substrate, sometimes pseudodichotomously divaricate. Cells irregular, with distance from one another, bigger than in epilithic part, elongated toward endolithic apex, apical cells club shaped. Sheaths in epilithic parts are usually more delimited, colored, and lamellate than in endolithic parts, where they are colorless, more or less homogeneous. Cell division irregular in three or more planes, or by baeocytes. All known species occur on splashing zone of the sea, on/in limestone rocks.

Dalmatella violacea—violet thallus (rest of all yellow-brown); *Dalmatella polyformis*—calcium incrusted thallus; *Dalmatella buaensis*—epilithic cells in rows, thallus up 500 µm, cells 3–7 µm wide and 4–15 µm long; *Dalmatella litoralis*—epilithic cells irregularly arranged, endolithic part to 100 µm, cells 15 µm wide and 23 µm long; *Dalmatella anomala*—epilithic cells irregularly arranged, endolithic part longer than 100 µm, cells wide 4–10 µm and 6–12 µm long [16].

***Entophysalis* Kützing (Chroococcales, Enthophysalidaceae)**
Colonies microscopic to macroscopic, attached to the substrate, young often polarized, later composed of subcolonies, which are sometimes radially or serially disposed. Cells or their groups in colonies more or less in radial or erect rows, with their

own, usually layered sheath. Cells spherical to irregular, often of variable size in a colony. Cell division in three or more planes. All genus contain more than 30 epilithic species with not so clear delimitation.

Marine species: *Entophysalis deusta*—3–6 µm, yellow-brown sheath, subcolonies, Mediterranean coasts; *Entophysalis granulosa*—2–5 µm, yellow-brown sheath, no subcolonies, cold seas; *Entophysalis maior*—cells 6–8 µm, sheath colorless or slightly yellowish at the edge, warm seas [3].

Nonmarine: *Entophysalis atroviolacea*—5–9 µm, sheath dark violet or black, Europe; *Entophysalis willei*—1.5–2.5 µm, sheath violet, Puerto Rico, China [4, 17]; *Entophysalis arboriformis*—macroscopic colonies, red, South America [14, 18].

Epilithia Ercegović (Pleurocapsales, Xenococcaceae)

Colonies microscopic, irregular, basal part forms "Gloeocapsa-like" agglomerations of more or less spherical or hemispherical cells, upper part is pseudofilamentous, with short rows of oval to cylindrical cells. Sheaths of colony colorless, umlamellated, open at the apex, cells usually with own colorless unlamellated sheaths. Little known genus; only one species, *Epilithia adriatica*, from calcareous supratidal rocks from Croatian marine coast [19].

Ercegovicia De Toni (Pleurocapsales, Hyellaceae)

Colonies microscopic, composed from short irregular pseudofilaments with basal groups of irregular rounded cells; larger cells are at the base of the pseudofilaments, a few apical cells smaller, mostly droplike.

Ercegovicia litoralis—splashing (marine) zone of Mediterranean Sea [20].

Eucapsis Clements and Shantz (Synechococcales, Merismopediaceae)

Colonies microscopic, more or less cubic (three-dimensional) or irregular, common sheaths of colony diffluent and colorless. Cells spherical or slightly oval arranged into perpendicular rows, later sometimes slightly irregular, sheaths nonlamellated or slightly lamellated.

Eucapsis alpina—cyanobacterium, which morphologically fits to this species is quite common on wet rocks, mostly on sandstone, in tropical to boreal zone, over the whole world, identity with typical *E. alpina*, which is originally known from clear freshwaters, needs molecular evaluation [3].

Gloeobacter Rippka, Waterbury and Cohen-Bazire (Gloeobacterales, Gloeobacteraceae)

Solitary or in irregular groups, rod shaped or widely oval, with thin, fine and indistinct sheaths. Cells with almost colorless content and solitary granules, in masses violet. Thylakoids and phycobilisomes absent.

Gloeobacter violaceus (incl. syn. *Aphanothece caldariorum*, *Gloeothece coerulea*, *Gloeothece linearis* [21] on rocks, mainly calcareous, Europe; *Gloeobacter kilaueensis*—cryptic species, Hawaii [22].

Gloeocapsa Kützing (Chroococcales, Microcystaceae)

Colonies microscopic, later macroscopic, gelatinous, spherical to amorphous, composed of irregular groups of cells, usually with wide and concentrically lamellated sheaths, colored or colorless. Cells spherical, after division hemispherical, with their own wide and sometimes lamellated sheaths. During vegetation, cycles develop morphologically different stages, depending on ecological conditions (e.g., arthrospores, resting stages with firm cell wall). Cell division in three perpendicular planes in successive generations; nanocytic cell division known.

Exist more than 80 taxa, almost all aerophytic. Most common species are as follows [3]:

Colorless sheaths: nonlamellated sheath: *Gloeocapsa punctata*—noncalcareous rocks, 1.8–3 µm; *Gloeocapsa coracina*—sandstone, 3–6 µm; *Gloeocapsa atrata*—mostly calcareous rocks, 2–5 µm, cells blue-green (Fig. 2.2.1E); *Gloeocapsa bituminosa*—calcareous rocks, 2–6 µm, cells olive green, violet or pinkish; *Gloeocapsa minutula*—1.7–2 µm, atmophytic.

Lamellated sheaths: *Gloeocapsa aeruginosa*—calcareous rocks, 2–3 µm; *Gloeocapsa caldariorum*—3–4.5 (8) µm.

Gloeocapsa gelatinosa—atmophytic species in hot springs.

Sheaths blue or violet: *Gloeocapsa compacta*—sheaths almost transparent, outer layer wide, usually almost colorless, 2–2.5 (3.5); *Gloeocapsa nigrescens*—lowland to mountain, 3–5 µm; *Gloeocapsa alpina*—high mountain, 4–6 (8) µm; *Gloeocapsa violascea*—sheaths almost not transparent, 3–4 (7) µm.

Sheaths red: *Gloeocapsa thermalis*—atmophytic in hot springs.

Gloeocapsa ralfsii—outer envelopes very widened, only slightly colored, noncalcareous high mountains (Fig. 2.2.1D); *Gloeocapsa rupicola*—2–3 (3.5) µm, noncalcareous; *Gloeocapsa sanguinea*—cells pale blue-green, 4–6.5 µm, noncalcareous high mountain; *Gloeocapsa novacekii*—3.5–9 µm, olive green cells, serpentinic or noncalcareous rocks; *Gloeocapsa multisphaerica*—3–3.8 µm, spherical colonies, young colony colorless, later pink to brick red, China [17].

Sheaths orange: *Gloeocapsa reicheltii*—2.5–3.5 µm; *Gloeocapsa shuttleworthiana*—6–8 µm;

Sheaths yellow or brown: *Gloeocapsa deusta*—marine calcareous splashing zone; *Gloeocapsa fusco-lutea*—1.5–2 µm, noncalcareous; *Gloeocapsa biformis*—0.8–3 µm, calcareous; *Gloeocapsa kuetzingiana*—3–5 µm, noncalcareous, high mountain; *Gloeocapsa rupestris*—(5) 6–8 (11) µm (Fig. 2.2.1F), lowland to mountain [3].

Gloeocapsopsis Geitler ex Komárek (Chroococcales, Chroococcaceae)

Colonies microscopic, later macroscopic, composed by groups of irregular packets of cells. Sheaths of individual cells are packets of the whole colony, mucilage firm, lamellated or not lamellated, and usually colored. Cells hemispherical, spherical, later polygonal. Cell division in different directions in succeeding generations, the cells do not grow into the original form before the next division. Occasional reproduction by spherical nanocytes.

Gloeocapsopsis crepidinum—sheath yellow–yellow-brown, supralittoral of sea or inland saline, 4–8 µm; *Gloeocapsopsis pleurocapsoides*—sheath yellow–yellow-brown, 5–11 (20) µm, dry or wet rocks; *Gloeocapsopsis dvorakii*—sheath rusty red or orange-red, occasionally lamellated, xenotherm, on serpentinite, sandstone, or volcanic rocks; *Gloeocapsopsis magma*—bloody red, wet, mostly acidic rocks, high mountain; *Gloeocapsopsis polyedrica*—inner sheath intensely violet, limestone (known only from Croatia); *Gloeocapsopsis chroococcoides*—sheath colorless or blackish (Fig. 2.2.1G); *Gloeocapsopsis cyanea*—sheath colorless, cells blue, sandstone, known only from Crete [3].

Gloeothece Nägeli (Chroococcales, Aphanothecaceae)
Unicellular, or in micro- or small macroscopic gelatinous colonies, common sheaths colorless or colored, with irregularly arranged cells. Cells widely oval, oval or rodlike, cells or group of cells with own individual sheaths, usually concentrically lamellated, colorless or colored. Cell division (binary fission) crosswise, perpendicular to the long axis of elongated cells in successive generations.

Exist more than 50 taxa, majority aerophytes. Most common or characteristic species [3]:

Rod-like cells (*Gloeothece linearis*—rocks with low pH; *Gloeothece violacea*—caves or shady places; *Gloeothece vibrio*—curved cells, New Caledonia; *Gloeothece filiformis*—cells 1.8–2.5 µm wide and 5–20 µm long, lamellated sheath, Brazil) [23].

Oval cells: colorless, lamellated sheaths (*Gloeothece tepidariorum*—in glasshouses in Europe, Fig. 2.2.1I; *Gloeothece membranacea*—rare on wet rocks, mainly submerged); colorless, nonlamellated sheaths (*Gloeothece krigeri*—1.5 µm wide, Papua New Guinea [24]; *Gloeothece confluens*—1.6–2.3 µm; *Gloeothece palea*—2.5–4.5 µm); blue or violet (*Gloeothece cyanochroa*—noncacareous rocks, 1.4—4 µm long; *Gloeothece ustulata*, 15—50 µm long); cells dark violet individual sheath, sheath of whole colony colorless—*Gloeothece endochromatica*, Central America [4]; orange, limestone—*Gloeothece dubia*; yellow-orange—brown *Gloeothece fusco-lutea* (width 4.5–5.5 um); *Gloeothece rupestris* (width 3.6–6 (10) µm) (Fig. 2.2.1L).

Hormathonema Ercegović (Pleurocapsales, Hydrococcaceae)
Colonies microscopic, mainly epilithic (pseudofilamentous, irregular uni- or moreseriate rows or irregular groups of spherical cells (more or less), enveloped by their own mucilaginous, sometimes widened and slightly lamellated envelopes); smaller endolithic part is pseudofilamentous and in limestone substrate in the form of uniseriate rows of cells, which are distant one from another (with thick, mucilaginous, usually crosswise lamellated sheath). Cells spherical, irregular or polygonal, slightly elongated in endolithic pseudofilaments, usually more widened at their ends. Cells divide irregularly in different planes, nanocytes not observed.

All species occur in marine splashing zone mostly on coastal limestone rocks; *Hormathonema sphaericum*—colorless sheath, almost oval cells; *Hormathonema luteo-bruneum*—yellow–yellow-brown sheath, epilithic part wide to 20 µm, endolithic short to 9 µm length; *Hormathonema longicellulare*—yellow–yellow-brown sheath, epilithic part to 8 µm wide, endolithic long to 40 µm length; *Hormathonema epilithicum*—violet or dark violet sheath, cells 4–5 µm, distant, few celled pseudofilament only; *Hormathonema violaceo-nigrum*—violet or dark violet sheath, cells 2–5 µm, many celled pseudofilamentous [4, 25].

Hormothece Jao (Chroococcales, Aphanothecaceae)
Colonies microscopic, irregular in form of thick mucilaginous stalks, which are colorless, usually more or less transversally lamellated (additional growth zones), at the bases confluent and sometimes pseudodichotomously branched. Cells in stalks solitary, in pairs or loosely arranged in short rows, widely oval to cylindrical.

Hormothece banyolensis—wet rocks above water level in clear lake in Spain, 2–3 µm wide × 4–5 µm long cells, stalks with slight striation; *Hormothece cylindrocellularis*—tropical wet rocks, long stalks without striation, cells 2–3 × 6 µm; *Hormothece rupestris*—densely lamellated sheath, short thallus, calcareous rocks, China, 8–10 × 5–5.5 µm; *Hormothece jaoi*—stalks with transversal striation, cells 2.2–2.5 × 5 µm, China [3, 26].

Hyella Bornet and Flahault (Pleurocapsales, Hyellaceae)
Colonies pseudofilamentous, microscopic; irregular rows of cells creeping on the substrate or boring into it, clearly polarized. Initial (surface) stages are cell groups enveloped by colorless sheaths (chroococcalean stages). Pseudofilaments, growing into the stony substrate, uniseriate or polyseriate, laterally pseudobranched, with sometimes layered sheaths; terminal cells distinctly elongated. Cells divide irregularly in different planes, sometimes forming nanocytes (usually in the surface, oldest parts of thallus).

Terrestrial is only *Hyella terrestris*, mostly from soil surface, all freshwater species are submerged. Rest of species are from marine splashing zone, almost euendolithic, with unclear morphological differences and without modern molecular study. Most frequent species are *Hyella balani*—branches frequent and short (Fig. 2.2.1A); *Hyella tenuior*—pseudofilament more or less with the same width, cells in pseudofilament distinctly separate; *Hyella gigas*—cells 40 µm × 2–5 µm, pseudofilament to 400 µm long; *Hyella dalmatica*—cells 30 µm × 2–8 µm, pseudofilament to 300 µm long [3].

Inacoccus Gama, Rigonato, Fiore, and Sant'Anna (Chroococcales, Chroococcaceae)
Solitary or few celled. Cells rounded or hemispherical after cell division. Sheaths colorless or often intense red, homogeneous, or lamellated. Thylakoids fasciculate, reproduction by binary fission in three irregular planes and by nanocytes present ("Chroococcus—like cells with nanocyte and intensely red-colored sheaths" [27]).

Inacoccus carmineus—rocks and concrete, known only from South Brazil.

***Lightfootiella* Hašler, Pentecost, Jahodářová, Dvořák, and Poulíčková (Chroococcales, Chroococcaceae)**

Colonies microscopic (spherical, hemispherical, or irregularly lobate), but forming thick mucilaginous layers up to 1 cm in thickness. Colonies disintegrate into subcolonies after gelatinization of layered, orange, or red envelopes. Thalli appear dark red in color often with an irregular ropelike morphology, weakly attached or loose lying on the substratum. Sheaths wide, consisting of distinct layers closely and often with external layer of amorphous mucilage. Cells spherical, hemispherical, oval. Reproduction by binary fission in three planes.

Lightfootiella montana—aerophytic to subaerophytic, on quarzite or granitic stones, sometimes among mosses, in mountains of Scotland, Wales, Faroe Islands, probably in Scandinavia [28].

***Lithocapsa* Ercegović (Chroococcales, Enthophysalidaceae)**

Cells ellipsoid to oval, forming linear colonies, parallelly arranged colonies

Colonies microscopic, composed of parallel and fasciculated rows of cells. Cells more or less spherical, after division subspherical, distant from one another; enveloped by wide, slightly lamellated, colorless gelatinous envelopes around each cells; common mucilage around the colony is missing.

Lithocapsa fasciculata, epilithic on limestone rocks, marine splashing zone, Croatia [29]. Little known taxon.

***Lithococcus* Ercegović (Synechococcales, Synechococcaceae)**

Colonies microscopic, composed by irregular net like rows of cells in mucilaginous pseudofilament, endolithic. Sheaths homogeneous and colorless. Cells (hemi)spherical to ellipsoidal, distant from one another.

Lithococcus ramosus—endo- and epilithic on limestone rocks, marine splashing zone, Croatia [29]. Little known taxon.

***Merismopedia* Meyen (Synechococcales, Merismopediaceae)**

Colonies microscopic (in several nonaerophytic species also small macroscopic), table-like, flat, with cells situated in one plane, in rows more or less perpendicular one to another, square or rectangular, common envelopes almost invisible, colorless, homogenous. Cells spherical or widely oval (before division), after division hemispherical. Cell division in two planes in successive generations, perpendicular one to another.

Merismopedia minima— the only known aerophytic species from the genus, cells 0.4–1 μm in diameter, colonies few celled (4–32), among other algae in wetted localities [3].

***Myxosarcina* Printz (Pleurocapsales, Hyellaceae)**

Colonies microscopic to macroscopic, packet like, with subcolonies, cubic rounded or irregular. Common envelopes narrow, colorless or yellow-brown, sometimes slightly lamellated. Cells irregular spherical, later irregular, more or less polygonal. Division of cells in three or more directions in successive generations; the daughter cells do not grow into original shape of the mother cell; develops motile baeocytes in old colonies.

Myxosarcina tatrica—known from High Tatra Mountains only; *Myxosarcina gloeocapsoides*—coastal supralittoral rocks, Croatia; *Myxosarcina decolorata*—over concrete and bricks, India [3]; several unclear species as *Myxosarcina rubra, Endospora mellea, Endospora nigra*, and *Endospora olivacea* (probably *Myxosarcina* species) were described from Puerto Rico [4].

Nephrococcus Li (Chroococcales, Chroococcaceae)

Colonies micro- or macroscopic (with subcolonies), spherical. Cells oval to kidney shaped, with lamellated or nonlamellated sheath.

Nephrococcus shilinensis—cells 6.5–9 µm wide and 13–15 µm long, carbonate rocks, China; *Nephrococcus serbicus*—1.5–6.5 µm, limestone cave, Serbia; *Nephrococcus* sp.—5–8 µm wide, up to 10 µm long, Himalaya [30].

Palikiella Claus (Chroococcales, Cyanobacteriaceae)

Colonies microscopic, pseudofilamentous, in form of isopolar uniseriate rows of cylindrical cells. Cells distant from one another, with visible centro and chromatoplasma, with pairs of granules. Sheaths homogeneous, colorless, closed at both ends. Very unclear genus (three species), not found after description [31], Drouet [32] considers it as Rhodophyta.

Placoma Schousboe ex Bornet and Thuret (Chroococcales, Enthophysalidaceae)

Colonies micro- and macroscopic, spherical, irregular spherical, slightly lobate or vesiculose with waved surface, usually hollow. Cells more or less spherical, in outline irregular or elongated, in one or several layers near the surface of a colony, arranged irregularly, radially or in irregular rows or groups, with or without individual envelopes. Sheath of colony colorless, yellow-brownish or pinkish. Little known genus.

Placoma vesiculosum—splashing zone of European Atlantic and Mediterranean coast, envelopes yellow-brown, cells about 4 µm; *Placoma violaceum*—splashing zone of Pacific coast of the USA, cells about 4 µm in diameter, violet envelopes; *Placoma regulare*—macroscopic red colonies, cells to 30 µm, New Zealand; *Placoma willei*—cells 2–2.4 µm, Puerto Rico and China [3].

Pleurocapsa Thuret in Hauck (Pleurocapsales, Hyellaceae)

Pseudofilamentous; irregular groups of cells, from which grow uni- or multicellular, irregular, or radial rows of cells, sometimes irregularly branched, more or less short. Crust like colonies. Without distinct space between cells. Cells divide mostly perpendicularly or irregularly, produced already baeocytes, forming in old part of pseudofilament, sometimes in masses.

Nonmarine: *Pleurocapsa muralis*, cells 4–6 µm, baeocytes 3–4 µm, South America; *Pleurocapsa minor*—cells 3–9 (12) µm, baeocytes 0.8–3 µm.

Marine splashing zone: sheaths dark brown or blackish: *Pleurocapsa brevissima, cells 3–5* µm wide, basal cells not elongated; *Pleurocapsa minuta*—cells 3–7 µm, basal cells elongated.

Sheath yellow-brown: *Pleurocapsa mucosa*—pseudofilament to 80 μm; *Pleurocapsa hansgirgiana*—pseudofilament to 150 μm and sometimes with violet or yellow cell content.

Sheath colorless: *Pleurocapsa fuliginosa*—cells 5–20 μm, yellowish, yellowredish, or dirty violet; *Pleurocapsa crepidinum*—cells 5–25 μm, blue-green; *Pleurocapsa magna*—cells 18–36 μm, tropical Asia [3].

Podocapsa Ercegović (Pleurocapsales, Hyellaceae)

Short irregular pseudofilaments (4–6 cells), larger cells at apices of the pseudofilament: narrowed toward the "basal" end and more widened in the upper part, terminal cells sometimes irregular spherical, pear shaped, droplike, or club shaped.

Podocapsa pedicellata—splashing zone on Mediterranean Sea, on limestone [33].

Pseudocapsa Ercegović (Chroococcales, Chroococcaceae)

Colonies micro- or later macroscopic, cells forming packetlike aggregates with tightly, radially and "fanlike" oriented cells (with the "conical" part oriented into the center of a colony), later irregular. Sheaths thin, colorless, usually firm, sometimes slightly lamellated.

Pseudocapsa dubia—wet limestone rocks, especially shaded [29].

Rhabdogloea Schröder (Synechococcales, Synechococcaceae)

Colonies microscopic, often few celled. Cells elongated, narrowed to both ends, distant from one another, sometimes oriented more or less in one direction in the colony. Envelopes of colony colorless, not lamellated.

Rhabdogloea brasilica—on rocks splashing zones of the sea, known only from south of Brazil [34].

Sinocapsa Wang and Li (Sinocapsaceae, _incertae sedis_)

Cells irregularly aggregated or form dense colonies with rigid colorless or golden common envelopes. Cells spherical or irregular rounded, without individual envelopes. Thylakoids irregular. Cell division in three or more planes by binary fission.

Sinocapsa zengkensis—crust samples in an exposed depression of rough concrete surface from a rural house roof. Known only from the type locality [35].

Solentia Ercegović (Pleurocapsales, Hyellaceae)

Pseudofilamentous, with epilithic and endolithic part. Epilithic part simple, short uniseriate or multiseriate rows of spherical or slightly elongate, irregularly disposed cells, intensely calcified. Endolithic pseudofilaments common, long, with cells enclosed within widened mucilaginous stalks, within which is an uniseriate row of solitary, usually elongate cells, which are distinctly distant one from another; at the ends of pseudofilaments one enlarged, oval, obovoid, club-shaped, or irregular terminal cell. Cell division in different planes, usually crosswise apical cells sometimes divide into baeocytes.

All species in marine splashing zones. *Solentia sanguinea*—red cells, tropical; *Solentia intricata*—thin pseudofilament, 3–9 μm wide; *Solentia achromatica*—terminal cells 7–16 μm wide and 15–90 μm long; *Solentia paulocellulare*—terminal cells maximum 45 μm long, endolithic pseudofilament 3–25 μm wide; *Solentia stratosa*—terminal cells maximum 45 μm, endolithic pseudofilament to 40 μm wide; *Solentia foveolarum*—terminal cells maximum 45 μm, endolithic pseudofilament to 60 μm wide [3].

***Synechococcus* Nägeli (Synechococcales, Synechococcaceae)**
Cells solitary or agglomerated (not in common colony), cylindrical or rodlike, sometimes curved, in suboptimal conditions with filament-like involution cells. Cell division by perpendicular binary fission, in the same plane in successive generations.

Synechococcus elongatus—on the wet walls and soil close to water level (or in the water), 1.2–3 μm width and 2–9 μm length; *Synechococcus sciophilus*—on calcium carbonate (calcareous rocks), 3.5–5.5 μm width and 6–14 length μm; *Synechococcus brunneolus*—noncalcium carbonate substrata, 5–11 μm width and more than 30 μm length [3].

***Synechocystis* Sauvageau (Synechococcales, Merismopediaceae)**
Cells solitary or agglomerated (not in common colony), spherical, or slightly widely oval before division, after division hemispherical. Individual sheaths very fine, colorless, almost invisible. Cells often with visible chromatoplasma (parietal thylakoidal position). Cell division (pinching) always in two perpendicular planes in successive generations.

Synechocystis primigenia—on limestone rock, 0.7–1 μm in diameter; *Synechocystis pevalekii*—marine splash zone in Croatia, 2.3–3.5 μm in diameter[3].

***Thermosynechococcus* Katoh, Itoh, Shen, and Ikeuchi (Synechococcales, Synechococcaceae)**
(nom. inval: its taxonomic description is invalid according to all codes).

This genus differ from *Synechococcus* only by molecular markers and partly with ecology. Cells solitary or agglomerated, cylindrical, or rodlike, sometimes curved. Cell division by perpendicular binary fission, in the same plane in successive generations.

Thermosynechococcus elongatus—atmophytic, near hot springs [36].

***Tryponema* Ercegović (Pleurocapsales, Hydrococcaceae)**
Pseudofilamentous, endolithic, with cells irregularly elongated, oval, or cylindrical, thallus consist with distinct rows of cells, in mucilaginous envelopes. Cells divide by perpendicular binary fission. Cells sometimes divaricated and forming later branches from pseudofilaments. Terminal cells rounded.

Tryponema endolithicum—limestone in marine splashing zone [25].

2.2.4 Simple trichal taxa

***Blennothrix* Kützing ex Anagnostidis and Komárek (Oscillatoriales, Oscillatoriaceae)**
Trichome long, 8–40 (60) μm wide, uniseriate, more or less waved, isopolar, cells several times shorter than wide, slightly constricted or unconstricted at the cross

walls, not attenuated toward ends, terminal cells widely rounded, sometimes with a narrow calyptra.

Filaments solitary or in colonies or mats, (1)2 to several trichomes within the common sheath. Sheath colorless, sometimes slightly lamellated, open at the apex; filaments (not trichomes), isopolar or more or less polarized, sometimes specifically branched.

All taxa era from marine splash zone. *Blennothrix lyngbyacea*—trichomes 8–16 µm wide; *Blennothrix glutinosa*—trichomes 16–19 µm wide, gelatinous thallus, terminal cells widely rounded, and capitate; *Blennothrix vermicularis*—trichomes 14–20 µm wide, fasciculate thallus, terminal cells widely conical, and truncate [37].

Borzia Cohn ex Gomont (Oscillatoriales, Borziaceae)

Trichome short, maximally 8–16 celled, 0.5–7 µm wide, isopolar, uniserial, more or less straight or slightly curved, unbranched, constricted or not constricted at cross walls, not attenuated at the end, and terminal cell rounded. Filaments solitary or small groups of trichomes. Cells isodiametric or slightly shorter or longer than wide, cylindrical up to barrel shaped. Sheaths mostly lacking, sometimes with fine mucilage. Immotile, without necridia.

Borzia periklei—5–6.5 µm wide, strictly constricted, chasmoendolith on Parthenon or epilith calcareous supralittoral rocks on Aeagean Sea shore; *Borzia susedana*—3.5 µm wide, almost not constricted, calcareous rocks from Croatia, maybe belong to this genus (or *Hormoscilla*) also *Lyngbya saxicola*—6–9 µm wide short trichomes in layered colorless sheath, living endogloeic in colonies of aerophytic *Aphanocapsa* from Pieniny Mountains in Slovakia [38].

Homoeothrix (Thuret ex Bornet and Flahault) Kirchner (Oscillatoriales, Homoeothrichaceae)

Trichomes long, heteropolar, with widely rounded base (up to 15 µm) and continually attenuated to the apex with thin cellular hair, usually attached by basal part to the substrate. Straight or slightly curved, unbranched, or slightly and rarely with false branches, solitary filament or small group of filaments. Sheaths firm, thin, homogenous, open at the apex, colorless or colored, only one trichome in the sheath. Cells shorter than wide up to ± isodiametric, discoid in basal and central part, at the terminal part elongated and hyaline.

Homoeothrix rubra—marine splashing zone [38]; *Homoeothrix muscicola*—endogloeic in mucilage of cyanobacteria on wet rocks, Korea [39].

Jaaginema Anagnostidis and Komárek (Synechoccoccales, Pseudanabaenaceae)

Trichomes long, 0.5–3 µm wide, isopolar, usually waved or coiled, unbranched, slightly constricted or unconstricted at the cross walls, usually not narrowed to the ends, terminal cells rounded to conical rounded. Filaments solitary or in small colonies, always immotile (including hormogonia). Cells usually longer than wide, cylindrical (up to several times), not or slightly constricted at the cell walls. Sheaths always lacking.

Jaaginema neglectum—rounded apical cell, slightly constricted, 0.8–1.3 µm wide × 1–2 µm long; *Jaaginema pseudogeminatum*—rounded apical cells, not constricted, 1.3–2.7 µm wide × 1–2.6 µm long; *Jaaginema kuetzingianum*— attenuated apical cell, 1.8–2 µm wide × 2–3 µm long; *Jaaginema litorale*—marine intertidal zone [40]; *Jaaginema geminatum*—atmophytic around hot springs, also glasshouses [38].

Kamptonema Strunecký, Komárek, and Šmarda (Oscillatoriales, Microcoleaceae)
Trichomes long (2.5) 3–5 (5.3) µm wide, isopolar, unbranched, motile, slightly constricted or unconstricted at cross walls. Terminal cells rounded, bent and hooked, without calyptra. Cells isodiametric or slightly longer or shorter than wide. Filaments solitary without sheaths, or with very fine, facultative, and diffluent sheaths. Arrangement of thylakoids is parietal or semiparietal.

Kamptonema animale—wet soil, glasshouses, rarely wet walls; *Kamptonema gebhardtianum*—cave in Hungary; *Kamptonema laetevirens*—marine splashing zones [41].

Komvophoron Anagnostidis and Komárek (Oscillatoriales, Gomontiellaceae)
Trichome short or rarely long, wide (1) 2–7 (10) µm, isopolar, moniliform, straight or slightly waved, unbranched, deeply and widely constricted, not attenuated toward the ends, terminal cells rounded or with conical outer cell walls. Filaments solitary or agglomerated in clusters or in fine, mucilaginous colonies (mats). Cells more or less spherical or barrel shaped. Trichomes without sheath, but sometimes in diffluent slime, with indistinct motility.

Komvophoron jovis—3.4—4.5 µm wide, thermal springs, also atmophytic; *Komvophoron tenuis*—2–3 µm wide, thermal springs, also atmophytic [42].

Leptolyngbya Anagnostidis and Komárek (Synechococcales, Leptolyngbyaceae)
Trichome long, 0.5–3.5 µm wide, isopolar, straight or slightly curved, unbranched or (rarely) falsely branched, constricted or unconstricted at cross walls, with terminal cells rounded, rarely attenuated, without thickened cell walls or calyptras. Filaments solitary or coiled into clusters and fine mats. Cells shorter to longer than wide. Sheaths thin but firm, usually colorless, opened at the apical end, mainly homogenous. Immotile, hormogonia motile [42].

Unconstricted species: *Leptolyngbya nana*—steel blue or violet, calcareous crusts; *Leptolyngbya cebenensis*—brown; *Leptolyngbya carnea*—red; *Leptolyngbya fallax*—blue-green, wide sheath 3–6 µm, yellow-brown mats; *Leptolyngbya gracillima*—blue-green, long cells, wide sheath 2–4 µm; *Leptolyngbya minus*—pink rose sheath, Puerto Rico; *Leptolyngbya muralis*—false branching 1.5–1.8 µm wide firm sheath, Puerto Rico; *Leptolyngbya tenuissima*—1.4–1.6 µm, sheath hardly visible [38].

Constricted species, rounded terminal cell: *Leptolyngbya norvegica*—marine splash zone, often false branching; *Leptolyngbya leptotrichiformis*—1 µm, older sheath funnel shaped; *Leptolyngbya perforans*—cells 0.8–2 (2.5) µm wide, calcareous rocks, epi- or endolith; *Leptolyngbya compacta*—forming thick mats, short cells; *Leptolyngbya cataractarum*—thick mats, long cells, constriction often indistinct; *Leptolyngbya subtilissima*—solitary filaments or clusters 1–2 µm wide; *Leptolyngbya schmidlei*—solitary filaments or clusters, 2–3 µm; *Leptolyngbya nigricans*—black sheath, maritime Antarctica [38].

Constricted species, plus minus conical terminal cell: *Leptolyngbya scotii*—clear conical terminal cell; *Leptolyngbya henningsii*—rounded conical, mats, 1.7–2.5 µm wide; *Leptolyngbya maius*—solitary filaments or clusters 2.8–3.3 µm wide [38].

Note: In last decades, a number of "Leptolyngbyoid" genera that are distinguishable from each other have been described practically only using molecular methods (cryptic taxa). Aerophytic among them are as follows:

Albertania Zammit (*A. skiophylla*—limestone in catacombs, Malta); *Cartusia* Mai, Johansen, and Pietrasiak (*C. fontana*—wet walls, Europe), *Kaiparowitsia* Mai, Johansen, and Bohunická (*K. implicata*—wet stone, Utah), *Komarkovaea* Mai, Johansen, and Pietrasiak (*K. angustata*—rock close waterfall, Puerto Rico); *Kovacikia* Miscoe, Pietrasiak, and Johansen (*K. muscicola*—caves, Hawaii); *Lusitaniella* Ramos, Brito, and Kaštovský (*L. coriacea*—marine splashing zone, Portugal); *Nodosilinea* Perkerson and Casamatta (four species living on rocks: *N. epilithica*—wet stones, Europe; *N. chupicuarensis*—stones of historical monuments, Mexico; *N. nodulosa*—rocks in splashing marine zone, USA; *N. signensis*—rocks on Signy Island, Antarctica); *Pegethrix* Mai, Johansen, and Bohunická (*P. bostrychoides* and *P. olivacea*—both sandstone seep wall, Utah, USA; *P. convoluta* and *P. indistincta*—both large seep wall and waterfalls, Utah, USA); *Tildeniella* Mai, Johansen, and Bohunická (*T. nuda*—wet stone, Switzerland; *T. torsiva*—limestone wall, Slovakia); *Timaviella* Sciuto and Moro (*T. circinata* and *T. carstica*—both in limestone cave, Italy; *T. obliquedivisa* and *T. radians*—both in seep wall, Utah, USA).

Lyngbya Agard ex Gomont (Oscillatoriales, Oscillatoriaceae)

Trichomes long, 5.5–60 um wide, isopolar, straight or slightly waved, unconstricted or constricted at the cross walls, not attenuated or attenuated to the ends, end cell with thickened outer cell wall or with prominent calyptra. Filaments thick, in mats (macroscopic, up to several cm or dm in diameter), free clusters or rarely solitary. Cells very short, discoid tells. Sheaths wide, firm, hyaline or lamellated, colorless or yellow-brown, opened at the ends. Sporadically false branching, short branches. Not motile or slightly motile (hormogonia motile).

Lyngbya calcarea—travertine forming species, not constricted trichome with width 5–6 µm, Europe, North America; *Lyngbya thermalis*—atmophytic; *Lyngbya kwangsiensis*—trichomes 5.5–6 µm, not constricted at cross walls, hyaline thin sheath, calcareous rocks, China; *Lyngbya koreana*—trichomes 8–11.5 µm, constricted at cross walls, lamellated colorless sheath, wet rocks, Korea; *Lyngbya patrickiana*— trichomes 5–10 µm, not constricted at cross walls, latter lamellated wide sheath, wet rocks, USA;

Lyngbya meneghiniana—marine splash zone, constricted at cross walls, trichome width 6.5–8 μm, volcanic rocks; *Lyngbya semiplena*—marine splash zone, not constricted at cross walls, trichome (5) 7–10 (12) μm, slightly narrowed end with calyptra; *Lyngbya confervoides*—marine splash zone, not constricted at cross walls, (8) 10–16 (25) μm trichome width, slightly narrowed end; *Lyngbya majuscula*—marine splash zone (and benthos), unconstructed at cross walls, 15–60 (80) width [38].

Microcoleus Desmazières ex Gomont (Oscillatoriales, Microcoleaceae)

Trichome long, wide 3—12 μm, isopolar, straight or slightly curved, unbranched, rarely simply branched, unconstricted or slightly constricted at cross walls, trichomes gradually narrowed toward ends, terminal cell attenuated, mostly with calyptra. Filaments solitary or joined in fine, thin strata or mats. Cells plus minus isodiametric. Trichomes motile, one in sheath or several trichomes enclosed in common sheath, with or without individual sheaths, sheath not tapering toward ends, colorless. Motile, reproduction by hormogonia with necridic cells [43].

Several trichomes in common sheath: *Microcoleus paludosus* cells isodiametric or longer; *Microcoleus vaginatus*—cells isodiametric or shorter, wide 6–10 μm, terminal cell conical or rounded; *Microcoleus subtorulosus*—cells isodiametric or shorter, wide 2.5–9 μm, terminal cell capitate or calyptrate.

One trichome in sheath: *Microcoleus crustaceus*—cells 2.8–3.7 μm wide, terminal cell narrow, with calyptra, Ukraine only; *Microcoleus wallrothii*—cells 4–5.7 μm wide, terminal cell narrow, with calyptra [38].

Oculatella Zammit, Billi, and Albertano (Synechococcales, Oculatellaceae)

Trichomes long, 1–3 μm wide, isopolar, straight or slightly wavy, unbranched or (rare) with false branching, slightly constricted at cross walls, not attenuated toward the ends. Terminal cells conical rounded with prominent rhodopsin granules (orange) at apical end. Filaments solitary or in reddish, sometimes greenish compact mats. Cells longer than wide. Sheaths thin but firm, colorless, opened at the apical end, homogenous. Reproduction by hormogonia, mostly without necridic cells. All species cryptic, not recognizable morphologically [44].

Oculatella subterranea—Mediterranean caves and catacombs [45]; *Oculatella kauaiensis*—sea cave, Hawaii; *Oculatella cataractarum*—splashing zone close to desert waterfall, USA.

Oscillatoria Vaucher ex Gomont (Oscillatoriales, Oscillatoriaceae)

Trichomes long, (4) 10–80 μm wide, isopolar, straight or slightly waved, not or slightly constricted at the cross walls, not attenuated or attenuated to the ends, end cell sometimes with thickened outer cell wall (calyptra). Filaments thick, in mats (macroscopic, up to several centimeter or decimeter in diameter), free clusters, or rarely solitary. Cells very short, discoid tells. In optimal condition without sheaths. Highly motile.

Taxonomical note: According to recent studies (e.g., [46]) the definition of genus *Oscillatoria* is narrower (for example, thin species are not in cluster with "true" Oscillatoria), but studies covering all *Oscillatoria* species are missing and this question is still open.

Oscillatoria nigro-viridis—marine splash zone, trichomes 7–12(13) µm wide, end slightly narrowed, constricted at the cross walls; *Oscillatoria funiformis*—marine splash zone, trichomes 11–16 µm wide, very loosely, but ± regularly screwlike coiled, not constricted at the cross walls; *Oscillatoria bonemaisonii*—marine splash zone, trichome (14) 23–30(40) µm wide constricted at the cross walls; *Oscillatoria rupicola*—trichome (3) 4–6 (9) µm wide, not constricted at the cross walls, Europe, Hawaii (sometimes is recorded similar *Oscillatoria tenuis* at wet walls too—constricted at cross walls, trichome (4) 6–11 (15) µm wide); *Oscillatoria vizagapatensis*—trichomes 8–10 µm wide, apical cell hyaline, slightly thickened cell wall of terminal cell, subtropical and tropical [38].

Phormidiochaete Komárek in Anagnostidis (Oscillatoriales, Homoeothrichaceae)
Trichomes long, heteropolar, creeping on the substrate by bases, plus minus erected, straight or curved, sometimes false branched, not or constricted at cross walls, rounded at the base, continually narrowed toward ends, up into hairlike formations. Cells at the base shorter than wide, later isodiametric, and longer than wide from the middle part. Sheaths distinct, ±wide, open at the end, sometimes lamellated. Irregular thylakoids.

Aerophytic epilithic: *Phormidiochaete balearica*—bases 2.5–9 µm wide, calcareous rocks, ± cosmopolitan [37].

Phormidium Kützing ex Gomont (Oscillatoriales, Oscillatoriaceae)
Trichomes long, wide 2–12(14) µm, isopolar, straight or slightly curved, unbranched, constricted or not constricted at cross walls, terminal cell usually rounded, less frequently with calyptra. Filaments solitary, in small groups or in mats. Cells always shorter than wide, with or without sheaths (usually facultative or diffluent). Motile, hormogonia separated mostly with necridic cells. Thylakoids radial or irregular.

There exist more than 200 taxa, taxonomical revision are still in the process. Most typical species are as follows:

Trichome gradually narrowed, terminal cell narrow, rounded, without calyptra: *Phormidium thwaitesii*—marine splashing zone, cosmopolitan [47].

End of trichome shortly narrowed, bent, without calyptra: *Phormidium crassivaginatum*—trichomes 3.5–4 µm wide, cave, Hungary; *Phormidium dudicsianum*—trichomes 7.5–11 µm wide, cave, Hungary.

Whole trichome cylindrical, only terminal cell is distinctly conical, without calyptra: *Phormidium papyraceum*—whole filament to 6 µm wide, cosmopolitan; *Phormidium pachydermaticum*—whole trichome cylindrical, only terminal cell is distinctly conical, filament 6–10 wide, tropical.

Whole trichome cylindrical, only terminal cell is rounded, without calyptra, sometimes thickened cell wall: *Phormidium jenkelianum*—cells shorter than wide, atmophytic; *Phormidium numidicum*—cells isodiametric or longer, atmophytic.
Normal rocks: *Phormidium kuetzingianum*—trichomes 3.5–4 µm wide; *Phormidium interruptum*—trichomes 5–6.5 µm wide, trichome yellow-green, fine granulation at cross walls; *Phormidium lividum*—trichomes 5.5–7 µm wide, blue-green, without granulation at cross walls; *Phormidium calcicola*—trichomes 5.8–6.4 µm wide, terminal cell with thickened wall, India, Puerto Rico; *Phormidium durum*—sheath colorless, latter red, trichomes 12–14 µm wide, Puerto Rico.

Trichome shortly or gradually narrowed, with calyptra: *Phormidium litorale*—trichomes 3.5–4.5 µm wide, marine splash zone, Adriatic Sea; *Phormidium submembranaceum*—trichomes 5 µm wide, marine splash zone, subtropical and temperate; *Phormidium nigro-viride*—trichomes 7–11 µm wide, marine splash zone; *Phormidium schroeteri*—atmophytic; *Phormidium calcareum*—3.7–4 µm wide, cells plus minus isodiametric, limestone; *Phormidium hieronymusii*—7 µm wide, short cells, all rocks.

Cells distinctly shorter than wide, shortly attenuated toward ends, terminal cell calyptrate: *Phormidium lucidum*—trichomes 6–8 µm wide, atmophytic; *Phormidium irriguum*—trichomes 6–11 µm wide purple-gray, wet rocks [38].

Porphyrosiphon Kützing ex Gomont (Oscillatoriales, Microcoleaceae)
Trichome long, wide 3–20 µm, isopolar, usually slightly coiled, unbranched, with or without constrictions at the cross walls, terminal cells widely rounded, sometimes with thickened outer cell wall. Filaments solitary or in microscopic to macroscopic clusters or compact mats. Cells isodiametric or slightly longer or shorter than wide. Sheaths obligatory thick, lamellated, colored, rarely colorless, open at the apex, containing one or rarely two trichomes; sheaths often characteristically widened and coiled after hormogonia release. Immotile, hormogonia motile, necridic cells present.

Porphyrosiphon luteum—marine splashing zone; *Porphyrosiphon lomniczensis*—trichomes 3 µm wide, filament 6 µm, sheath yellow-brown, constricted at cell walls, European Mountains; *Porphyrosiphon calciferus*—trichomes 4 µm wide, filament 9 µm, sheath brown, mats macroscopic to 2 mm thickness, incrusted by calcium carbonate, India; *Porphyrosiphon fuscus*—trichomes 4–7 µm, filament 6–10 µm wide, sheath colorless to brownish-violet, tropical or subtropical cosmopolitan; *Porphyrosiphon latissimum*—trichomes 4.5–6 (7) µm wide, filament 30–40 (50) µm, sheath yellow-brown, Venezuela [48]; *Porphyrosiphon aureofulvus*—trichome 8.5–9.5 µm, filament 10–15 µm wide, sheath yellow to brown, India, Angola; *Porphyrosiphon samoensis*—trichomes 10–11 µm, filament 11–12 µm wide, sheath red to brown, Samoa; *Porphyrosiphon robustus*—trichomes 16–20 µm, filament 30–36 µm wide, sheath colorless to rose-pink, Puerto Rico; *Porphyrosiphon notarisii*—trichomes 8–23 µm, filament up to 28 µm wide, sheath colorless, pink, orange, red, or cosmopolitan [38].

Pseudanabaena Lauterborn (Synechococcales, Pseudanabaenaceae)

Trichome usually short, rarely long, 0.8–3 µm wide, isopolar, straight or slightly waved or arcuate, unbranched, constricted at cross walls, terminal cells rounded or sometimes conical, without calyptras. Filaments solitary or agglomerated in fine, mucilaginous mats. Cells cylindrical, longer, or shorter than wide. Without firm sheath, sometimes with fine, colorless, diffluent, narrow slime envelope. Motility usually indistinct and facultative.

Pseudanabaena spelaea—cells 0.8–1.5 µm wide, 2–3 µm long, apical cells rounded, caves, Greece; *Pseudanabaena skujae*—1.2 µm—2.3 3–7.5 µm, apical cell conical, caves, Hungary.

Epilithic way of life have been described in these two species only, but probably some edaphic species are able live as an epilith (*Pseudanabaena curta* from Russia, Ukraine, and central Asia; *Pseudanabaena minima* from Korea). Several times were recorded as an aerophyte some undescribed types, similar to water species as *Pseudanabaena catenata* or *Pseudanabaena lonchoides* (marine splash zone) [38].

Pseudophormidium (Forti) Anagnostidis and Komárek (Oscillatoriales, Microcoleaceae)

Trichome short or long, 1–18 µm wide, isopolar, irregularly curved and coiled, richly falsely branched, constricted or unconstricted at cross walls, not attenuated or slightly attenuated to the end, terminal cell rounded or conical rounded, without calyptra. Cells isodiametrical or slightly longer or shorter than wide, barrel shaped, or cylindrical. Sheaths firm, irregularly tubelike, colorless, homogeneous or lamellated, sometimes two trichomes in a sheath. Immotile, hormogonia motile, necridic cells present.

Pseudophormidium golenkinianum—marine rocks, reddish (mostly submerged); *Pseudophormidium spelaea*—0.8–1.5 µm wide, 2–3 µm long, cave; *Pseudophormidium skujae*—1.2–2.3 µm wide, 3–7.5 µm long, cave

Pseudophormidium hollerbachianum filament 2.5–6 µm, sheath colorless (mainly soil species, but recorded as epi- or endolith too); *Pseudophormidium tenue*—filament 5–10 µm, golden yellow sheath (mainly from streams etc., but recorded subaerophytic too) [38].

Pycnacronema Martins and Branco (Oscillatoriales, Coleofasciculaceae)

Trichome long, wide 4.8–8 µm, isopolar, straight or slightly curved, unbranched, not constricted or constricted at the cross walls, not attenuated toward ends, terminal cells rounded or conical rounded, with thickened cell wall. Cell content finely granular or with scattered larger granules. Filaments densely entangled, forming mats attached to the substrate. Cells ± isodiametric or shorter or longer than wide; sheaths facultative, firm, thin, colorless, and hyaline; motile.

Six species were described, from tree bark and soil crust. One not named member of this genus (designated as "Phormidium B-Tom") is described from wet rock near Atlantic coast of Brazil [49].

Schizothrix Kützing ex Gomont (Synechococcales, Schizotrichaceae) (Fig. 2.2.2A, B)
Trichome long, 1–4 µm wide, filaments isopolar, but their fascicles joined often to the substrate and heteropolar in form of erect fascicles, straight or slightly curved, unbranched (filaments branched often divaricated), not constricted or slightly constricted at cross walls, terminal cells mainly conical or rounded, without calyptra. Cells cylindrical, usually distinctly longer than wide, rarely isodiametrical. Several trichomes are enclosed in common sheath, which is narrowed and bluntly pointed, usually tapering toward end, usually closed, sometimes divaricated branched, homogenous, rarely lamellated, colorless or colored, with or without individual sheaths. Whole thallus microscopic to macroscopic mats. Hormogonia motile, separating by necridic cells.

The taxonomic situation in this genus is extraordinary complicated, from approximately 150 described species in literature, 40 are almost epilithic. The differences between these species are disputable, and genus *Schizothrix* needs special modern revision [38]. The most characteristic for rocks are as follows:

Filaments solitary or in clusters, mats or fascicles, but not incrusted (subgenus *Schizothrix*): *Schizothrix septentrionalis*—marine splash zone (cold seas), cells shorter than wide; *Schizothrix fragilis*—cells shorter than wide or isodiametric, very wetted rocks; *Schizothrix calcicola*—cells 1–2 µm wide, 1.5–6 µm long, limestone; *Schizothrix cuspidata*—cells 1.9–2.3 (3.5) µm wide, 5–8 µm long, distinct constriction on very visible cell walls; *Schizothrix lardacea*—terminal cell slightly narrowed, 1.5–2 (2.5) µm wide and 2–3.5 (5) µm long, often with granulation at the cross walls; *Schizothrix delicatissima*—terminal cell acute conical, trichome width 0.5–0.8 (1.8) µm, cosmopolitan; *Schizothrix arenaria*—terminal cell acute conical, trichome width 1.3–3 µm, cosmopolitan; *Schizothrix heufleri*—terminal cells conical, sheath blackish-blue or reddish, cells 1.5–3 µm wide and 3–8 µm long; *Schizothrix fusescens*—terminal cells rounded, sheath yellow to yellow-brown, cells 2–3 µm wide and (6)8–13(15) µm long; *Schizothrix epilitica*—terminal cells rounded, cells 0.5–1.3 (1.5) µm wide and 1–2.5(3.5) µm long, south of temperate Europe.

Filaments in macroscopic hemispherical or flattened thallus, partly or totally encrusted by calcium carbonate (subgenus *Inactis*): *Schizothrix cresswellii*—marine splashing zone, sandstone; *Schizothrix lateritia*—terminal cell rounded, sheath colorless, cells 1.3–1.6 µm wide, 2–5 (12) µm long; *Schizothrix vaginata*—terminal cell rounded, sheath colorless, cells 2–3 µm wide, 1–2 µm long; *Schizothrix incrustans*—terminal cell rounded, sheath blackish, cells 1–1.5 (2) µm wide, 1.5–2.5 µm long, south of Europe (e.g., Plitvice Lakes); *Schizothrix coriacea*—terminal cells conical and pointed, sheath colorless, cells 1–2.5 µm wide, 3–6 µm long; *Schizothrix braunii*—terminal cells obtuse conical, sheath dark steel blue, cells 1.7 µm wide, 2–5 µm long [38].

Scytolyngbya Song and Li (Synechococcales, Leptolyngbyaceae)
Trichome long, wide 2–2.6 µm wide, isopolar, bent, repeated false branching, branches mostly narrower than the main filaments, distinctly constricted at cross walls, not attenuated at the ends. Apical cells rounded. Filaments solitary or in mats. Cells

longer than wide, cylindrical. Sheaths initially thin, colorless, later yellow-brown, widened, homogenous. Immotile.

Scytolyngbya timoleontis—on stones in a water treatment facility, China [50].

Symploca Kützing ex Gomont (Oscillatoriales, Microcoleaceae)

Filaments form erected fascicules "grass-like" mats. Trichome long, 1–8 (14) µm wide, isopolar, straight or slightly coiled, unbranched, usually not constricted at cross walls, terminal cells rounded, without calyptra. Fascicles (not filaments) are sometimes branched. Cells isodiametric, shorter or longer than wide. Sheaths firm, thin. Thylakoids radial.

Symploca aeruginosa—marine splashing zone, North America; *Symploca lacrimans*—cells 1.8–2 µm wide, dark cave, Hungary; *Symploca symbiotica*—endogloeic in sheath of *Microcoleus* on wet rocks; *Symploca cavernarum*—atmophytic in shady zone, cells 2.8—3.4 µm wide, Yellowstone.

Symploca thermalis—atmophytic, cells 1.2–2 µm wide, 1.6–6 µm long, cosmopolitan; *Symploca dubia*—atmophytic or subaerophytic, cells 1.5–2.5 wide, 3–8(12) long, trichome pale blue-green, cosmopolitan; *Symploca meneghiniana*—1.5–2.5 µm wide, 3–4.5 µm long, thick mats up to 3 cm high [38].

Common rocks: *Symploca roseola*—cells 1–1.3 (2.5) µm wide, plus minus isodiametric, pink or red sheath, wider Caribbean region; *Symploca elegans*—1.3–2.5 µm wide, (1) 2–4 µm long, calcareous substrates, cosmopolitan; *Symploca parietina*—1.8–3 µm wide, 3.5–9 µm long, trichome pale yellow-green, tropical area and greenhouses in temperate; *Symploca erecta*—2.7–3 µm wide, 3.5–4 µm long, erected fascicules, calcareous substrates, Croatia; *Symploca muralis*—(2) 3.4—4 (5) µm wide, (1.5) 2.5—5.5 µm long, cosmopolitan; *Symploca muscorum*—4.5—8 µm wide, 5–11 (13) µm long, cosmopolitan [38].

Tapinothrix Sauvageau (Oscillatoriales, Homoeotrichaceae)

Filaments solitary or in small fascicules. Trichomes relative long, wide 1–5 (7) µm heteropolar, attached to substrate with wider basis and tapering toward ends, sometimes with hyaline hairs with elongated cells, straight or slightly curved, unbranched, not constricted or constricted at cross walls. Cells isodiametric, shorter or longer than wide. Sheaths thin, firm, hyaline, or lamellated and yellowish. Thylakoids parietal, centro- and chromatoplasma often distinguishable.

Tapinothrix janthina—trichome at base 1–2.3 µm, submersed or subaerophyt, mostly silicates; *Tapinothrix bornetii*—trichome at base 2.8–3.2 µm; *Tapinothrix clintonii*—trichome at base 3–4 µm, known from sandstone desert seep walls, Utah only; *Tapinothrix fusca*—whole filament 7.5 µm wide, trichome at the base 4 µm, sheath yellow or violet, submersed or subaerophytic on calcareous and silicate substrates close to water of clear mountain lakes [51].

Trichocoleus Anagnostidis (Synechococcales, Trichocoleusaceae)

Filaments solitary or rarely densely aggregated in mats. Trichomes long, 0.5—2.5 (3) µm wide, isopolar, straight or slightly curved, unbranched, constricted or not constricted at cross walls, terminal cell acute conical, obtuse or rounded, without calyptra.

Cells usually longer than wide, cylindrical. Several trichomes enclosed in common sheath, not tapering toward end, colorless, homogenous or lamellated, with or without individual sheaths. Necridic cells probably presents, probably absent.

Trichocoleus tenerrimus—marine splash zone; *Trichocoleus acutissimus*—filaments with 15–30 trichome, on wet rocks in tropical and subtropical countries; *Trichocoleus minor*—filaments up to 75 trichomes, Puerto Rico, Brazil [38].

Yonedaella Umezaki (Synechococcales, Pseudanabaenaceae)
Filaments form subspherical colonies, 30–200 μm in diameter, are concentrated freely inside of spherical mucilaginous. Trichomes short or long, 1–2.5 μm wide, isopolar, slightly curved, unbranched, constricted at cross walls, terminal cell rounded. Cells ±isodiametric, barrel shaped, ellipsoid, or almost spherical. Sheaths of the whole colony hyaline, colorless, or yellowish to brownish, trichomes without individual sheaths. Immotile.

Yonedaella lithophila—on calcareous rocks in splash zone in sea coasts—Mediterranean, southern Japan [52].

2.2.5 Heterocytous taxa

Adrianema DeToni (Nostocales, Symphyonemataceae)
Solitary filaments in irregular flattened colonies spreading on substrate, and endolithic or epilithic. Trichomes isopolar, uniseriate, ± cylindrical, unconstricted or only slightly constricted at cross walls, sometimes with widened both ends, with frequent true branching. Sheaths not present, but trichomes enclosed in wide, diffluent, sometimes indistinct, mucilaginous envelopes, which are colorless to brownish, not stratified. Cells ± isodiametric to longer than wide, ± barrel shaped to cylindrical. Heterocytes not known.

Adrianema adriaticum—on limestone rocks in marine splash zone; Mediterranean region [53].

Albrightia Copeland (Nostocales, Hapalosiphonaceae)
Solitary filaments sometimes forming macroscopic colonies (mats). Trichomes isopolar, uniseriate, cylindrical, not tapering in any part, strongly constricted at cross walls, with true branching. Sheaths clearly delimited, usually not stratified, yellow, yellow-brownish to brown, always with just one trichome inside. Cells longer than wide, cylindrical, oval, or sometimes citriform. Heterocytes not known. Reproduction by hormocytes.

Albrightia roraimae—trichomes 4–8 μm wide, nonaquatic type known from sandstone rocks in Venezuela [54].

Baradlaia Palik (Nostocales, Hapalosiphonaceae)
Solitary filaments spreading on substrate. Trichomes isopolar, uniseriate, cylindrical, not constricted at cross walls, with true branching. Terminal parts of trichomes sometimes bent upward. Sheaths very thin, with strong calcification on surface. Cells

longer than wide, cylindrical, cross walls are not visible without staining. Heterocytes located at ends of short lateral braches, never intercalary, spherical to ellipsoid, wider than vegetative cells.

Baradlaia speluncicola—on walls in dark part of a limestone cave; known from one cave in Hungary [55].

Borzinema De Toni (Nostocales, Borzinemataceae)

Solitary filaments spreading on substrate by their bases groups, small colonies, straight, later distorted. Trichomes heteropolar, ± attached to the substrate, uniseriate, cylindrical, not or only very slightly tapering toward their apical end, distinctly constricted at cross walls, with frequent false branching. Sheaths clearly delimited, usually not stratified, yellowish to yellow-brownish, always with just one trichome inside. Cells ± isodiametric or slightly longer than wide, ± barrel shaped. Heterocytes basal on branches, exceptionally intercalary, solitary, oval to almost spherical, with ± same width as the vegetative cells.

Borzinema rupicola—on volcanic rocks, Italy [56].

Brachytrichia Zanardini ex Bornet and Flahault (Nostocales, Symphyonemataceae)

Solitary filaments to brownish, gelatinous, flat to hemispherical macroscopic colonies attached to substrate, hollow when old. Filaments sometimes arranged into horizontal layers with erected ends. Trichomes of different width along their length, uniseriate, ± cylindrical in central parts, distinctly lonely tapering in apical parts, distinctly constricted at cross walls, richly branched. Sheaths distinct and delimited in young filaments, colorless to yellowish, older filaments enclosed in diffluent mucilaginous envelopes. Cells shorter than wide, isodiametric or slightly longer than wide (particularly in hair like ends and in oldest, creeping parts), cylindrical to barrel shaped. Heterocytes intercalary, solitary, ± spherical or barrel shaped, with ± same width as the vegetative cells. On alkaline rocks in marine splash zones; probably worldwide in warmer regions.

Brachytrichia quoyi—colonies up to 5 cm in diameter, width of filaments in different parts changing 4–9 µm [57].

Brachytrichiopsis Jao (Nostocales, Hapalosiphonaceae)

Fasciculated or irregularly arranged filaments forming to macroscopic, black-brownish colonies attached with bases to the substrate. Trichomes ± heteropolar, uniseriate, cylindrical, constricted at cross walls, with true branching, not narrowed toward ends. Sheaths very thick, distinctly lamellate, yellow-brown. Cells ± isodiametric, barrel shaped. Heterocytes intercalary, spherical to ellipsoidal, wider than vegetative cells. Reproduction by ensheathed hormogonia.

Brachytrichiopsis filiformis on wet, limestone walls exposed to direct sunlight; known from China only [58].

Brasilonema Fiore et al. (Nostocales, Scytonemataceae)

Solitary filaments to feltlike, fasciculated, brownish, macroscopic colonies. Trichomes isopolar, uniseriate, cylindrical, not or slightly tapering toward their end, not constricted or constricted at cross walls, with not frequent false branching, especially geminate. Sheaths clearly delimited, usually not stratified, colorless, yellowish to yellow-brownish, always with just one trichome inside. Cells of variable length, cylindrical, with ± violet content. Heterocytes intercalary, short discoid, solitary, cylindrical to subspherical, of ± same width as the vegetative cells, with golden-like color. Epilithic on rocks and bigger stones, on bark of trees, and on rocks or in soil in warm regions, or in warm humid habitats outside these regions, e.g., glasshouses, industrial plants; ± worldwide.

Brasilonema sennae—trichomes 6–12.5 µm wide; *Brasilonema terrestre*—trichomes 9–15 µm wide; *Brasilonema tolantongensis*—trichomes 12.5–20 µm wide, *Brasilonema roberti-lamyi*—trichomes 9.5–11.5 µm wide, terminal cell rounded to barrel shaped [59].

Camptylonemopsis Desikachary (Nostocales, Nostocaceae)

Solitary filaments, joint into smaller clusters, attached to the substrate. Apical parts of filaments bent in upright position. Trichomes heteropolar (seemingly isopolar), uniseriate, ± cylindrical, sometimes with widened apical part, constricted cross walls, with ± common false branching. Sheaths relatively thin, distinctive, homogenous to stratified, hyaline, yellow-brownish or brown, always with just one trichome inside. Cells ± isodiametric, cylindrical. Heterocytes intercalary, after trichome break basal, spherical to barrel shaped. Akinetes in central part of trichomes, but known in some species only.

Epilithic on rocks or stones, many species in soil or epiphytic.

Camptylonemopsis boldii—3–8 µm wide in central part, underneath rocks in Himalaya;

Camptylonemopsis maior—15–19 µm wide in central part, tidal zone in south China;

Camptylonemopsis semiaquatica—1.6–3.2 µm wide in central part, on moistened pebbles, India [60].

Calothrix Agardh ex Bornet and Flahault (Nostocales, Rivulariaceae)

Solitary filaments sometimes agglomerated to flat, felt-like macroscopic colonies attached to substrates. Trichomes heteropolar, uniseriate, tapering from the basal parts toward their ends in some species up to terminal hair, unconstricted or distinctly constricted at cross walls, not or very seldom falsely branched, sometimes with basal heterocytes in branches. Sheaths delimited, sometimes stratified, colorless, yellow to yellow-brownish. Cells usually shorter than wide, less frequently isodiametric or slightly longer than wide, discoid, cylindrical, barrel shaped, or conical. Heterocytes mainly basal, solitary, or few in small groups (rows), oval to almost spherical, ± kidney shaped or conical, with ± same width as the vegetative cells at the trichome base or narrower. On rocks, plants, or soil particles in both marine and nonmarine environments; worldwide. Most of the terrestrial species will be probably transferred to recently established, morphologically similar genus *Dulcicalothrix* [61].

Calothrix aequalis—trichomes 7–10 µm wide in basal part, tapering very gradually toward apical ends, no akinetes; *Calothrix parietina*—trichomes 10–15 µm wide in basal part, sheath frayed at apical ends, no akinetes; *Calothrix parva*—trichomes 6.5–11 µm wide in basal part, no akinetes; *Calothrix tenella*—trichomes 10–12 µm wide in basal part, no akinetes [60].

Colteronema Copeland (Nostocales, Hapalosiphonaceae)

Colonies in form of leathery biofilm, up to 1 mm thick, or ± filamentous, attached to the substrate. Trichomes isopolar, composed from basal, creeping filaments and ± divaricated branches, uniseriate, ± cylindrical, not tapering toward their ends, distinctly constricted at cross walls, often perpendicularly branched. Sheaths delimited, in younger parts distinctly striated, golden to yellow-brown. Cells usually longer than wide, oval to long barrel shaped, with prominent polar granules. Heterocytes nor akinetes were observed.

Colteronema funebre—atmophytic on soil and rocks near thermal springs, in contact with steam in temperature 36–43°C, known only from Yellowstone National Park [62].

Croatella Ercegović (Nostocales, Tolypothrichaceae)

Solitary filaments, joint into crust. Trichomes heteropolar, uniseriate, cylindrical, distinctly widened in terminal part, unconstricted to slightly but clearly constricted at cross walls, with ± common single false branching. Sheaths very thick, homogenous to stratified, yellow-brownish always with just one trichome inside. Cells longer than wide in basal parts to shorter than wide apical parts, cylindrical. Heterocytes basal, ± spherical. Akinetes or arthrospores not known.

Croatella lithophila—on limestone walls and cave entrances, Croatia [29].

Cylindrospermum Kützing ex Bornet and Flahault (Nostocales, Nostocaceae)

Solitary filaments, later forming dark green, leathery macroscopic colonies (mats). Trichomes isopolar, symmetric, uniseriate, cylindrical, not tapering to the ends, distinctly constricted at cross walls, not branched, freely crimped. Sheaths absent, but common mucilage around trichomes may be present. Cells ± isodiametric or slightly longer than wide, cylindrical to barrel shaped. Heterocytes always terminal at both ends (but often develop successively) of the trichome with adjacent akinetes, conical or oval.

On wet rocks in Spain and Australia, several soil species secondary on rocks.

Cylindrospermum toledii—verrucose, brown exospore; *Cylindrospermum rectangulare*—smooth, colorless exospore; *Cylindrospermum majus*—warty brown exospore, soil species; *Cylindrospermum muscicola*—smooth, brown exospore, soil species [60].

Cyanocohniella Kaštovský et al. (Nostocales, Nostocaceae)

Solitary filaments forming blue-green or dark green macroscopic colonies, laying on substrate. Trichomes isopolar, uniseriate, cylindrical, slightly attenuated toward terminal parts, not constricted or constricted at cross walls, not branched, freely crimped. Sheaths lacking or fine, colorless to yellowish. Cells shorter than wide, isodiametric or longer than wide, cylindrical, oval or ± spherical, terminal cells longer than wide, oval to conical. Heterocytes intercalary, solitary, ± spherical. *Cyanocohniella calida*—atmophytic near thermal springs; known only from Czech Republic [63].

Dactylothamnos **Fiore et al. (Nostocales, Tolypothrichaceae)**
Solitary filaments, sometimes joint into small clusters. Trichomes heteropolar, uniseriate, cylindrical, slightly attenuated toward terminal part, unconstricted to slightly but clearly constricted at cross walls, with ± common false branching. Sheaths thin, colorless to brownish always with just one trichome inside. Cells ± isodiametric or slightly shorter or longer than wide, cylindrical to barrel shaped. Heterocytes both basal or intercalary, solitary or in pairs, spherical to slightly barrel shaped or elongated. Akinetes or arthrospores not known.

Dactylothamnos antarcticus—on wet rocks in Antarctica [64].

Desmonostoc **Hrouzek and Ventura (Nostocales, Nostocaceae)**
Solitary filaments to pale to dark green or brown, gelatinous, diffluent macroscopic colonies (mats), laying on substrate. Trichomes isopolar, uniseriate, cylindrical, not tapering in any part, distinctly constricted at cross walls, not branched, freely crimped. Sheaths diffluent, colorless to yellow-brown, without firm periderm of the whole colony. Cells shorter than wide to isodiametric or slightly longer than wide, usually barrel shaped. Heterocytes both intercalary and terminal, solitary, spherical to barrel shaped. Sometimes oval akinetes formed in rows.

Desmonostoc muscorum—on alkaline and acidic rocks among other algae; worldwide [65].

Dichothrix **Zanardini ex Bornet and Flahault (Nostocales, Rivulariaceae)**
Solitary filaments to bushlike macroscopic colonies attached to the substrate, sometimes moderately incrusted with calcium carbonate. Trichomes heteropolar, uniseriate, tapering toward their ends in some species up to terminal hair, unconstricted or distinctly constricted at cross walls, commonly falsely branched. Sheaths delimited, sometimes distinctly stratified, colorless, yellow to yellow-brownish. Cells shorter than wide in basal parts, isodiametric or slightly longer than wide, discoid, barrel shaped or conical, particularly in terminal parts. Heterocytes mainly basal, particularly in branches solitary, conical to hemispherical, with ± same width as the vegetative cells at the trichome base.

On both alkaline and acidic rocks, or soil; worldwide.

Dichothrix gypsophila—trichomes 6–8 µm wide in basal part, sheath lamellated and intensely frayed at apical ends; *Dichothrix chungii*—trichomes 9–12 µm wide in basal part, very long terminal hair; *Dichothrix interrupta*—trichomes 2–3 µm wide in basal part, cylindrical along almost whole length [60].

Fischerella **(Bornet and Flahault) Gomont (Nostocales, Hapalosiphonaceae)**
Solitary filaments often forming mats, main basal filaments wider, multiseriate, spreading on substrate, commonly branched. Branches thinner, uniseriate, ± erect. Main axes cylindrical or irregularly cylindrical, branches cylindrical walls. Trichomes in older parts of thallus constricted at cross walls, often not constricted in cylindrical young parts, especially at terminal parts of branches. Sheaths relatively thin, colorless, yellowish to yellow-brown, sometimes stratified, especially in creeping basal

parts. Cells irregular, spherical, barrel shaped or cylindrical. Heterocytes intercalary, in both main axes and branches, spherical to cylindrical, in branches cylindrical.

On wide range of substrates, often on steam exposed sites (thermal springs).

Fischerella indica—sheath thick, not lamellated, hyaline; *Fischerella major*—sheath thick, lamellated, yellow-brown [60].

Fortiea De Toni (Nostocales, Fortieaceae)

Solitary filaments to small, greenish-grayish, feltlike colonies. Trichomes heteropolar, uniseriate, cylindrical, slightly, but form thin bases distinctly widened to the apex, distinctly constricted at cross walls in apical parts, slightly constricted or unconstructed at basal parts, unbranched or with single branches. Sheaths clearly delimited, usually not stratified, colorless to yellow-brownish, always with just one trichome inside. Cells cylindrical and longer than wide in basal parts and ± barrel shaped and shortening toward apical end. Heterocytes basal or intercalary, solitary, mostly spherical to hemispherical if basal, cylindrical if intercalary.

Fortiea caucasica—trichomes 2–4 µm wide in middle part, dripping limestone walls [20]; *Fortiea subaiana*—trichomes approximately 2 µm wide in middle part, caves [66].

Geitleria Friedmann (Nostocales, Geitleriaceae)

Solitary, ±erect filaments forming feltlike or mildew-like macroscopic colonies attached to the substrate. Trichomes heteropolar, uniseriate, cylindrical, not or only very slightly tapered or widened toward their apical end, distinctly constricted at cross walls, with ± frequent true branching. Terminal cells widened to slightly tapered. Sheaths clearly delimited, sometimes slightly stratified, usually heavily incrusted with calcium carbonate, always with just one trichome inside. Cells isodiametric or shorter or longer than wide, cylindrical, spherical to discoid. Heterocytes not sufficiently approved, akinetes not known.

In dimly lit parts of limestone caves [67, 68].

Geitleria appalachiana—cell length variable along the trichome, from wet temperate climate; *Geitleria calcarea*—cell length variable along the trichome, from desert climate (Fig. 2.2.2H); *Geitleria floridana*—cells distinctly longer than wide mostly along the whole trichome, terminal cell usually distinctly inflated.

Goleter Miscoe and Johansen (Nostocales, Nostocaceae)

Solitary filaments to gelatinous, light blue-green macroscopic colonies. Trichomes heteropolar, uniseriate, clearly tapering toward their ends, distinctly constricted at cross walls, not branched, freely crimped. Sheaths thin, indistinct. Cells ± isodiametric in basal parts, longer than wide in apical parts, barrel shaped. Heterocytes basal, spherical, solitary. Akinetes solitary or in small groups in basal parts of trichomes.

Goleter apudmare—known from volcanic cave in island of Kauai only [12].

Handeliella Skuja (Nostocales, Hapalosiphonaceae)

Filaments often forming crustose, layered dense mats, with main filaments spreading on substrate and ± erect branches. Trichomes heteropolar, uniseriate, constricted at cross walls, with both true and false branching. Sheaths relatively thin or thick, colorless, yellowish to yellow-brown, sometimes stratified. Cells spherical, short barrel shaped to almost cylindrical. Heterocytes both basal and intercalary, spherical to cylindrical, of ± the same width as the vegetative cells. On wet calcareous rocks, known from China [58, 69].

Handeliella sparsa—trichomes 7–10 μm wide, filaments among other organisms; *Handeliella stockmayeri*—trichomes 10–22 μm wide, macroscopic, blackish stratified mats.

Hassallia Berkeley ex Bornet and Flahault (Nostocales, Tolypothrichaceae) (Borzinemataceae)

Solitary filaments feltlike, brownish, macroscopic colonies. Trichomes heteropolar, uniseriate, cylindrical, not or only slightly tapering toward their apical end, distinctly constricted at cross walls, with frequent false lateral branching. The daughter trichomes are bent to the same directions as the mother trichomes. Sheaths clearly delimited, usually not stratified, yellowish to yellow-brownish, always with just one trichome inside. Cells always distinctly shorter than wide, ± discoid. Heterocytes basal, exceptionally intercalary, solitary, oval to almost spherical, of ± same width as the vegetative cells.

On alkaline and acidic rocks, worldwide. Certain species occur in marine splash zones [70].

Hassallia bouteillei—trichomes 4–5 μm wide, forms ±circular mats; *Hassallia byssoidea*—trichomes 8–11 μm wide, forms irregular mats.

Herpyzonema Weber van Bosse (Nostocales, Symphyonemataceae)

Solitary filaments joined into larger groups, which sometimes form mats. Trichomes ± heteropolar, uniseriate, tapering slightly toward their ends, constricted at cross walls, with reverse Y-true branching. Sheaths relatively thick, yellow-brown, sometimes with colorless parts, often stratified. Cells ± spherical, short barrel shaped to almost cylindrical. Heterocytes intercalary, spherical, oval to almost cylindrical, of ± same width as the vegetative cells.

On limestone in splash zone of seas in Southeast Asian region, one species described from caves in Spain [71].

Herpyzonema intermedium—trichomes ca. 4 μm wide, marine splash zone, *Herpyzonema pulvelurentum*—trichomes 10–20 μm wide, limestone caves.

Iphinoe Lamprinou and Pantazidou (Nostocales, Symphyonemataceae)

Solitary filaments, later joined into aggregates spreading on substrate, whitish to purple. Trichomes uniseriate, cylindrical, distinctly constricted at cross walls, with both true and false branching. Sheaths distinct and delimited, colorless, incrusted with calcium carbonate Cells shorter than wide, isodiametric or slightly longer than wide, cylindrical to barrel shaped. Heterocytes mainly intercalary, solitary, ± spherical or barrel shaped, with ± same width as the vegetative cells.

I. spelaea—in a limestone cave, Greece [72].

Isocystis Borzì ex Bornet and Flahault (Nostocales, Nostocaceae)

Gelatinous, macroscopic colonies. Trichomes isopolar, uniseriate, tapering toward their ends, distinctly constricted at cross walls, not branched, freely crimped. Sheaths diffluent. Cells ± barrel shaped, oval, ellipsoidal or spherical, ± isodiametric in central parts, longer than wide in terminal parts. Heterocytes lacking. Akinetes solitary or in small groups (rows) intercalary, spherical to oval, with unstructured or structured cell wall. Maybe Pseudanabaenaceae.

Isocystis messanensis—epilithic type known from a wet stony wall in Sicily [73].

Kyrtuthrix **Ercegović (Nostocales, Rivulariaceae)**

Solitary filaments to bright greenish, gelatinous, macroscopic colonies attached to substrate, some species endolithic. Trichomes isopolar, uniseriate, cylindrical, tapering mainly in their apical parts, distinctly constricted at cross walls, not or very seldom branched, U-bent in their middle part and attached with this part to the substrate. Sheaths delimited, sometimes stratified, in upper parts yellow to yellow-brownish. Cells shorter than wide, isodiametric, or slightly longer than wide, barrel shaped. Heterocytes intercalary, solitary, oval to almost spherical, with ± of the same width as the vegetative cells.

On alkaline and acidic rocks in marine splash zones; probably worldwide.

Kyrtuthrix dalmatica—trichome ends attenuated to a long hyaline hair (may be broken), colonies greenish [53] (Fig. 2.2.2C); *Kyrtuthrix maculans*—trichome ends attenuated, but not to a long hyaline hair, colonies blue-green to blackish colonies [74].

Leptopogon **Borzì (Nostocales, Hapalosiphonaceae)**

Solitary filaments usually forming macroscopic mats, freely coiled, attached to the substrate with upright bent ends. Trichomes isopolar when young, later heteropolar, always uniseriate, ± cylindrical, unconstricted, later constricted at cross walls, with irregular, repeated true branching, branches joined into erect fascicles. At terminal parts of branches often develop the hormocytes in rows. Sheaths thin or slightly widened, delimited, colorless to brown. Cells isodiametric to longer than wide, in older parts of trichomes oval to barrel shaped, in young parts and branches cylindrical. Heterocytes intercalary, ± spherical, barrel shaped to almost oval, of ± same width as the vegetative cells.

Leptopogon intricatus—known from glasshouses only [75].

Loriella **Borzì (Nostocales, Hapalosiphonaceae)**

Filaments usually forming macroscopic fasciculated mats, freely crimped, erect, frequently branched. Trichomes heteropolar, always uniseriate, ± cylindrical, constricted at cross walls, with irregular, repeated true branching (T). Sheaths widened, delimited, funnel-like structured, colorless, incrusted with calcium carbonate. Cells ± isodiametric, cylindrical to barrel shaped, of ± same shape within the thallus. Heterocytes intercalary, exceptionally also basal or terminal, ± spherical to barrel shaped, of ± same width as the vegetative cells. Akinetes intercalary, in rows, with brownish cell walls.

Loriella osteophila—on bones and calcareous rocks in Papua New Guinea [60].

Loriellopsis Hernândez-Marine and Canals (Nostocales, Symphyonemataceae)
Solitary filaments, later joined into aggregates spreading on substrate or partially endolithic. Trichomes uniseriate, cylindrical, distinctly constricted at cross walls, true branched. Sheaths distinct and delimited, colorless, in basal parts incrusted with calcium carbonate. Cells ± isodiametric or slightly shorter or longer than wide, cylindrical to barrel shaped. Heterocytes intercalary, solitary, ± spherical or barrel shaped, with ± same width as the vegetative cells.

Loriellopsis cavernicola—in limestone cave; Spain [72].

Macrochaete Berrendero et al. (Nostocales, Rivulariaceae)
Solitary filaments, sometimes joint into microscopic groups. Trichomes heteropolar, uniseriate, tapering toward their ends, constricted at cross walls, not or very seldom false branched, terminated usually with long hyaline hair. Sheaths delimited, sometimes slightly stratified, colorless, yellow to yellow-brown. Cells shorter than wide, isodiametric or slightly longer than wide, discoid, barrel shaped. Heterocytes mainly basal, in young trichomes single, in older trichomes in pairs, hemispherical to spherical, slightly narrower than the vegetative cells at the trichome base or narrower.

Macrochaete psychrophila—epilithic species described from rocks in Sao Paulo state, Brazil [76].

Mastigocoleus Lagerheim ex Bornet and Flahault (Nostocales, Hapalosiphonaceae) (Nostochopsidaceae)
Solitary filaments spreading on substrate. Trichomes isopolar, uniseriate, ± cylindrical, constricted at cross walls, with true branching—branches of two types: (1) morphologically similar to main filament ending with heterocytes and (2) tapering toward end without heterocytes. Sheaths firm, thin, not structured. Cells longer than wide or isodiametric, cylindrical. Heterocytes located at ends of short lateral braches or lateral in main trichomes, never intercalary, spherical to ellipsoid, often wider than vegetative cells.

Mastigocoleus testarum—on rocks in marine splash zone, worldwide outside cold regions [77] (Fig. 2.2.2D).

Nostoc Vaucher ex Bornet and Flahault (Nostocales, Nostocaceae)
Solitary filaments are aggregated to bright or dark green, gelatinous, macroscopic colonies enveloped by gelatinous periderma, with firm periderm of the whole colony, sometimes attached to substrate. Trichomes isopolar, uniseriate, cylindrical, not tapering in any part, distinctly constricted at cross walls, not branched, freely coiled. Sheaths are developed sometimes around individual trichomes in central part of colonies, diffluent, in peripheral parts delimited are yellow to yellow-brownish. Cells ± isodiametric or slightly longer than wide, barrel shaped. Heterocytes mainly terminal, less frequently intercalary, solitary, spherical to oval.

On both alkaline and acidic rocks; worldwide.

Note: Genera *Compactonostoc* [78] and *Minunostoc* [79] have been recently published, which differ from *Nostoc* sensu stricto mostly on molecular level.

Nostoc calcicola—cells 2.5–3 µm wide, trichomes freely entangled, colonies irregular flat; *Nostoc ellipsosporum*—cells 4–5 µm wide, trichomes freely entangled, colonies irregular; *Nostoc intestinale*—cells 2.8–3.5 µm wide, trichomes freely entangled, colonies microscopic, soft; *Nostoc microscopicum*—cells 5–9 µm wide, trichomes densely entangled in young colonies, colonies ± spherical, firm [60].

Nostochopsis Wood (Nostocales, Hapalosiphonaceae)

Solitary filaments or to macroscopic, hemispherical to almost spherical gelatinous colonies attached to substrate, epilithic. Trichomes isopolar, uniseriate, ± cylindrical, sometimes narrowed toward ends, unconstricted or constricted at cross walls, with true lateral branching. Sheaths thick, diffluent, not structured. Cells longer than wide, cylindrical to long barrel shaped. Heterocytes located at ends of short lateral braches, or intercalary, spherical to ellipsoid, often wider than vegetative cells.

Nostochopsis hansgirgii—on rocks near streams, nonaquatic type known from India [80].

Nunduva González-Resendis et al. (Nostocales, Rivulariaceae)

Solitary filaments, usually joint in fascicles forming macroscopic tuft colonies. Trichomes isopolar when young (hormogonia), later heteropolar, uniseriate, not tapering or tapering toward their ends, terminated sometimes with semihyaline or hyaline hair, unconstricted or constricted at cross walls, not or very seldom false branched. Sheaths narrow or widened, delimited, diffluent at their apical ends, sometimes slightly stratified, colorless, yellow to yellow-brown. Cells of variable shape and length. Heterocytes intercalary, solitary, or in groups (rows) up to 3, ± hemispherical, conical or cylindrical, of ± same width as the vegetative cells.

Nunduva fasciculata—on rocks in marine intertidal and supratidal zones, known from Mexico [81].

Pelatocladus Johansen and Vacarino (Nostocales, Hapalosiphonaceae)

Solitary filaments spreading on substrate brown to blue-green tuft colonies. Trichomes isopolar, uniseriate, cylindrical, strongly constricted at cross walls, with mostly unilateral true branching. Sheaths lacking. Cells ± spherical in main axis and basal parts of branches, longer than wide, barrel shaped to cylindrical in branches. Heterocytes intercalary in both main axes and branches, solitary, spherical, barrel shaped to cylindrical, of ± the same width as the vegetative cells.

Pelatocladus maniniholoensis—known from volcanic cave in island of Kauai [12].

Petalonema Berkeley ex Correns (Nostocales, Scytonemataceae)

Solitary filaments or forming gelatinous, yellowish, macroscopic colonies. Trichomes isopolar, uniseriate, cylindrical, not tapering toward their end, rarely a little widened and in some species narrowed in central parts, not constricted or constricted at cross walls, with false branching, especially geminate. Sheaths very wide and distinct, clearly delimited, rarely not stratified, colorless to yellow or brownish, always with just one trichome inside, sometimes funnel-like widened at the ends. Cells of variable length, barrel shaped. Heterocytes intercalary, solitary, cylindrical to subspherical, usually under than trichomes, rarely only with ± same width as the vegetative cells.

On limestone rocks usually in shady places or shaded by other organisms or crumbled substrate; worldwide.

Petalonema alatum—funnel-like stratified sheath, heterocytes ± spherical (Fig. 2.2.2E), *Petalonema densum*—trichomes distinctly narrowed in central parts, heterocytes oval to barrel shaped; *Petalonema incrustans*—trichomes 7–10 μm wide, forming loops before branching, usually thick sheath [60] (Fig. 2.2.2G).

***Rexia* Casamatta et al. (Nostocales, Tolypothrichaceae) (Borzinemataceae)**

Solitary filaments attached to substrate, sometimes erect. Trichomes isopolar, uniseriate, cylindrical, not or slightly tapering toward their apical end, indistinctly to distinctly constricted at cross walls, with frequent false branching. Sheaths clearly delimited, usually not stratified, thin, colorless, in cultures may contain more than one trichome inside. Cells shorter than wide, to ± isodiametric, discoid, barrel shaped to cylindrical, able of division in two planes. Heterocytes intercalary, but with single pore, ± hemispherical, of the same width as the vegetative cells.

Rexia erecta—known from one rock in Great Smoky Mountains, North Carolina [82].

***Rivularia* Agardh ex Bornet and Flahault (Nostocales, Rivulariaceae)**

Solitary filaments, aggregated to hemispherical macroscopic colonies attached to the substrate, often distinctly layered and heavily incrusted with calcium carbonate. Trichomes heteropolar, uniseriate, tapering toward their ends in some species up to terminal hair, unconstricted or distinctly constricted at cross walls, commonly falsely branched. Sheaths firm, delimited, sometimes distinctly stratified, colorless, yellow to yellow-brownish. Cells shorter than wide, isodiametric or slightly longer than wide, discoid, barrel shaped or conical, in apical parts elongated. Heterocytes mainly basal in both main filament and branches, solitary, oval to almost spherical, of ± same width as the vegetative cells at the trichome base or slightly wider. Akinetes absent.

On rocks, plants, or soil in both marine and nonmarine environment; worldwide.

Note: Genus *Cyanomargarita* is morphologically similar with species *Cyanomargarita calcarea*, as recently described by Shalygin et al. [83], but differs on genetic level.

Rivularia atra—colonies hard, marine; *Rivularia calcarata*—colonies incrusted, but relatively soft, up to 1 cm in diameter, nonmarine; *Rivularia haematites*—colonies incrusted, hard, up to initially hemispherical, later forming continuous crust, nonmarine; *Rivularia mesenterica*—colonies soft, marine [60].

***Sacconema* Borzì ex Bornet and Flahault (Nostocales, Rivulariaceae)**

Solitary filaments, radially aggregated to amorphous to lobate macroscopic colonies attached to the substrate. Trichomes heteropolar, uniseriate, tapering toward their ends to a terminal hair, distinctly constricted at cross walls, commonly falsely branched. Sheaths very widened, delimited, closed or open, lamellated, brown. One sheath may contain more than one trichome. Cells isodiametric or slightly longer or shorter than wide. Heterocytes, solitary or in groups up to three, oval to almost spherical, of ± same width as the vegetative cells at the trichome base or slightly wider. Akinetes spherical at base of trichomes.

Sacconema rupestre—on wet rocks, known from Italy only [60].

Schmidleinema De Toni (Nostocales, Hapalosiphonaceae)

Solitary filaments usually forming macroscopic mats, freely coiled, attached to the substrate by thin central parts with upright bent ends. Trichomes isopolar when young, later heteropolar, always uniseriate, ± cylindrical, unconstricted, later mostly constricted at cross walls, with both rare true and common false branching, branches sometimes joined into erect fascicles. At terminal parts of branches often develop the hormocytes. Sheaths thin or widened, delimited, colorless to yellow-brown, stratified. Cells ± isodiametric to shorter or longer than wide, in older parts of trichomes oval to barrel shaped, in young parts and branches ± cylindrical. Heterocytes intercalary, ± hemispherical to cylindrical, of ± the same width as the vegetative cells.

Schmidleinema indicum—known from wet walls and tree trunks in India [80].

Scytonema Agardh ex Bornet and Flahault (Nostocales, Scytonemataceae)

Solitary filaments to feltlike, brownish, macroscopic colonies (mats). Trichomes isopolar, uniseriate, cylindrical, not tapering toward their ends, but in some species slightly narrowed in central parts, not constricted or constricted at cross walls, with frequent false branching, especially geminate. Sheaths clearly delimited, usually not stratified, colorless, yellowish to brown, always with just one trichome inside. Cells of variable length, cylindrical. Heterocytes intercalary, solitary, cylindrical to subspherical, with ± same width as the vegetative cells.

On rocks, plants or soil; worldwide.

Note: Komárek [60] separated a group of *Scytonema* species with narrowed central part of trichomes into section *Myochrotes* (Fig. 2.2.21) Formal promotion of the section to generic level was not performed yet.

Scytonema drilosiphon—trichomes 5.5–10 µm wide, filaments usually densely encrusted with $CaCO_3$, sheath thin, not lamellated; *Scytonema hofmanii*—trichomes 4.5–9.5 µm wide, sheath thin, not or slightly lamellated; *Scytonema myochrous*—cells in central part of trichome narrowed, 6–12 µm wide, sheath distinctly layered; *Scytonema ocellatum*—trichomes 6–14 µm wide, sheath thin, usually lamellated [60].

Scytonematopsis Kiseleva (Nostocales, Scytonemataceae)

Solitary filaments to feltlike, brownish, macroscopic colonies (mats). Trichomes heteropolar when young, later isopolar, uniseriate, distinctly tapering toward their end (both ends in isopolar trichomes) up to a terminal hair, usually constricted at cross walls, with frequent false branching, especially geminate. Sheaths clearly delimited, usually not stratified, colorless, yellowish to brown, always with just one trichome inside. Cells of variable length, sometimes very short, isopolar up to long cylindrical in apical parts, or barrel shaped or cylindrical. Heterocytes intercalary or basal, solitary, cylindrical to subspherical, with ± same width as the vegetative cells. On rocks, plants or soil in both marine and nonmarine environment; worldwide.

Scytonematopsis contorta—filaments 14–20 µm wide, Hawaii [84]; *Scytonematopsis starmachii*—filaments 9–18 µm wide, granitic mountains in Europe [85].

Seguenzaea Borzì (Nostocales, Tolypothrichaceae)
Solitary filaments aggregated to fasciculate, brownish macroscopic colonies or flat mats. Trichomes isopolar, later heteropolar, uniseriate, cylindrical, not or only slightly tapering toward their apical ends, unconstricted (particularly in young parts) to distinctly constricted at cross walls, with infrequent true to false branching. Trichomes/filaments of two types: torulous (i.e., composed of ± spherical cells) creeping on substrate, and cylindrical branches. Sheaths firm, clearly delimited, thin, not stratified, colorless. Cells shorter than wide, isodiametric or up to longer than wide, cylindrical, or barrel shaped, in basal parts of branches ± spherical. Heterocytes both basal and intercalary, solitary or in short rows, oval, almost spherical to almost cylindrical, of ± same width as the vegetative cells. On rocks among mosses [56] and on wet wall in a botanical garden [84]; temperate and tropical regions. Certain species occur in marine splash zones.

Seguenzaea minor—creeping filaments 6–8 µm wide; *Seguenzaea sicula*—creeping filaments 12–14 µm wide [86].

Spelaeonaias Lamprinou et al. (Nostocales, Symphyonemataceae)
Solitary filaments joined into small tufts or grayish wooly mats. Trichomes isopolar, uniseriate, cylindrical, unconstricted in young branches to constricted at cross walls in older parts of thallus, with both true (Y) and false branching. Sheaths distinct and delimited, hyaline. Cells shorter than wide, isodiametric or longer than wide, cylindrical to barrel shaped. Heterocytes intercalary or basal in germinating hormogonia, solitary, barrel shaped or hemispherical, with ± the same width as the vegetative cells.

Spelaeonaias floccida—known from one limestone cave in Greece [87].

Spelaeopogon Borzì (Nostocales, Hapalosiphonaceae)
Solitary filaments, often joined into fascicles forming small tufts or flat, blue-green, reddish, brownish to black mats. Trichomes isopolar, uniseriate, ± cylindrical, unconstricted or constricted at cross walls, with both true and false branching, branches do not differ morphologically from the main axis. At terminal parts of branches often develop the hormocytes. Sheaths thin or widened, delimited, not structured or lamellated, colorless or yellow to brown. Cells ± isodiametric to shorter or longer than wide, cylindrical, or barrel shaped. Heterocytes intercalary, solitary, cylindrical, ± barrel shaped or hemispherical, of ± the same width as the vegetative cells.

On wet walls and stones and among mosses in caves; known from Italy [56] and Japan [60].

Spelaeopogon cavarae—old filaments 16–18 µm wide, sheath thin, colorless; *Spelaeopogon koidzumianum*—filaments 15–18 µm wide; *Spelaeopogon lucifugus*—no heterocytes, sheath thin; *Spelaeopogon sommieri*—filaments 8–10 µm wide, sheath thin, colorless.

Stigonema **Agardh ex Bornet and Flahault (Nostocales, Stigonemataceae)**
Solitary filaments to feltlike, brownish, macroscopic colonies. Trichomes heteropolar when young, later isopolar, uniseriate or multiseriate, usually constricted at cross walls, with frequent true branching, in some species branches differ morphologically from the main axes. Sheaths clearly delimited, not stratified or lamellated, sometimes with lengthwise surface cracks, colorless, yellowish to brown. Cells ± hemispherical, of variable length, barrel shaped, ± spherical or cylindrical. Heterocytes intercalary, solitary, cylindrical to subspherical, with ± same width as the vegetative cells.

On rocks, plants or soil in; worldwide.

Stigonema hormoides—trichomes mostly uniseriate, cells 5.2–8.8 μm wide; *Stigonema informe*—trichomes mostly multiseriate (up to 4–6), cells 11–15 μm wide; *Stigonema minutum*—trichomes mostly multiseriate (1–4), cells 15–18 μm in diameter; *Stigonema panniforme*—trichomes mostly uniseriate, cells 9.2–24 μm wide; *Stigonema multipartitum*—trichomes mostly multiseriate (up to 8), cells 8–16 μm wide [60].

Streptostemon **Sant'Anna et al. (Nostocales, Tolypothrichaceae)**
Solitary filaments later joined to dense, erect fascicles, forming brownish tufts. Trichomes heteropolar to isopolar, uniseriate, cylindrical, not or slightly narrowed in their central parts, unconstricted to slightly constricted at cross walls, with very rare geminate false branching. Sheaths clearly delimited, sometimes slightly lamellated, colorless to yellow-brownish. Cells shorter than wide, isodiametric or longer than wide, cylindrical to widely rounded. Heterocytes basal or intercalary, solitary, oval, almost spherical to almost cylindrical, of ± the same width as the vegetative cells.

Streptostemon lutescens—on stones in tropical rain forest, known from Brazil only [88].

Symphyonema **Jao (Nostocales, Symphyonemataceae)**
Solitary filaments to macroscopic, cushion-shaped colonies, attached to substrate. Filaments sometimes arranged into fasciculate groups. Trichomes uniseriate, with cylindrical central parts, unconstricted to slightly constricted at cross walls, richly true branched (T- or reverse V-form). Sheaths distinct and delimited, hyaline, in older filaments indistinctly lamellated, brownish. Cells shorter than wide, isodiametric or slightly longer than wide, cylindrical to barrel shaped. Heterocytes intercalary, solitary, cylindrical to subspherical, with ± the same width as the vegetative cells.

On limestone rocks and in caves; known from China [58] and Spain [89].

Symphyonema cavernicola—trichomes constricted on cross walls, chasmolithic in limestone caves; *Symphyonema sinense*—trichomes usually not constricted at cross walls, on wet limestone walls

Thalpophila **Borzì (Hapalosiphonaceae)**
Solitary filaments, often joined into fascicles forming flat colonies. Trichomes isopolar, always uniseriate, ± cylindrical, slightly constricted in young parts to distinctly constricted at cross walls in older parts, with true lateral branching, branches do not differ morphologically from the main axes, when more developed growing in the

same direction like the main axes. Sheaths firm, slightly widened, delimited to diffluent, lamellated. Cells ± isodiametric to longer than wide, cylindrical, or barrel shaped. Heterocytes intercalary, solitary, cylindrical, or ± barrel shaped, of ± the same width as the vegetative cells.

Thalpophila cossyrensis—on wet walls near thermal springs; known only from Sicily [56].

Tolypothrix **Kützing ex Bornet and Flahault (Nostocales, Tolypothrichaceae)**
Solitary filaments to feltlike, brownish, macroscopic colonies. Trichomes heteropolar, uniseriate, cylindrical, not or only very slightly tapering toward their apical ends, unconstricted to distinctly constricted at cross walls, with frequent false branching. The daughter trichomes are bent to both, i.e., the same and the opposite, directions as the mother trichomes. Sheaths clearly delimited, usually not stratified or slightly stratified, colorless, yellowish to yellow-brownish, sometimes may contain more than one trichome inside. Cells shorter than wide, isodiametric, or longer than wide, cylindrical to barrel shaped. Heterocytes basal, exceptionally intercalary, solitary or in short rows, oval, almost spherical to almost cylindrical, of ± the same width as the vegetative cells.

On ± wet alkaline and acidic rocks; worldwide. Certain species occur in marine splash zones. Recent taxonomic study published by Hauer et al. [90].

Tolypothrix elenkinii—trichomes 5–7 µm wide, filaments up to approximately four times wider, sheath very thick, yellow (Fig. 2.2.2F); *Tolypothrix fasciculata*—trichomes ±8 µm wide, sheath thin; *Tolypothrix pseudorexia*—trichomes 10–20 µm wide, sheaths pale brown.

Toxopsis **Lamprinou et al. (Nostocales, Godleyaceae)**
Solitary filaments to yellow-green to dark green macroscopic colonies. Trichomes isopolar when young, heteropolar when old, uniseriate, ± cylindrical, not or only very slightly tapering toward their apical ends, distinctly constricted at cross walls, with frequent false lateral branching. Sheaths thin, clearly delimited, usually not stratified. Cells shorter than wide, shortly barrel shaped. Heterocytes basal (especially in branches), exceptionally intercalary, solitary, oval or bluntly conical, of ± same width as the vegetative cells.

Toxopsis calypsus—known only from one cave in Greece [91].

Voukiella **Ercegović (Nostocales, Symphyonemataceae)**
Solitary filaments to hemispherical, firm, yellow-brown, macroscopic colonies, in peripheral part of colonies filaments arranged ± radially. Trichomes heteropolar, always uniseriate, ± cylindrical, constricted at cross walls, with true branching (V or T-type), branches do not differ morphologically from the main axes. Sheaths wide, delimited, slightly lamellated. Cells isodiametric to longer than wide, barrel shaped. Heterocytes basal or intercalary, ± spherical, of ± the same width as the vegetative cells.

Voukiella rupestris—known from limestone rocks in Croatia only [29].

Westiella Borzì (Nostocales, Hapalosiphonaceae)

Solitary filaments, freely coiled. Trichomes isopolar, always uniseriate, ± cylindrical, unconstricted or constricted at cross walls, with irregular true branching (T-type); branches do not differ morphologically from the main axis. The terminal parts of branches often develop the hormocytes, sometimes in short rows. Sheaths are finely thin, delimited, not structured. Cells ± isodiametric to longer than wide, cylindrical. Heterocytes intercalary, cylindrical, of ± the same width as the vegetative cells.

Westiella intricata—epilithic species from rock wall facing the steam from a thermal spring [56].

References

[1] Hauer T, Mühlsteinová R, Bohunická M, Kaštovský J, Mareš J. Diversity of cyanobacteria on rock surfaces. *Biodivers Conserv.* 2015; 24:759–79.

[2] Komárek J, Kaštovský J, Mareš J, Johansen JR. Taxonomic classification of cyanoprokaryotes (cyanobacterial genera) 2014 using a polyphasic approach. *Preslia.* 2014;86(4):295–35.

[3] Komárek J, Anagnostidis K. Cyanoprokaryota. Part 1: Chroococcales. In: Ettl H, Gärtner G, Heynig H, Mollenhauer D, editors. Süsswasserflora von Mitteleuropa 19/1, Jena-Stuttgart-Lübeck-Ulm, Germany: Gustav Fischer; 1998. p. 1–548.

[4] Gardner NL. New Myxophyceae from Porto Rico. In: Memoirs of the New York Botanical Garden. Vol. 7; 1927. p. 1–144.

[5] Komárek J, Komárková-Legnerová J. Taxonomic evaluation of the cyanobacterial microflora from alkaline marshes of northern Belize. 1. Phenotypic diversity of coccoid morphotypes. *Nova Hedwigia.* 2007;84(1–2):65–111.

[6] Komárek J. Validation of the genera *Gloeocapsopsis* and *Asterocapsa* (Cyanoprokaryota) with regards to species from Japan, Mexico and Himalayas. *Bul Nat Sci Mus Tokyo Ser B.* 1993;19(1):19–37.

[7] Borzì A. Genri nuovi di Chroococcacee. *Nuovo Notar.* 1905;16:20–1.

[8] Roldán M, Ramírez M, del Campo J, Hernández-Mariné M, Komárek J. *Chalicogloea cavernicola* gen. nov., sp. nov. (Chroococcales, Cyanobacteria), from low-light aerophytic environments: combined molecular, phenotypic and ecological criteria. *Int J Syst Evol Microbiol.* 2012;63(6):2326–33.

[9] Wood HC. Prodromus of a study of the freshwater algae of North America. *Proc Am Philos Soc.* 1869;11:119–145.

[10] Krieger W. Algenassoziationen von den Azoren und aus Kamerun. *Hedwigia.* 1931;70:140–56.

[11] Dor I, Carl N, Baldinger I. Polymorphism and salinity tolerance as a criterion for differentiation of three new species of *Chroococcidiopsis* (Chroococcales). *Arch Hydrobiol/Algolog Stud.* 1991;64:411–21.

[12] Miscoe LH, Johansen JR, Vaccarino MA, Pietrasiak N, Sherwood AR. The diatom flora and cyobacteria from caves on Kauai, Hawaii. II. Novel cyanobacteria from caves on Kauai, Hawaii. *Bibl Phycol.* 2016;123:75–152.

[13] Wang CZ. *Cyanoarbor*, a new genus of the Entophysalidaceae (Cyanophyta). *Acta Phytotax Sin.* 1989;27(2):129–131.

[14] Gama WA, Laughinghouse, HD, SantAnna CL. How diverse are coccoid cyanobacteria? A case study of terrestrial habitats from the Atlantic Rainforest (Sao Paulo, Brazil). *Phytotaxa.* 2014;178(2):61–97.

[15] Komárek J, Kopecký J, Cepák V. Generic characters of the simplest cyanoprokaryotes *Cyanobium, Cyanobacterium* and Synechococcus. *Cryptogam Algol.* 1999;20(3):209–22.

[16] Ercegovic A. *Dalmatella*, nouveau genre des Cyanophycées lithophytes de la côtes adriatique. *Acta Botanica Instituti Botanici Universitatis Zagrebensis*. 1929;4:35–41.
[17] Gardner NL. A new species of *Entophysalis* from China, and notes on other species of the genus. *Univ Calif Public Botany*. 1927;13:369–72.
[18] Kaštovský J, Fučíková K, Hauer, T, Bohunická M. Microvegetation on the top of Mt. Roraima, Venezuela. *Fottea*. 2011;11(1):171–186.
[19] Ercegovic A. Ekoloske i socioloske studije o litofitskim cijanoficejama sa jugoslovenske obale Jadrana. Rad Jugoslovenske Akademije Znanosti i Umjetnosti, Zagreb. 1932;244:129–220.
[20] De Toni G. Noterelle di nomenclatura algologica. Terzo elenco di Missoficee omonime. 1936;8:1–6.
[21] Mareš J, Hrouzek P, Kaňa R, Ventura S, Strunecký O, Komárek J. The primitive thylakoid-less Cyanobacterium Gloeobacter is a common rock-dwelling organism. *PLOS One*.2013;8(6), e66323.
[22] Saw JHW, Schatz M, Brown MV, Kunkel DD, Foster JS, Shick H, Christensen S, Hoa S, Wan X, Donachie SP. Cultivation and complete genome sequencing of Gloeobacter kilaueensis sp. nov., from a lava cave in Kilauea Caldera, Hawaii. *PLOS One*. 2014;8(10):1–12.
[23] Sant'Anna CL, Silva SMF, Branco LHZ. Cyanophyceae da Gruta-que-chora, Município de Ubatuba, Estado de Sao Paulo, Brasil. *Hoehnea*. 1992;18(2):75–97.
[24] Frémy P. Cyanophycées de la Nouvelle-Guinée. *Annal Cryptogam Exotiq*. 1932;5:190–7.
[25] Ercegovic A. Sur quelques nouveaux types des Cyanophycées lithophytes de la côte adriatique. *Arch Protist*. 1929;66:164–74.
[26] Ley SH. New Myxophyceae from Northern Kwangtung. *Bot Bull Acad Sinica*. 1947;1:77–9.
[27] Gama WA, Rigonato J, Fiore MF, Sant'Anna, CL. New insights into *Chroococcus* (Cyanobacteria) and two related genera: *Cryptococcum* gen. nov. and *Inacoccus* gen. nov. *Eur J Phycol*. 2019;54(3):315–25.
[28] Hašler P, Pentecost A, Jahodářova E, Dvořák P, Poulíčková A. Taxonomic revision of *Ulva montana* (Lightfoot 1777) and description of a new genus of *Lightfootiella* (Cyanophyceae, Chroococcaceae). *Phytotaxa*. 2018;362(2):173–86.
[29] Ercegovic A. Litofitska vegetacija vapnenaca i dolomita u Hrvatskoj. La végétation lithophytes sur les calcaires et les dolomites en Croatie. *Acta Botanica Instituti Botanici Universitatis Zagrebensis*. 1925;10:64–114.
[30] Watanabe M, Komárek J. Blue-green algae from Kathmandu. In: Watanabe M, Malla SB, editors. Cryptogams of the Himalayas, Vol. 1. The Kathamandu Valley. Tsukuba, Japan; 1988. p. 1–20.
[31] Claus G. Betitage zur Kenntnis der Algenflora der Abaligeter Hole. *Hydrobiologia*. 1962;19:192–216.
[32] Drouet F. Revision of the classification of the Oscillatoriaceae. *Proc Natl Acad Sci USA*. 1968;15:1–370.
[33] Ercegovic A. *Podocapsa* et *Brachynema*, deux genres nouveaux Chamaesiphonales de la côte adriatique de Dalmatie. *Acta Botanica Instituti Botanici Universitatis Zagrebiensis*. 1931;6:33–7.
[34] Azevedo MT de P, Kováčik L. *Rhabdogloea brasilica* sp. nov. (Chroococcales, Synechococcaceae): morphological and morphometric variability under cross-gradient cultures. *Arch Hydrobiol/ Algolog Stud* 1996;83:83–92.
[35] Wang Y, Cai F, Jia N, Li R. Description of a novel coccoid cyanobacterial genus and species *Sinocapsa zengkensis* gen. nov. sp. nov. (Sinocapsaceae, incertae sedis), with taxonomic notes on genera in Chroococcidiopsidales. *Phytotaxa*. 2019;409(3):146–60.
[36] Katoh H, Itoh S, Shen JR, Ikeuchi M. Functional analysis of psbV and a novel c-type cytochrome gene psbV2 of the thermophilic cyanobacterium *Thermosynechococcus elongatus* strain BP-1. *Plant Cell Physiol*. 2001;42(6):599–607.
[37] Anagnostidis K. Nomenclatural changes in cyanoprokaryotic order Oscillatoriales. *Preslia*. 2001;73:359–75.
[38] Komárek J, Anagnostidis K. Cyanoprokaryota. Part 2: oscillatoriales. In: Ettl H, Gärtner G, Heynig H, Mollenhauer D, editors. Süsswasserflora von Mitteleuropa 19/2. München, Germany: Elsevier Spektrum Akademischer Verlag; 2005. p. 1–759.

[39] An GS. Validation of new species names of blue-green algae (Cyanophyta) from Korea. *Fragmenta Floristica et Geobotanica*. 1992;37(2):417–424.
[40] Brito A, Ramos V, Mota R, Lima S, Santos A, Vieira J, Vieira CP, Kaštovský J, Vasconcelos VM, Tamagnini P. Description of new genera and species of marine cyanobacteria from the Portuguese Atlantic coast. *Mol Phylogenet Evol*. 2017;111:18–34.
[41] Strunecký O, Komárek J, Šmarda J. *Kamptonema* (Microcoleaceae, Cyanobacteria), a new genus derived from the polyphyletic Phormidium on the basis of combined molecular and cytomorphological markers. *Preslia*. 2014;86:193–207.
[42] Anagnostidis K, Komárek J. Modern approach to the classification system of cyanophytes. 3. Oscillatoriales. *Arch Hydrobiol*. 1988(Suppl. 80):327–472.
[43] Strunecký O, Komárek J, Johansen JR, Lukešová A, Elster J. Molecular and morphological criteria for revision of the genus *Microcoleus* (Oscillatoriales, cyanobacteria). *J Phycol*. 2013;49(6):1167–80.
[44] Osorio-Santos K, Pietrasiak N, Bohunická M, Miscoe LH, Kovácik L, Martin MP, Johansen JR. Seven new species of *Oculatella* (Pseudanabaenales, Cyanobacteria): taxonomically recognizing cryptic diversification. *Eur J Phycol*. 2014;49(4): 450–470.
[45] Zammit G, Billi D, Albertano P. The subaerophytic cyanobacterium *Oculatella subterranea* (Oscillatoriales, Cyanophyceae) gen. et sp. nov.: a cytomorphological and molecular description. *Eur J Phycol*. 2012;47(4):341–54.
[46] Mühlsteinová R, Hauer T, De Lay P, Pietrasiak N. Seeking the true *Oscillatoria*: a quest for a reliable phylogenetic and taxonomic reference point. *Preslia*. 2018;90:151–69.
[47] Umezaki I, Watanabe M. Enumeration of the Cyanophyta (blue-green algae) of Japan 1. Chroococcales and Oscillatoriales. *Jap J Phycol*. 1994;42:175–219.
[48] Kaštovský J, Veselá J, Bohunická M, Fučíková K, Štenclová L, Brewer-Carias C. New and unusual species of cyanobacteria, diatoms and green algae, with a description of a new genus *Ekerewekia* gen. nov. (Chlorophyta) from the table mountain Churí-tepui, Chimantá Massif (Venezuela). *Phytotaxa*. 2016;247(3):153–80.
[49] Martins MD, Machado De Lima NM, Branco LHZ. Polyphasic approach using multilocus analyses supports the establishment of the new aerophytic cyanobacterial genus *Pycnacronema* (Coleofasciculaceae, Oscillatoriales). *J Phycol*. 2019;55:146–59.
[50] Song GF, Jiang, YG, Li, R. *Scytolyngbya timoleontis*, gen. et sp. nov. (Leptolyngbyaceae, Cyanobacteria): a novel false branching Cyanobacteria from China. *Phytotaxa*. 2015;224(1): 72–84.
[51] Bohunická M, Johansen JR, Fučíková, K. *Tapinothrix clintonii* sp. nov. (Pseudanabaenaceae, Cyanobacteria), a new species at the nexus of five genera. *Fottea*. 2011;11:127–40.
[52] Umezaki I. *Yonedaella* nom. nov. *Taxon*. 1962;10:204.
[53] Ercegović A. Sur la valeur systématique et la ramification des genres *Brachytrichia* Zan. et *Kyrtuthrix* Erceg. et sur un nouveau type d'algue perforante. *Annal Protistol*. 1929;2:127–38.
[54] Kaštovský J, Fučíková K, Hauer T, Bohunická M. Microvegetation on the top of Mt. Roraima, Venezuela. *Fottea*. 2011;**11**(1):171–186.
[55] Palik P. A new blue-green alga from the cave Baradla near Aggtelek. Annales Universitatis Scientiarum Budapestinensis de Rolando Eötvös Nominatae. *Sectio Biologica*. 1960;3:275–85.
[56] Borzì A. Studi sulle Mixoficee. II (continuaz.). Nuovo Giornale Botanico Italiano, Nuova serie 1917;24:17–25.
[57] Bornet E, Flahault C. Revision des Nostocacées hétérocystées contenues dans les principaux herbiers de France (deuxième fragment). Annales des Sciences Naturelles, Botanique, Septième Série 1886 '1887';4:343–73.
[58] Jao CC. Studies on the fresh-water algae of China. XIII. New Myxophyceae from Kwangsi. *Sinensia*. 1944;15:75–90.
[59] Sant'Anna CL, Azevedo MTP, Fiore MF, Lorenzi AS, Kaštovský J, Komárek J. Subgeneric diversity of *Brasilionema* (Cyanobacteria, Scytonemataceae). *Rev Bras Botan*. 2011;34(1):51–62.

[60] Komárek J. Cyanoprokaryota, Part 3: Heterocytous genera. In: Büdel B, Gärtner G, Krienitz L, Schagerl M, editors. Süsswasserflora von Mitteleuropa 19/3. Berlin, Germany: Springer Spectrum; 2013. p. 1–1130.

[61] Saraf A, Suradkar A, Dawda HG, Gaysina LA, Gabidullin Y, Kumat A, Behere I, Kotulkar M, Batule P, Singh P. Phylogenetic complexities of the members of Rivulariaceae with the re-creation of the family Calotrichaceae and description of *Dulcicalothrix necridiiformans* gen nov., sp nov., and reclassification of *Calothrix desertica*. *FEMS Microbiol Lett.* 2019;366, fnz219.

[62] Copeland JJ. Yellowstone thermal Myxophyceae. *Ann N Y Acad Sci.* 1936;36:1–232.

[63] Kaštovský J, Berrendero EG, Hladil J, Johansen JR. *Cyanocohniella calida* gen. et sp. nov. (Cyanobacteria: Aphanizomenonaceae) a new cyanobacterium from the thermal springs from Karlovy Vary, Czech Republic. *Phytotaxa.* 2014;181(5):279–292.

[64] Komárek J, Genuário DB, Fiore MF, Elster J. Heterocytous cyanobacteria of the Ulu Peninsula, James Ross Island, Antarctica. *Polar Biol.* 2015;38:475–92.

[65] Hrouzek P, Lukešová A, Mareš J, Ventura S. (2013). Description of the cyanobacterial genus *Desmonostoc* gen. nov. including *D. muscorum* comb. nov. as a distinct, phylogenetically coherent taxon related to the genus Nostoc. *Fottea.* 2013;13(2):201–13.

[66] Claus G. Data on the ecology of the algae of Peace Cave in Hungary. *Nova Hedwigia.* 1962;4: 55–79.

[67] Friedmann I. *Geitleria calcarea* n. gen. et n. sp. A new atmophytic lime-incrusting blue-green alga. *Bot Notiser.* 1955;108(4):439–45.

[68] Kilgore Ch, Johansen JR, Mai T, Hauer T, Casamatta DA, Sheil CA. Molecular characterization of *Geitleria appalachiana* sp. nov. (Nostocales, Cyanobacteria) and formation of Geitleriaceae fam. nov.. *Fottea.* 2018;18(2):150–63.

[69] Skuja H. Algae. In: Handel-Mazzetti H., ed. Symbolae Sinicae. Botanische Ergebnisse der Expedition der Akademie der Wissenschaften in Wien nach Südwest-China 1914/1918. Vol. 1. Wien, Austria: Springer; 1937. p. 1–106.

[70] González-Resendiz L, León-Tejera H, Díaz-Larrea J, Alba-Lois L, Segal-Kischinezky C. *Hassallia littoralis* sp. nov. (Cyanobacteria, Microchaetaceae) from Mexico's marine supralittoral based on morphological and molecular evidence. *Phytotaxa.* 2013;137(1):35–47.

[71] Hernández-Mariné M, Canals T. *Herpyzonema pulverulentum* (Mastigocladaceae), a new cavernicolous atmophytic and limestone incrusted cyanophyte. *Algol Stud.* 1994;75:123–36.

[72] Lamprinou V, Hernández-Mariné M, Canals T, Kormas K, Economou-Amili A, Pantazidou, A. Morphology and molecular evaluation of *Iphinoe spelaeobios* gen. nov., sp. nov. and *Loriellopsis cavernicola* gen. nov., sp. nov., two stigonematalean cyanobacteria from Greek and Spanish caves. *Int J Syst Evol Microbiol.* 2011;61:2907–15.

[73] Borzì A. Note alla morfologia e biologia della Alghe Ficocromacee. *Nuovo Giornale Botanico Italiano.* 1878;10:236–88.

[74] Umezaki I. Revision of *Brachytrichia* Zanard. and *Krythuthrix* Erceg. Memoirs of the College of Agriculture, Kyoto University; 1958. p. 55–67.

[75] Borzì A. I generi delle Stigonemacee. In: Anon. Eds. Atti del Congresso dei naturalisti italiani: promosso dalla Società italiana di scienze naturali. Milano, Italy: Società italiana di scienze naturali; 1907. p. 370–75.

[76] Berrendero EG, Johansen JR, Kaštovský J, Bohunická M, Čapková K. *Macrochaete* gen. nov. (Nostocales, Cyanobacteria), a taxon morphologically and molecularly distinct from Calothrix. *J Phycol.* 2016;52(4):638–55.

[77] Vondrášková A, Fibich P, Lepš J, Kaštovský J. Determinants of cyanobacterial species composition in the splash zone of two Croatian islands. *Eur J Phycol.* 2017;52(12):179–89.

[78] Cai F, Li X, Yang Y, Jia N, Huo D, Li, R. *Compactonostoc shennongjiaensis* gen. et sp. nov. (Nostocales, Cyanobacteria) from a wet rocky wall in China. *Phycologia.* 2019;58(2):200–10.

[79] Cai F, Li X, Geng R, Peng Z, Li R. Phylogenetically distant clade of *Nostoc*–like taxa with the description of *Minunostoc* gen. nov. and *Minunostoc cylindricum* sp. nov. *Fottea*. 2019;19(1):13–24.

[80] Schmidle W. Algologische Notizen. Einige neue von Professor Dr. Hansgirg in Vorderindien gesammelte Süsswasseralgen. Allgemeine botanische Zeitschrift für Systematik, Floristik, Pflanzengeographie. 1900;VI (5):77–9.

[81] González-Resendiz L, Johansen JR, Alba-Lois L, Segal-Kischinevzky C, Escobar-Sánchez V, Jimenez Garcia LF, Hauer T, León-Tejera H. *Nunduva*, a new marine genus of Rivulariaceae (Nostocales, Cyanobacteria) from marine rocky shores. *Fottea*. 2018;18(1):86–105.

[82] Casamatta DA, Gomez SR, Johansen JR. *Rexia erecta* gen. et sp. nov. and *Capsosira lowei* sp. nov., two newly described cyanobacterial taxa from the Great Smoky Mountains National Park (USA). *Hydrobiologia*. 2006;561:13—26.

[83] Shalygin S, Shalygina R, Johansen JR, Pietrasiak N, Berrendero, EG, Bohunická M, Mareš J, Sheil CA. *Cyanomargarita* gen. nov. (Nostocales, Cyanobacteria): convergent evolution resulting in a cryptic genus. *J Phycol*.2017;53:762–77.

[84] Vaccarino MA, Johansen JR. *Scytonematopsis contorta* sp. nov. (Nostocales), a new species from the Hawaiian Islands. *Fottea*. 2011;11(1):149–61.

[85] Kováčik L, Komárek J. *Scytonematopsis starmachii*, a new cyanophyte species from the High Tatra Mts. (Czechosolvakia). *Algol Stud*. 1988;50–3:303–14.

[86] Geitler L. Cyanophyceae. In: Rabenhorst L., ed. Kryptogamen-Flora von Deutschland, Österreich und der Schweiz. 2nd ed. Vol. 14. Leipzig, Germany: Akademische Verlagsgesellschaft; 1932. p. 673–1196.

[87] Lamprinou V, Christodoulou M, Hernández-Mariné M, Parmakelis A, Economou-Amili A. *Spelaeonaias* gen. nov., a new true-branched cyanobacterium from Cave Vlychada (Diros, Peloponnese, Greece). *Phytotaxa*. 2016;282(3):171–18.

[88] Sant'anna CL, Azevedo TMP, Kaštovský J, Komárek J. Two form-genera of aerophytic heterocytous cyanobacteria from Brasilian rainy forest 'Mata Atlântica'. *Fottea*. 2010;**10**(2):217–28.

[89] Asencio AD, Aboal M, Hofmann L. A new cave-inhabiting blue-green alga: *Symphyonema cavernicolum* sp. nova (Mastigocladaceae, Stigonematales). *Arch Hydrobiol/Algol Stud*. 1996;83:73–82.

[90] Hauer T, Bohunická M, Johansen JR, Mareš J, Berrendero-Gómez E. Reassessment of the cyanobacterial family Microchaetaceae and establishment of new families Tolypothrichaceae and Godleyaceae. *J Phycol*.2014;50:1089–1100.

[91] Lamprinou V, Skaraki K, Kotoulas G, Economou-Amilli A, Pantazidou A. *Toxopsis calypsus* gen. nov., sp. nov. (Cyanobacteria, Nostocales) from cave 'Francthi', Peloponnese, Greece: a morphological and molecular evaluation. *Int J Syst Evol Microbiol*. 2012;62:2870–77.

Patrick Jung and Burkhard Büdel

2.3 Lichens as pioneers on rock surfaces

Lichens can be understood as multiorganismic complex organisms with a symbiotic character that represents a self-sustaining ecosystem formed by the interaction of an exhabitant fungus (sometimes more) and an extracellular arrangement of one or more photosynthetic partners, as well as an indeterminate number of other microscopic organisms. Lichens are among the first microbial organisms colonizing barren surfaces as, for example, rocks, where they have various benefits that make them superior to other organisms and thus they are considered pioneers. Although the greatest proportion of lichens colonize the rock's surfaces (saxicolous lichens), others prefer to colonize deeper structures underneath the rock's surface (endolithic lichens), which interact with the lithomatrix. Here, we discuss the ecology of lichens as pioneers from various biomes, their adaptations to a life on rocks, and the succession of lichen communities and give insights into their interactions with the lithomatrix.

2.3.1 Lichens: a self-sustaining ecosystem

Carl Linnaeus described lichens as "rustici pauperimi," which could be translated as miserable mob of vegetation [1]. At that time, he did not know that lichens represent an obligate mutualistic ectosymbiosis between at least one photosynthetic eukaryotic green algal or prokaryotic cyanobacterial species (photobiont) and one (in most cases) fungal ascomycete or, more rarely, basidiomycete species (mycobiont). During this relationship, the primary photobiont does not need to be restricted to a single strain of algae, but rather a coexistence of multiple strains may contribute to the resilience of lichens [2]. These multiorganismic interactions are so complex that it took until 2016 to find out that the "one lichen–one fungus" paradigm, which lasted over 140 years, seemed to be wrong or is at least not the full story about lichens [3]. Instead of being composed of the known ascomycete and the photosynthesizing partner, unexpectedly specific basidiomycete yeasts embedded in the cortex seem to play a crucial but rather unknown role within this intimate association. Meanwhile, a plethora of heterotrophic bacteria, protists, and even viruses have been discovered in association with lichens [4, 5], highlighting the complex character of lichens of any kind. As early as 1976, Farrar [6] suggested that lichens might have evolved as open systems, sometimes with special structures to facilitate gaseous exchange, and thus might be interpreted as miniature ecosystems, including a variety of organisms operating at different trophic levels. This progressive concept has only recently been accepted and lead to a new definition of lichens. According to that, the lichen symbiosis should be understood as "a self-sustaining ecosystem formed by the interaction of an exhabitant

https://doi.org/10.1515/9783110646467-005

fungus and an extracellular arrangement of one or more photosynthetic partners and an indeterminate number of other microscopic organisms" [7], finally making a point of a lichen being a multiorganismic complex, a microecosystem in itself.

2.3.2 Ecology and diversity of rock lichens

Lichens grow naturally on all substrates, including very nutrient-poor ones such as rocks or even artificial surfaces. Their ability to take up most nutrients from the air or ambient water enables them to colonize even rather hostile microhabitats that are mostly devoid of other vegetation. Enormous amplitudes of variation in water availability, surface temperature, or wind speed over very short time spans are just some of the diverse stresses that an organism conquering rock or mineral surfaces faces. Because of their poikilohydric nature, they are only active during periods of water availability in their environment, and thus, they are poor competitors compared with vascular plants and tend to favor open habitats such as bare soil and stony substrates. Besides pro- and eukaryotic algae, as well as few fungi, the majority of the biodiversity on rocks usually consists of lichens except for submerged or heavily shaded rocks. Besides those growing as epiphytes on other plantlike organisms, the majority of lichens seem to be saxicolous (rock inhabiting), living in most parts of the world. Despite lichens on natural rocks, we encounter them every day on artificial substrates such as concrete, cobblestones, or walls of houses wherever we go. Some might even grow on glass surfaces. Most of them appear to be as colorful as one could think. They do this on purpose just to excessively demonstrate how resiliently they can live on pure rock surface. Saxicolous lichens include members of the three main traditional morphological groups showing crustose, fruticose, and foliose growth forms (Fig. 2.3.1), each having a different way of attaching themselves to the substrate, beside other traits. Fruticose lichens appear to be bushlike and are only slightly attached to the substrate such as the *Ramalina* species (Fig. 2.3.1A), whereas crustose thalli form a tight crustlike structure that is inseparably bound to the lithic surface such as *Lecidea* or some *Acarospora* species (Figs. 2.3.1B and 2.3.3B, yellow). The leaflike flat or slightly wrinkled thalli of foliose lichens are only partly attached to the substrate via rootlike structures called rhizines (laciniate lichens) or with a central umbilicus emerging from their underside (umbilicate lichens) such as *Lassalia pustulata* (Fig. 2.3.1C) or *Umbilicaria cylindrica* (Fig. 2.3.1D).

In contrast to those saxicolous lichens that are visibly colonizing the rock surfaces, endolithic lichens are hidden a few millimeters underneath the rock surface in the zone of light penetration. They are not easily recognizable because their color is very similar to the rock and their thalli completely penetrate the lithomatrix of the stone and weaken the stone structure by altering the substratum. These endolithic lichens can withstand extreme environmental conditions by colonizing the inside of lithic substrata; thus, they can especially be found in areas where the environmental

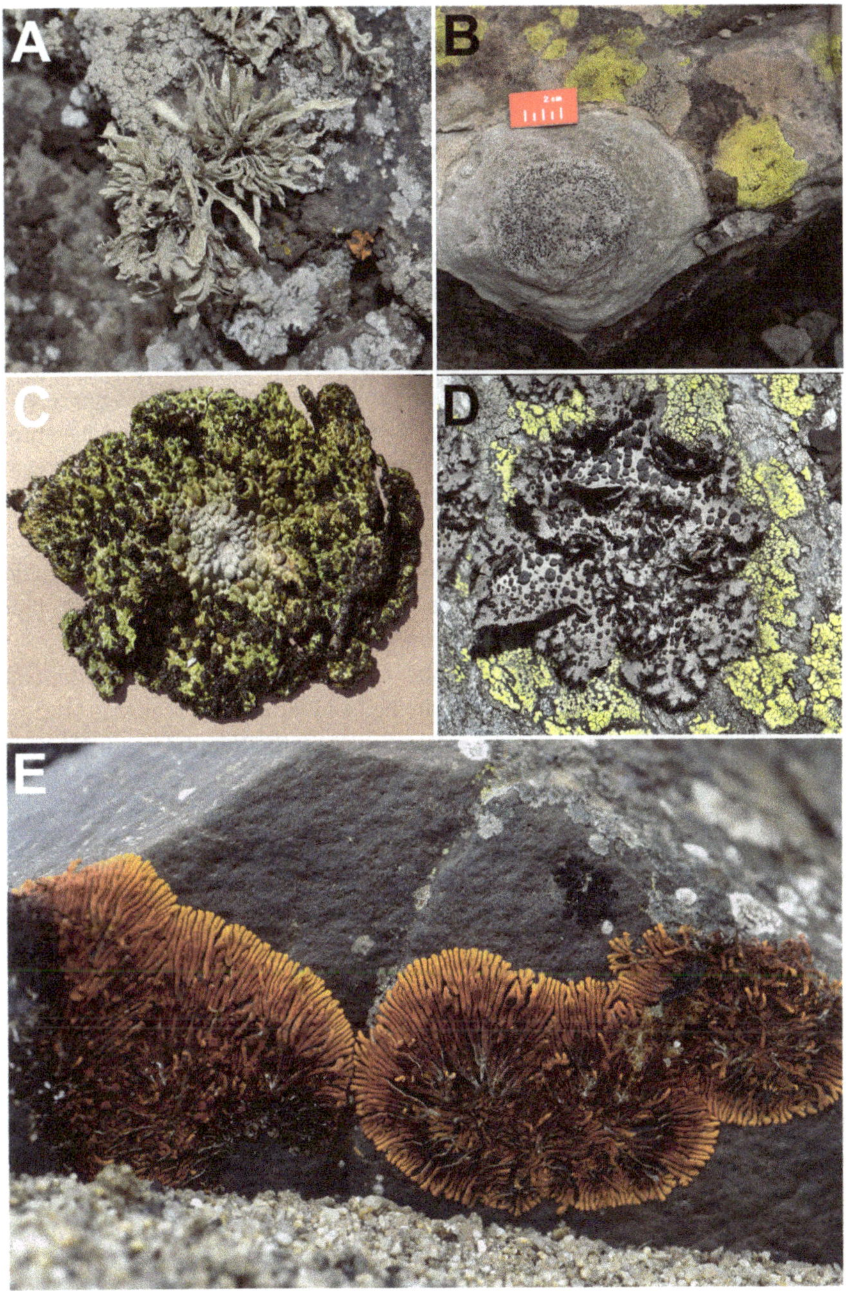

Fig. 2.3.1: Growth forms of rock lichens. (A) Pioneer *Ramalina* sp. from La Palma, Canary Island. (B) Crustose lichen *Lecidea* sp. (whitish) and *Acarospora* sp. (yellow) on Clarens Sandstone from South Africa. (C) The umbilicate lichen *Lassalia pustulata* as a pioneer on silicate rocks in temperate zones (wet thallus). (D) The umbilicate lichen *Umbilicaria cylindrica* (gray) and the crustose *Rhizocarpon geographicum* on silicate rock from the alpine mountains (Hochtor), Austria, European Alps. (E) *Caloplaca elegantissima*, a pioneer on basaltic rock at Wlotzkasbaken, Namib Desert.

conditions exacerbate the colonization of rock surfaces such as, e.g., Antarctica or the Sonora Desert [8, 9].

Growth rates and specific thallus structures

One of the most abundant saxicolous lichens is *Rhizocarpon geographicum* (Fig. 2.3.1B, yellow), a frequent pioneer colonizer of newly exposed rock surfaces [10]. The thallus comprises yellow-green lichenized areolae, which develop and grow on the surface of a nonlichenized fungal hypothallus, the latter extending beyond the edge of the areolae to form a marginal ring [11]. Because of its broad geographic distribution, *R. geographicum* has been the major subject of several studies recording its radial growth rates ranging from 0.05 mm year^{-1} in West Greenland [12], 0.5 mm year^{-1} in Switzerland [13], to over 0.94 mm year^{-1} in North Wales [14]. All of these measurements are possibly still considerably higher than the slowest radial growth rate yet measured for a crustose lichen by Benedict [15] in a study over 16 years of the related *Rhizocarpon superficiale* from the Front Range in Colorado, a site at which an overall average of 0.006 mm year^{-1} was recorded. The granite rock-inhabiting *Buellia frigida* (Fig. 2.3.4C) in the Dry Valleys of Antarctica has growth rates (0.01 mm year^{-1}) that match or slightly exceed those of *R. superficiale* [16].

Another easily recognizable saxicolous lichen is *Caloplaca elegantissima* (Fig. 2.3.1E), with its slightly dichotomous branched and radial vivid scarlet to orange-red thallus colonizing, e.g., NE-exposed sides of rocks in the Namib Desert [17]. This species prefers to grow on whitish quartz stones, and it has been shown that it has a comparably low light compensation point (the light intensity at which carbon fixation during photosynthesis equals carbon release during respiration), resulting in a longer phase of net photosynthesis that can have striking benefits in extreme ecosystems such as deserts [18]. This beneficial ecophysiological performance might also be explained by specific physiomorphological adaptations, such as algal cells that occur in thick vertical stacks (algal stacks), which are separated by vertical channels of light-transferring fungal hyphae (fungal stacks) that were also detected in *C. elegantissima* [19]. Lichens such as *C. elegantissima* with algal stacks can have both a higher photosynthetic capacity and higher dark respiration than species without algal stacks [19]. Fungal and algal stacks are not a rare feature but were also found in lichens that occur on rocks of deserts to steppe regions with high solar radiation, all over the world, especially among saxicolous lichen communities [19]. Other saxicolous representatives with these morphological features are, for example, *Diplotomma atacamae* from the Atacama Desert, *Aspicilia* species from Spain, or *Acarospora nodulosa* from Australia [19].

Endolithic lichens

By contrast, endolithic lichens have developed even more sophisticated ways of avoiding the harsh environmental conditions that can result in morphological adaptations.

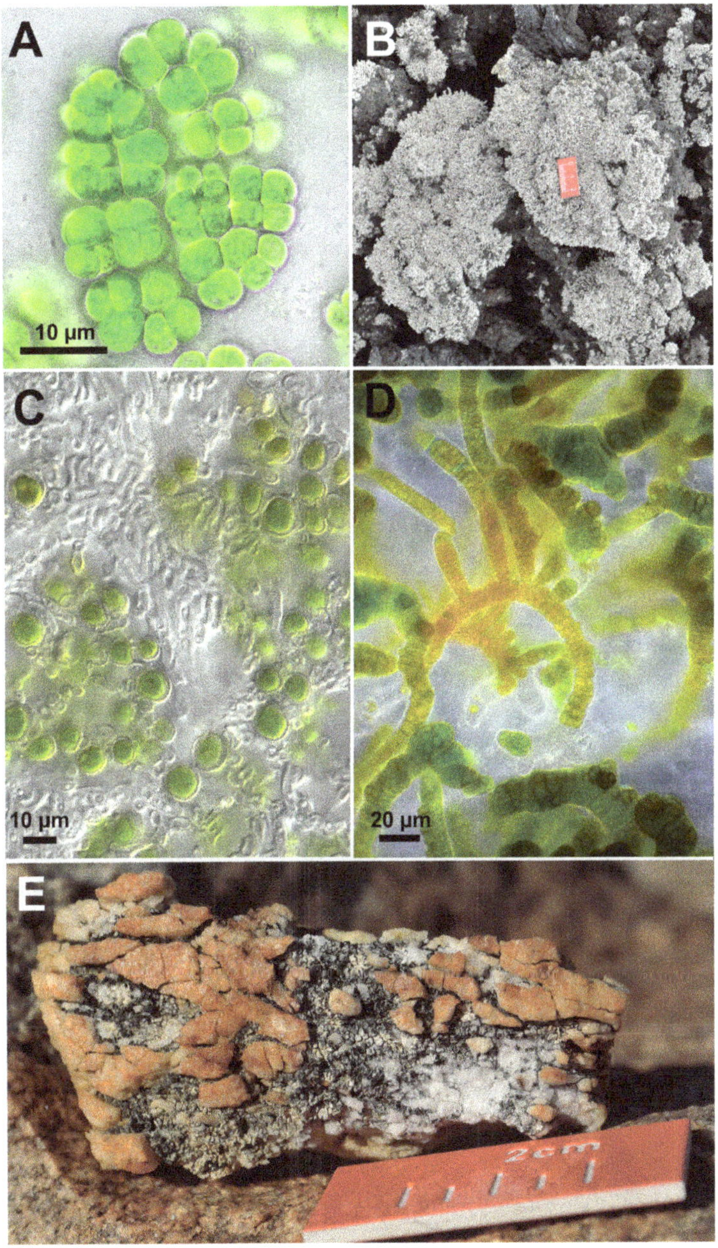

Fig. 2.3.2: Rock lichens and photobionts. (A) The cyanobacterium *Chroococcidiopsis*, a frequent photobiont of many cyanolichens. (B) *Stereocaulon* sp. from La Palma, Canary Island, a fruticose lichen and pioneer on cooled lava flows; the species has green algae as primary and cyanobacteria as secondary photobionts. (C) Microscopic cross section of a saxicolous *Buellia* lichen from the Atacama Desert with green algal photobionts of the genus *Trebouxia*. (D) *Stigonema* sp., a frequent cyanobacterial symbiont of some cyanolichens. (E) *Lecidea cancriformis*, an endolithic lichen on granitoid rock from the Darwin Glacier region, Antarctica.

Caliche plates commonly found in the Sonoran Desert, for example, were found to be colonized by the endolithic, whitish lichen *Verrucaria rubrocincta* [9, 20]. Here, the surface of the caliche reaches temperatures that regularly exceed 60°C during the summer and approach 0°C in the winter. Incident light intensities are high, with photosynthetically active radiation levels typically up to 2,600 µmol m^{-2} s^{-1} during the summer [20]. The endolithic growth of *V. rubrocincta* with its rounded patches on the rock surfaces is an adaptation to both rapid evaporation on the rock surface and high light intensities [9]. Cross sections of rocks inhabited by *V. rubrocincta* showed an anatomical zonation comprising an upper inorganic micrite layer that acts as a cap to the lichen that helps to trap moisture [20]. At the same time, the micrite layer is highly reflective and reduces light intensity to the algae below and acts as an efficient sunscreen that blocks harmful ultraviolet radiation [20].

Pioneer lichens and their photobionts of different climatic regions

As demonstrated in various studies on rock colonizing lichens across the globe, a specific pattern seems to emerge: besides rock type (acidic or alkaline), climate regime of the geographic region defines the rock-inhabiting community of lichens with a few dominant species for each habitat. Fully exposed rock surfaces of all types of tropical biomes, for example, such as table mountains in the Guyana Uplands of South America or Inselbergs in rainforests or humid and dry savannas have been shown to be home to a surprisingly diverse community of lichens with cyanobacterial photobionts, known as cyanolichens (reviewed by Büdel [21]). Several studies revealed that the characteristic color of Inselbergs is due to a dense colonization of mainly lichens of the genus *Peltula* (Fig. 2.3.4D) and members of the cyanolichen family Lichinaceae [21]. Those lichens appear blackish due to ultraviolet-absorbing substances like scytonemin and mycosporine-glycine that are produced by their unicellular cyanobacterial photobionts such as *Chroococcidiopsis* (Fig. 2.3.2A) [22]. Some of these cyanobacterial strains have been shown to be able to fix nitrogen [23–25], which can potentially contribute to the nitrogen budged of the soils at the close surroundings of such inselbergs. This is caused by rainfall runoff and desquamation of rock material, thus supporting the growth of gallery forests around inselbergs [21].

Jackson [26] described that *Stereocaulon vulcani* (comparable to Fig. 2.3.2B) together with species of the genera *Cladonia* and *Parmelia* were among the pioneer lichen community as the initial stage of vegetation on lava flows in Hawaii up to an elevation of 1000 m. He detected that *S. vulcani* preferred regions of higher rainfall, whereas all species of *Parmelia* and *Cladonia*, together with an unidentified crustose lichen, were found only in areas of lower rainfall, demonstrating the clear ecological preference of species within a lichen community. The dominance of *S. vulcani* in such a harsh and poor environment of lava was ascribed to its ability to invade vesicles and narrow recesses in the rock, its ability to accelerate the chemical weathering of the rock, and its rapid rates of dispersal, establishment, and growth [26]. In addition,

S. vulcani is a green algal lichen with *Trebouxia irregularis* (comparable to Fig. 2.3.2C) as the main photobiont and the cyanobacterium *Stigonema* sp. (Fig. 2.3.2D) as the second photobiont located in cephalodia, specialized structures of some lichens. Generally, the ability of the cyanobacterial photobiont to fix N_2 from the air enables the lichen consortium being an ideal pioneer on bare rock with the advantage of nitrogen supply, which could make it superior to the competing *Parmelia* and *Cladonia* species. Cyanobacterial photobionts in dominant rock pioneer lichens might also be the key for the dominance of *Placopsis gelida* and *Stereocaulon vesuvianum*, which were found to be among the first lichens on lava rock on the Island of Surtsey, about 30 km off the southern coast of Iceland, that newly emerged in late 1963 [27]. The green algal lichen *P. gelida* is characterized by a central cephalodium, harboring the filamentous, branched, and heterocytous cyanobacterium *Stigonema* sp. as a second photobiont, whereas the cyanobacteria (*Nostoc* or *Stigonema*) hosting cephalodia of *S. vesuvianum* are several per thallus. In general, the role of the photobionts seems to be striking for the establishment of rock lichens also determining their ecological preference. Recently, 232 crustose lichen specimens, collected along an elevational gradient (171–959 m a.s.l.) within the McMurdo Dry Valleys at Antarctica, were screened for their mycobiont-photobiont networks in relation to abiotic factors [28]. It turned out that the mycobionts were highly dependent on the availability of climatically adapted photobionts with elevation positively associated with water availability as the key factor explaining most of the distribution patterns of the lichens [28]. *Lecidea cancriformis* (Fig. 2.3.2E) and *Rhizoplaca macleanii*, for example, were found to be significantly more common at higher elevations and *Carbonea vorticosa* and *Lecidea polypycnidophora* at lower elevations [28]. This goes along with the clear ecological preferences and niche differentiations of lichen photobionts because some were mostly restricted to the higher elevated habitats (cold and humid) with a clearly dividing threshold at 600 m a.s.l. dividing higher and lower sites [28].

Making use out of growth and age: lichenometry

The intimate relation between long living lichens and their rock substrate has been a focal point of interest for lichenometry, a technique to date, e.g., late-Holocene terminal moraines that record glacier fluctuations. Traditionally, it relies on dating curves that relate diameters of the largest lichens in a population to surface ages and can give insights into past and contemporary climate change. The inferred rates at which landscapes respond, both biologically and physically, to these climatic changes are hence dependent on the accuracy and precision of lichenometric dates, which is still a controversial debate. The basic physiology of at least some lichens might be understood, but applications of lichenometry is greatly hampered by exceptional slow growth rates of the lichens, the resemblance of dead lichens to living ones, and by the difficulty of documenting colonization by sexual or asexual propagules [29]. However, calibrated dating curves are the basis of lichenometry that relates diameters of the largest—and

Fig. 2.3.3: Diversity of rock lichens. (A) Microscopic cross section of an apothecium of a saxicolous *Buellia* species from the Atacama Desert with black (melanized) spores. (B) *Acarospora schleicheri* (yellow) on silicate rock and *Xanthoria elegans* (orange) on slate from the Namib Desert, South Africa. (C) *Buellia frigida* on granite from Taylor Valley, Antarctica. (D) *Peltula euploca* as pioneer species on silicate rich Amphibolite, Switzerland (courtesy of Karl Bürgi-Meyer).

presumably oldest—lichens in a series of populations with the known ages of surfaces where those populations reside. Such approaches revealed, for example, that the lichenometrically "dated" moraines of Southeast Iceland to the second half of the 19th century may actually predate this time by several decades (30–100 years), thus throwing doubt on the exact timing of maximum glaciation during the "Little Ice Age" [30]. The method has also been used to suggest ages for surfaces that were initially lichen-free, such as the moai of Easter Island, rock surfaces exposed by tool stone quarrying or other applications in archeology [reviewed in 31]. However, the method had and still has to face valid criticism and without agreement on range of utility, treatment of error, and methods of measurement, sampling, and data handling. It turned out that a major source of error seems to be the assumption that the largest lichen within a population colonized soon after deposition and will survive indefinitely (reviewed by Osborn et al. [32]). A review of literature of *Rhizocarpon* (Fig. 2.3.1D, yellow), one of the most frequently used lichens for lichenometry, summarized gaps in knowledge regarding early development, growth rate/size curve, mortality, regeneration, competitive effects, colonization, and succession on rock surfaces [33]. The authors demonstrate that these processes at least for this lichen may not be comparable on different rock surfaces, especially in regions where growth rates and thallus turnover are high.

2.3.3 Colonizing the rock surface

Of the many properties of a substrate, one might examine to determine what causes the particular behavior of a lichen; the texture, the water relations, and the chemistry are most important. Each rock where a lichen population occurs constitutes a well-defined unit of a patchily distributed habitat type to which lichens are attached for the greatest part of their life cycle. The smoothness, hardness, relative stability, and surface features of substrates have often been cited as factors causing the restriction of lichens to one substrate or another. This becomes clear when considered how lichens distribute: lichens can easily conquer new habitats via fragmentation, a process during which parts of the lichen containing hyphae from the mycobiont wrapped around a few photobiont cells break apart. Transported by wind or runoff water, or carried by animals or humans, those propagules are able to develop a full lichen thallus once reaching a final destination under suitable conditions. Experiments showed, for example, that thalli of *Parmelia conspersa*, *P. glabratula* ssp. *fuliginosa*, and *Buellia aethalea* are established in plots on undisturbed and newly exposed smooth slate grown from such propagules transported by runoff waters over a period of 6 years [34]. By contrast, the spore formation in, e.g., the apothecium of the mycobiont (Fig. 2.3.3A) as a second propagation strategy enables only the fungal partner of the symbiosis to reach out for new habitats. Once the fungal spore transported via wind, water, animals, or humans reaches an adequate surface, the germination of the ascospores is followed by the formation of a prothallus-like mycelium [35]. This purely hyphal stage might be able to persist on the surface of rocks or other substrates over long periods once contact is made by a free-living compatible photobiont. This appears to be a critical stage of the early life of a lichen because, e.g., mycobionts could be nourished by, e.g., carbohydrate-rich leachates of the rocks dissolved in rain [36]. However, it is obvious that rocks are a substrate that is easy to conquer because lichen diaspores can become trapped and begin to develop on rough surfaces more easily than on smooth surfaces.

Succession of lichen communities on deglaciated rocks

In most cases, a succession of lichen species colonizing such an exposed rock over time can be observed beginning with a very few pioneer species that dominate the rocks over several years up to decades before others arise. A study that assessed the pattern of primary succession on a chronosequence of five rock outcrops exposed during the past 140 years by the retread of Glaciar Frias in the Patagonian Andes, Argentina, for example, showed that the saxicolous lichen species *Placopsis perrugosa* can be such a pioneer species [37]. This species is not only considered a successful colonizer there, with high growth rate and dispersal ability, but also frequently found dominating recently deglaciated terrains in Chile, New Zealand, and Antarctica [38–41]. This crustose lichen formed pure stands on the study rock outcrops during the first 50 years after deglaciation on the Patagonian Andes [37]. With increasing terrain age, *P. perrugosa* centers disintegrate probably because of limitations in nutrient transport from the periphery to the center of the thallus until it completely disappears [42], indicating that it has little effect of later colonizers [37]. This was also supported by others, who considered that crustose lichens were able to colonize bare surfaces early because of their ruderal life cycle traits but do not have relevant positive effects on later colonizers [43, 44]. During mid-succession, the fruticose lichens *Stereocaulon* spp. (comparable to Fig. 2.3.2B), *Cladonia lepidophora*, and *Cladonia subchordalis* followed together with mosses, 80 years after deglaciation until vascular plants and ferns took over 140 years after deglaciation together with the lichen species *C. subchordalis*, *Neofuscelia plana*, and *Buellia* sp. [37]. This example shows that the development of vegetation on rocks requires at least 100 years longer on bedrock outcrops than on unconsolidated glacial deposits and is closely related to the time required for the formation of a cryptogamic carpet starting with lichens (and algae).

Each of those lichens has to face various extreme conditions owed to the properties of rock once established. Rapid temperature shifts of rocks and therewith also of the lichen thallus are commonly happening, but it has been shown that extreme high surface temperatures do not seem to determine the presence or absence of lichen coverage on stonework. Instead, average stone surface temperatures over the course of the year seems to play a critical role in determining whether or not surfaces are receptive to lichen colonization at least on stone walls [45].

However, a study about *Caloplaca sublobulata* growing on rocks on two different islands of the South Shetland Islands of maritime Antarctica, close to the front of large glaciers, found striking differences in its colonization behavior depending on environmental conditions [46]. On the moraine of Livingston Island, rock size played an important role in lichen development, explaining most of the differences observed in the diameter of *C. sublobulata*, the number of species, and the percentage of cover among the rocks studied while on Robert Island; the distance from the glacier front was associated with the lichen cover of the rocks but not with its diameter [46]. On Robert Island, the lichen development seemed to be drastically affected by fluctuations in the persistence of snow cover after glacier front retreat [46]. The authors state that tentative associations between ice retreat and colonization, on the one hand, and changes in snow cover duration and the dynamic processes of extinction and recolonization, on the other hand, are suggested from comparison of the two zones.

Influences of the rock chemistry on lichens

Lichens are also affected by the chemical milieu, and many aspects of their microdistribution are determined by the chemistry of the rock substrate. The underlying geochemistry and mineralogy of rocks plays an important role in the occurrence of individual lichen species and the assembly of lichen communities [47]. Investigations on saxicolous lichen communities of the New Idria serpentinite mass, San Benito County, California, for example, showed that only 4 out of 119 lichen species were shared between various sites containing ultramafic and nonultramafic rocks, depending on the chemistry of the substrate [48]. The diversity of a lichen community is determined by the rock's chemistry, and it can also cause the appearance of chemosyndromes in lichen species. Lichen acids built by the mycobiont, for example, affect the acidity tolerance of lichens and, thus, the choice of their substrate [49]. Other authors showed that even secondary metabolites of saxicolous crustose lichens varied on different rock types, e.g., depending on the concentration of Ca and metals in the substrate [50]. In detail, gyrophoric, lobaric, psoromic, rhizocarpic, and stictic acids were common in crustose lichens in metal-poor habitats, whereas species with anthraquinones and lichens without any secondary metabolites were most abundant on limestone (alkalic and metal-poor) [50].

Slope and altitude

However, most of the abovementioned characters seem to be set by a few superior features such as the slope of the rock surface or altitude. A study on divers saxicolous lichen communities comprising 54 lichen species at Jonas Rockslide, Jasper National Park, Alberta, showed that the variables that were found to be most highly correlated with species distributions were altitude on the rockslide and inclination from horizontal of the rock face, thus determining temperature and water availability at the rock surface [51]. Here, direct correlations between specific lichen species and slope were found, indicating that, for example, *Rhizocarpon eupetraeum* was found on steeper rocks, whereas *Rhizocarpon bolanderi* had a negative correlation with slope [51]. In addition, species preferring steep slopping such as *R. eupetraeum* receive little direct radiation and retain less precipitation, both rain and snow [52, 53]. Thalli of species in this group are often discrete and surrounded by uncolonized rock, which suggests that competition may be unimportant in these environments [51].

2.3.4 Colonizing the lithomatrix

In contrast to epilithic lichens that colonize rock surfaces, endolithic lichens colonize the lithomatrix of stones and rocks with a loose hyphal-photobiont network beyond the extremes to which the surface is exposed. These organisms are able to colonize existing cracks and fissures (chasmoendolithic) and internal pores (cryptoendolithic) or penetrate actively into the rock (euendolithic) [54]. All of them are subjected to the most severe conditions of the terrestrial environment and are especially abundant under extreme ecological conditions, showing an extraordinary viability even there. Unfortunately, it is very difficult to study these lichens, their biology, and their adaptation to the rigors of their habitat, and even their taxonomy is as yet poorly understood.

Living under extreme conditions might make endolithic lichens unrivaled, but this is also true for the availability of potential photobionts, which are strongly limited there as well. Genetic investigations of an endolithic *Lecidea* sp. and others from Antarctica, for example, unveiled only one genotype of the photobiont green algae *Trebouxia* (comparable to Fig. 2.3.2C) [8]. These results could indicate a generally low abundance of free-living *Trebouxia* species in these harsh conditions and/or poor specificity of the mycobiont [8]. However, mycobionts less specific in their choice of photobiont are also able to survive in conditions in which only some photobiont species exist [55].

Although the interactions between the various partners of the symbiosis, mainly the relation between the mycobiont and the photobiont, are similar to those of epilithic lichens, endolithic lichens show quite some differences. Some endolithic species such as *Verrucaria marmorea*, *Opegrapha saxicola*, or *Rinodina bischoffii* contain the so-called oil hyphae in the medullary layer of the thallus [56, 57]. These hyphae of unknown function appear in different forms and accumulate unusually large amounts of oil in the form of oil globules of various sizes and shapes, which were first described in endolithic lichens by Zukal [58] and named “Sphäroidzellen”. It appears that the formation of oil hyphae and lipid production is an intrinsic characteristic of the fungal partner of some endolithic species because this feature also remains when the isolated mycobionts grow on culture medium [56].

The endolithic lifestyle is also reflected in the ecophysiological performance of some endolithic lichens such as *Acrocordia conoidea*, *Petractis clausa*, and

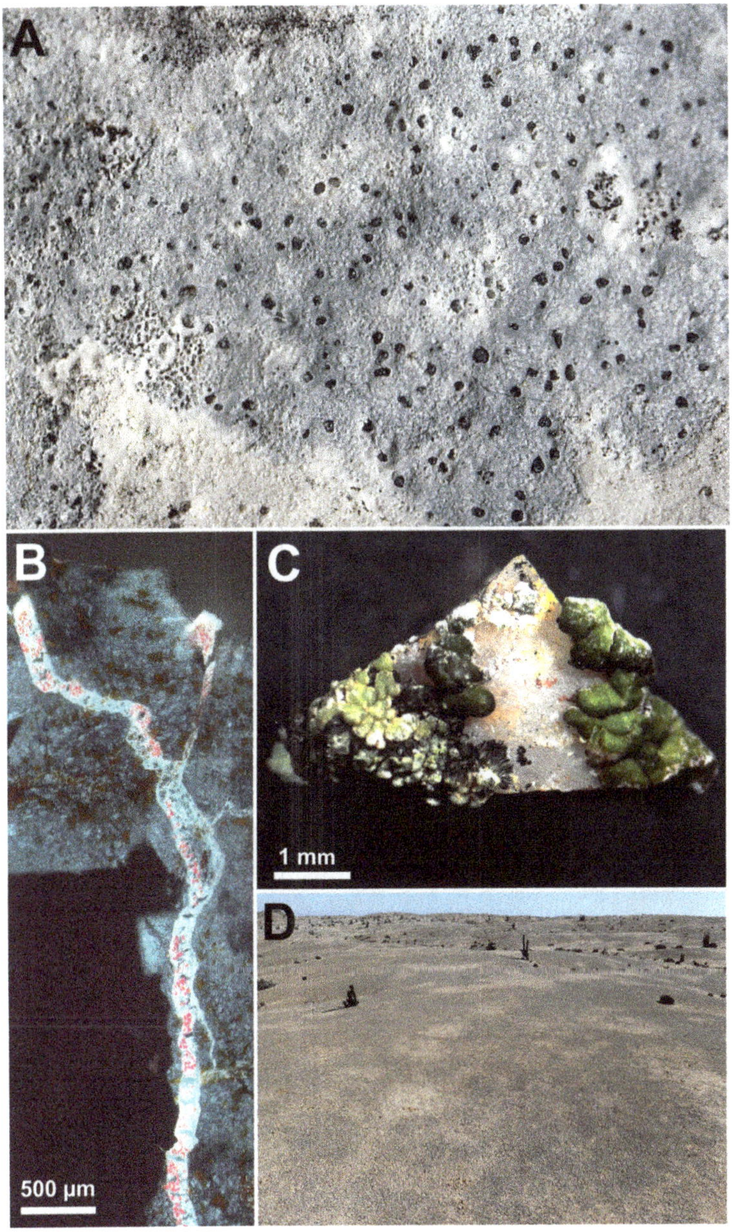

Fig. 2.3.4: Lichens and the lithomatrix. (A) The endolithic lichen *Hymenelia coerulea*, a pioneer of limestone of the European Alps, Austria. (B) Thin section of polycrystalline grit from the Atacama Desert with lichens (comparable to C) under fluorescence microscopy showing the penetration of the lichen (whitish autofluorescence of chitin) with its photobionts (red autofluorescence of chlorophyll) into deeper structures of the lithomatrix. (C) Stereomicroscopic image of a grit stone from the Atacama Desert colonized by various saxicolous lichens. (D) Landscape of the National Park Pan de Azúcar, Atacama Desert, showing blackish patterns of the grit crust, a biocenosis of lichens, algae, and fungi on and inside polycrystalline quartz stones (comparable to C).

Rinodina immersa from the Trieste Karst of Italy, which showed a rather small maximum photosynthetic rate ranging between 0.2 and 1.5 μmol CO_2 m^{-2} h^{-1} at optimal conditions compared with, e.g., fruticose lichens [59]. In addition, they have a high resistance to CO_2 diffusion with a saturation being reached only at a very large CO_2 concentration, indicating that endolithic lichens are slow-growing, stress-tolerant organisms, which are rather similar in their physiology to epilithic crustose lichens [59].

Endolithic lichens have for sure the most intimate contact with the lithomatrix compared with other lichens, and it is clear that they interact with it. Observations of such bioweathering patterns on, e.g., limestone triggered by lichens have been described for multiple cases, first in 1880, when Sollas [60] observed minute hemispherical pits on exposed limestone surfaces, which were produced by the apothecia of the endolithic lichen *Verrucaria rupestris* (=*Verrucaria muralis*) [61]. Simultaneous measurements of the oxygen uptake/release and pH shift of isolated green algal photobionts of the endolithic lichens *Hymenelia prevostii* and *Hymenelia coerulea* (Fig. 2.3.4A) revealed respiration-induced acidification of the medium in the dark followed by acidification effects at higher light intensities [61]. This mechanism allows both lichen species to conquer their endolithic niche, the lithomatrix of limestone of the eastern Alpine mountains, at the Untersberg [61].

2.3.5 Bioweathering

Several types of interactions between the lithic/mineral substrate and the lichens are well known as bioweathering processes that can significantly influence ecosystems from the micro- to the macroscale. Unlike the situation in soil, where there are more complex and often interacting factors, the zone of contact between saxicolous lichens and their rock substrate provides an ideal environment for studying the biological weathering of minerals. The lithomatrix is the place of physical and chemical activity, presenting a complicated heterogeneity in which primary and secondary minerals, organic acids and compounds, and living organisms, including the myco- and photobionts of lichens, free-living algae and fungi, and bacteria, are all involved [62]. Several studies have shown that lichens, even in some of the most extreme ecosystems as, for example, in Antarctica, mediate bioweathering actions. Here, various lichen to mineral actions have been documented that might serve as pathways for the initial pedogenesis of soils on Earth [63]. *Xanthoria elegans*, *Lecidea lapicida*, *R. geographicum*, and *Bacidia stipata*, for example, are cosmopolitan as well as endemic lichen species from maritime Antarctica that showed biomineralization processes as well as mineral transformations of significant meaning for such a little productive ecosystem dominated by cold and therewith slow rates of activity of any kind [64].

Bioweathering processes

One of the first bioweathering actions taking place once a lichen established on a rock surface is the penetration of the lithomatrix with fungal hyphae. These processes have been described in many studies (e.g., [65]), and often a significant pH reduction in the vicinity of cells on mineral surface attachment or a significant turgor pressure of around 10–20 MPa that is applied toward the lithomatrix can be observed [66]. This eventually leads to mineral mass loss at the interface of lithomatrices causing pits and tramlines, which was observed for ancient stones such as the Persepolis monuments, Iran [67], and also many other natural rock types (summarized by Farrar [6]). Such drillings and pits subsequently lead often to the establishment of the whole lichen in deeper parts of the stones within the lithomatrix (Fig. 2.3.2B), promoting further bioweathering processes. As mentioned before, lichens are poikilohydric and take up water from their surroundings. However, even during times without rain or fog, high air humidity or dew leads to frequent shrinking and swelling actions of single hyphae, photobiont cells, or even whole thalli within the lithomatrix. This has, for example, recently been observed within polycrystalline stones of only several millimeters in size in coastal regions of the Atacama Desert where fog leads to such swelling actions followed by the deterioration of the stones induced by the swelling action of the lichens [68]. In addition, freezing and thawing of such biological structures within stones can have similar effects [e.g., 69] as well as the swelling action of organic and inorganic salts as a by-product of other weathering processes [e.g., 70].

Besides physical forces, lichens can enhance bioweathering process also via a plethora of chemical actions. Their photosynthetic character, for example, that leads to the dissolution of respiratory CO_2 in water held by lichen thalli results in the generation of carbonic acid, which advances the solubilization processes by lowering the local pH values of the thallus and the related microenvironment. For example, Jackson and Keller [71] observed that the respiratory CO_2 of lichen-colonized lava flows on the Island of Hawaii effectively resulted in localized pH reduction in the microenvironment and, thus, promoted the chemical weathering rate.

In addition to respiratory CO_2, oxalic acid synthesized and secreted by the mycobiont plays a significant role in the deterioration of stone and rock material induced by lichens. Lichen plays a crucial role in the biochemical weathering of, e.g., granite via the production of oxalic acid and therewith in the formation of local soils in the cold temperate forest area of northeast China, such as northern Greater Hinggan [72]. During their experiments, the authors applied concentrations of oxalic acids similar to those within local lichens to granite powder and found great dissolving effects of granite, which significantly promoted the release of Na^+, K^+, Al^{3+}, Fe^{3+}, Mg^{2+}, Mn^{2+}, Ca^{2+}, and SiO_3^{2-} [72].

The above mentioned as well as several other mechanisms of bioweathering processes have been documented in the past, but most studies present single mechanisms. Recently, it could be shown that lichens can play significant roles from their microhabitat to the landscape in the coastal areas of the Atacama Desert. Here, various lichens together with algae and heterotrophic organisms colonize 6-mm-sized polycrystalline granitoid stones called *grits* (Fig. 2.3.4C), causing a blackish pattern on the ground (Fig. 2.3.4D) [73]. This unique biocenosis was termed *grit crust*, and the organisms involved mediate various of the abovementioned bioweathering processes simultaneous. Although the hyphae of the mycobionts penetrate the lithomatirx and embed the thallus into the lithomatrix, the photobiont gets in tight contact with mineral fragment and raises the pH, thus leading to the dissolution of quartz. In addition, fog and dew as the main regular water sources lead to the shrinking and swelling action of the lichens, which also colonize great parts of the internal structure of the grit stones, thus leading to a physical breakdown of the grits. The wetted lichen thalli on the surface of the grits can also trap dust and mineral particles from the air, which, together with the other effects, seems to lead to the accumulation of fine material as an initial stage of soil formation within the ecosystem of the coastal Atacama Desert. This is only one example highlighting the important biogeological and ecological role lichens can play as they are the first and persistent colonizers of exposed rock surfaces [68, 73].

References

[1] Linneaus, C. Species plantarum. Stockholm: Laurentius Salvius; 1753.

[2] Casano LM, del Campo EM, Garcia-Breijo FJ, Reig-Arminana J, Gasulla F, del Hoyo A, Guera A, Barreno E. Two *Trebouxia* algae with different physiological performances are ever-present in lichen thalli of *Ramalina farinacea*. Coexistence versus competition? *Environm Microbiol.* 2011; 13:806–18.

[3] Spribille T, Tuovinen V, Resl P, Vanderpool D, Wolinski H, Aime MC, Schneider K., Stabentheiner E, Toome-Heller M, Thor G, Mayrhofer H, Johannesson H, McCutcheon JP. Basidiomycete yeasts in the cortex of ascomycete macrolichens. *Science.* 2016;353:488–92.

[4] Wilkinson DM, Creevy AL, Kalu CL, Schwartzman DW. Are heterotrophic and silica-rich eukaryotic microbes an important part of the lichen symbiosis? *Mycology.* 2015;6:4–7.

[5] Petrzik K, Koloniuk I, Sehadova H, Sarkisova T. Chrysoviruses inhabited symbiotic fungi of lichens. *Viruses.* 2019;11:1120.

[6] Farrar JF. The lichen as an ecosystem: observation and experiment. In: Brown DH, Hawksworth DL, Bailey RH, editors. Lichenology: Progress and Problems. London (UK): Academic Press; 1976. p. 385–406.

[7] Hawksworth DL, Grube M. Lichens redefined as complex ecosystems. *New Phytol.* 2020;227(5):1281.

[8] De los Rios A, Sancho LG, Grube M, Wierzchos J, Ascaso C. Endolithic growth of two *Lecidea* lichens in granite from continental Antarctica detected by molecular and microscopy techniques. *New Phytol.* 2005;165(1):181–90.

[9] Bungartz F, Garvie LAJ, Nash TH III. Anatomy of the endolithic Sonoran desert lichen *Verrucaria rubrocincta* Breuss: implications for biodeterioration and biomineralization. *Lichenologist.* 2004;36:55–73.

[10] Armstrong RA. The biology of the crustose lichen *Rhizocarpon geographicum. Symbiosis.* 2011;55(2):53–67.

[11] Armstrong R, Bradwell T. Growth of crustose lichens: a review. *Geogr Ann A Phys Geogr.* 2010;92(1):3–17.
[12] Beschel RE. Dating rock surfaces by lichen growth and its application to the glaciology and physiography (lichenometry). In: Raasch GO, editor. Geology of the Arctic. Toronto: University of Toronto Press; 1961. p. 1044–62.
[13] Proctor MCF. Sizes and growth-rates of thalli of the lichen *Rhizocarpon geographicum* on the moraines of the Glacier de Valsorey, Valais, Switzerland. *Lichenologist.* 1983;15:249–61.
[14] Armstrong RA. Growth curves of four crustose lichens. *Symbiosis.* 2005;38:45–57.
[15] Benedict JB. Experiments on lichen growth III. The shape of the age–size curve. *Arct Antarct Alp Res.* 2008;40:15–26.
[16] Sancho LG, Green TGA., Pintado A. Slowest to fastest: Extreme range in lichen growth rates supports their use as an indicator of climate change in Antarctica. *Flora.* 2007;202:667–73.
[17] Schieferstein B, Loris K. Ecological investigations on lichen fields of the Central Namib. *Vegetatio.* 1992;98(2):113–28.
[18] Lange OL, Meyer A, Ullmann I, Zellner H. Mikroklima, Wassergehalt und Photosynthese von Flechten in der küstennahen Nebelzone der Namib-Wüste: Messungen während der herbstlichen Witterungsperiode. *Flora.* 1991;18(4):233–66.
[19] Vondrak J, Kubásek J. Algal stacks and fungal stacks as adaptations to high light in lichens. *Lichenologist.* 2013;45(1):115–24.
[20] Garvie LA, Knauth LP, Bungartz F, Klonowski S, Nash TH. Life in extreme environments: survival strategy of the endolithic desert lichen *Verrucaria rubrocincta. Naturwissenschaften.* 2008;95(8):705–12.
[21] Büdel B. Ecology and diversity of rock-inhabiting cyanobacteria in tropical regions. *Eur J Phycol.* 1999;34(4):361–70.
[22] Büdel B, Karsten U, Garcia-Pichel F. Ultraviolet-absorbing scytonemin and mycosporine-like amino acid derivates in exposed, rock inhabiting cyanobacterial lichens. *Oecologia.* 1997;112:165–72.
[23] Stewart WDP. Some aspects of structure and function in N_2-fixing cyanobacteria. *Ann Rev Microbiol.* 1980;34:497–536.
[24] Crittenden PD, Llimoa X, Sancho LG. Lichenized unicellular cyanobacteria fix nitrogen in the light. *Can J Bot.* 2007;87:1003–6.
[25] Mazhar S, Cohen JD, Hasnain S. Novel approach for the determination of nitrogen fixation in Cyanobacteria. *J Chem Soc Pakistan.* 2019;41(1):105.
[26] Jackson TA. Study of the ecology of pioneer lichens, mosses, and algae on recent Hawaiian lava flows. *Pac Sci.* 1971;25:22–32.
[27] Kristinsson H. Studies on lichen colonization in Surtsey 1970. *Surtsey Progress Report.* 1972;VI: 1.
[28] Wagner M, Bathke AC, Cary SC, Green TGA, Junker RR, Trutschnig W, Ruprecht U. Myco- and photobiont associations in crustose lichens in the McMurdo Dry Valleys (Antarctica) reveal high differentiation along an elevational gradient. *Polar Biol.* 2020: 1–17.
[29] Loso MG, Doak DF. The biology behind lichenometric dating curves. *Oecologia.* 2006;147(2): 223–9.
[30] Bradwell T. A new lichenometric dating curve for southeast Iceland. *Geogr Ann A Phys Geogr.* 2001;83(3):91–101.
[31] Benedict JB. A review of lichenometric dating and its applications to archaeology. *Am Antiq.* 2009: 143–72.
[32] Osborn G, McCarthy D, LaBrie A, Burke R. Lichenometric dating: Science or pseudo-science? *Quat Res*, 2015;83(1): 1–12.
[33] Armstrong RA. Lichenometric dating (lichenometry) and the biology of the lichen genus *Rhizocarpon*: challenges and future directions. *Geogr Ann A Phys Geogr.* 2016;98(3): 183–206.

[34] Armstrong RA. Field experiments on the dispersal, establishment and colonization of lichens on a slate rock surface. *Environ Exp Bot*. 1981;21(1):115–20.
[35] Smith AL. Lichens. Cambridge (UK): Cambridge University Press; 1921.
[36] Hale ME. The Biology of Lichens. London (UK): Edward Arnold; 1983.
[37] Garibotti IA, Pissolito CI, Villalba R. Vegetation development on deglaciated rock outcrops from Glaciar Frías, Argentina. *Arct Antarct Alp Res*. 2011;43(1):35–45.
[38] Orwin JF. Lichen succession on recently deposited rock surfaces. *N Z J Bot*. 1970;8:452–77.
[39] Lindsay DC. The role of lichens in Antarctic ecosystems. *Bryologist*. 1978;81:268–76.
[40] Galloway DJ. Lichens of Laguna San Rafael, Parque Nacional "Laguna San Rafael", Southern Chile: indicators of environmental change. *Glob Ecol Biogeogr*. 1992;2:37–45.
[41] Raggio J, Green TGA, Crittenden PD, Pintado A, Vivas M, Pérez-Ortega S, De los Ríos A, Sancho LG. Comparative ecophysiology of three *Placopsis* species, pioneer lichens in recently exposed Chilean glacial forelands. *Symbiosis*. 2012;56(2):55–66.
[42] Nash III TH. Photosynthesis, respiration, productivity and growth. In: Nash III TH, editor. Lichen Biology. Cambridge (UK): Cambridge University Press; 1996. p. 88–120.
[43] Longton RE. The role of bryophytes and lichens in terrestrial ecosystems. In: Bates JW, Farmer AM, editors. Bryophytes and Lichens in a Changing Environment. Oxford: Oxford University Press; 1992. p. 32–75.
[44] Kurina LM, Vitousek PM. Nitrogen fixation rates of *Stereocaulon vulcani* on young Hawaiian lava flows. *Biogeochemistry*. 2001;55:179–94.
[45] de la Rosa JPM, Porcel MC, Warke PA. Mapping stone surface temperature fluctuations: implications for lichen distribution and biomodification on historic stone surfaces. *J Cult Herit*. 2013;14(4):346–53.
[46] Valladares F, Sancho LG. Lichen colonization and recolonization of two recently deglaciated zones in the maritime Antarctic. *Lichenologist*. 1995;27(6):485–93.
[47] Purvis OW, Halls C. A review of lichens in metal-enriched environments. *Lichenologist*. 1996;28(6):571–601.
[48] Rajakaruna N, Knudsen K, Fryday A, O'Del RE, Pope N, Olday FC, Woolhouse S. Investigation of the importance of rock chemistry for saxicolous lichen communities of the New Idria serpentinite mass, San Benito County, California, USA. *Lichenologist*. 2012;44(5):695.
[49] Hauck M, Jürgens SR. Usnic acid controls the acidity tolerance of lichens. *Environ Pollut*. 2008;156:115–22.
[50] Paukov A, Teptina A, Morozova M, Kruglova E, Favero-Longo SE, Bishop C, Rajakaruna N. The effects of edaphic and climatic factors on secondary lichen chemistry: a case study using saxicolous lichens. *Diversity*. 2019;11(6):94.
[51] John E, Dale MRT. Environmental correlates of species distributions in a saxicolous lichen community. *J Veg Sci*. 1990;1(3):385–92.
[52] Creveld M. Epilithic lichen communities in the alpine zone of southern Norway. *Bibliotheca Lichenologica*. 1981;17.
[53] Daniels FJA. Vegetation of the Angmagassalik district of southeast Greenland. III. Epilithic microlichen communities. *Medd Gronl*. 1975;198(3):1–32.
[54] Golubic S, Friedmann I, Schneider J. The lithobiontic ecological niche, with special reference to microorganisms. *J Sediment Petrol*. 1981;51:475–8.
[55] De Los Rios A, Wierzchos J, Sancho LG, Green TGA, Ascaso C. Ecology of endolithic lichens colonizing granite in continental Antarctica. *Lichenologist*. 2005;37(5):383–95.
[56] Kushnir E, Tietz A, Galun M. "Oil hyphae" of endolithic lichens and their fatty acid composition. *Protoplasma*. 1978;97(1):47–60.
[57] Pinna D, Salvadori O, Tretiach M. An anatomical investigation of calcicolous endolithic lichens from the Trieste karst (NE Italy). *Plant Biosystems*. 1998;132(3):183–95.
[58] Zukal H. Über das Vorkommen von Reservestoffbehältern bei Kalkflechten. *Bot Zeitung*. 1886;44:761–70.

[59] Tretiach M, Pecchiari M. Gas exchange rates and chlorophyll content of epi-and endolithic lichens from the Trieste Karst (NE Italy). *New Phytol.* 1995;130(4):585–92.
[60] Sollas WJ. On the activity of a lichen on limestone. *Rep Br Ass Advmt Sci.* 1880:586.
[61] Weber B, Scherr C, Bicker F, Friedl T, Büdel B. Respiration-induced weathering patterns of two endolithically growing lichens. *Geobiology.* 2011;9(1):34–43.
[62] Chen J, Blume HP, Beyer L. Weathering of rocks induced by lichen colonization—a review. *Catena.* 2000;39(2):121–46.
[63] Mergelov N, Mueller CW, Prater I, Shorkunov I, Dolgikh A, Zazovskaya E, Shishkov V, Krupskaya V, Abrosimov K, Cherkinsky A, Goryachkin S. Alteration of rocks by endolithic organisms is one of the pathways for the beginning of soils on Earth. *Sci Rep.* 2018;8(1):1–15.
[64] Ascaso C, Sancho LG, Rodriguez-Pascual C. The weathering action of saxicolous lichens in maritime Antarctica. *Polar Biol.* 1990;11(1):33–9.
[65] Adamo P, Violante P. Weathering of rocks and neogenesis of minerals associated with lichen activity. *Appl Clay Sci.* 2000;16(5–6):229–56.
[66] Howard RJ, Ferrari MA, Roach DH, Money NP. Penetration of hard substrates by a fungus employing enormous turgor pressures. *PNAS.* 1991;88(24):11281–4.
[67] Mohammadi P, Krumbein WE. Biodeterioration of ancient stone materials from the Persepolis monuments (Iran). *Aerobiologia.* 2008;24(1):27–33.
[68] Jung P, Baumann K, Emrich D, Springer A, Felde VJ, Dultz S, Baum C, Frank M, Büdel B, Leinweber P. Lichens bite the dust—a bio-weathering scenario in the Atacama Desert. *iScience.* 2020:101647.
[69] Arino X, Gomez-Bolea A, Saiz-Jimenez C. Lichens on ancient mortars. *Int Biodeterior Biodegrad.* 1997;40:217–24.
[70] Sand W. Microbial mechanisms of deterioration of inorganic substrates—a general mechanistic overview. *Int Biodeterior Biodegrad*, 1997;40:183–90.
[71] Jackson TA, Keller WD. A comparative study of the role of lichens and "inorganic" processes in the chemical weathering of recent Hawaiian lava flows. *Am J Sci.* 1970;269:446–66.
[72] Song JF, Ru JX, Liu XP, Cui XY. Oxalic acid and succinic acid mediate the weathering process of granite in the cold-temperate forest regions of Northeast China. *Eurasian Soil Sci.* 2019;52(8):903–15.
[73] Jung P, Baumann K, Lehnert LW, Samolov E, Achilles S, Schermer M, Wraase LM, Eckhardt K-U, Baader MY, Leinweber P, Karsten U, Bendix J, Büdel B. Desert breath—how fog promotes a novel type of soil biocenosis, forming the coastal Atacama Desert's living skin. *Geobiology.* 2020;18(1):113–24.

Christoph Scheidegger

2.4 High alpine lichens

Mountain regions host an overproportionate fraction of global biodiversity, supporting an estimated one-third of terrestrial biological diversity. In alpine landscapes, lichen diversity has often been reported to follow a unimodal distribution along altitudinal gradients with a midaltitudinal peak. In the European Alps, the montane belt includes the highest number of lichen species, but the mid altitudinal peak for rocks with a low calcareous content (saxicolous intermediate), terricolous and bryophilous species, and the growth form of umbilicate species peak at the subalpine belt. Alpine landscapes are strongly influenced by glacial oscillations during the Pleistocene and most prominently by the last glaciation. We hypothesize that *Umbilicaria virginis*, the only macrolichen restricted to the nival belt in the Alps, is a candidate species that survived the last glacial maximum on central alpine nunataks in the Alps and possibly also on ice-free peaks in other mountain systems. Survival on central alpine nunataks is also a likely hypothesis for a number of other alpine lichen species, including the rare saxicolous *Lecanora diaboli* and *Lecanora freyi* on rocks with a low content in carbonates, *Lecanora concolor*, and several lecideoid lichens on siliceous rocks. High alpine regions, including the nival belt, provide key habitats for extremophile lichen species that depend on cold climates, which are unavailable at lower altitudes. Nival regions of mountain systems are therefore important "museums" for a considerable number of lichens, many of them with boreal arctic-alpine distributions across diverse mountain systems. However, more studies are needed to test if high alpine and nival environments also act as "cradles" for lichen diversity, and immediate studies on threats and a possible decline of high alpine and nival lichens is needed to avoid that alpine and nival environments turn into "graves" because of the dramatic retreat of glaciers and permanent snow due to climate change.

2.4.1 Introduction

Mountain regions host an overproportionate fraction of global biodiversity, supporting an estimated one-third of terrestrial biological diversity [1]. Steep mountain slopes and opposite expositions dramatically influence temperature over short distances, and mountain ranges may separate humid from dry areas at a few kilometer distance [2]. At high elevations, permanent snow and glaciers limit the available habitat for plants, bryophytes, and lichens, but most of the highest peaks in the Alps and in other high mountain systems also provide ice-free habitats for lichens under the current situation and even during the last glacial maximum (Fig. 2.4.1). Over geologically short time scales, mountain regions are intrinsically unstable

https://doi.org/10.1515/9783110646467-006

Fig. 2.4.1: Most of the highest peaks in the Alps (Weisshorn, canton of Valais, 4505 m a.s.l., A) provided ice-free habitats for lichens during the last glacial maximum (LGM). Also, lower peaks like Ränfenhorn (front, 3255 m a.s.l., B) were possibly ice-free, although the slightly higher Rosenhorn (right in the back, 3688 m a.s.l., B) was covered by ice during the LGM and their rock surfaces had to be colonized after the retreat of the ice.

systems, undergoing substantial changes in response to tectonic, erosional, and climatic processes that lead to splitting and subsequent isolation of species ranges, evolutionary adaptation to changing conditions, and consequently, population differentiation [3]. These biological processes create a shifting balance between speciation and extinction, in which mountains may act as "cradles" (areas of especially rapid species origination), "museums" (areas of especially long-term persistence of species), and "graves" (areas with especially high rates of extinction) for biodiversity [3–5].

2.4.2 Upper altitudinal limits of lichens

The upper altitudinal limit of higher plants is currently known from the north flank of Sagarmatha at 6400 m [6]. The highest records of lichens are known from Makalu (Nepal) and were reported from 7400 m [7–10], and lichen records from about 6000 m and above are known to occur in Africa, Asia, and South America (Tab. 2.4.1). In the Alps, there is no altitudinal limit for lichens from where lichens have been reported from the permanently ice-free rocks at Dufourspitze (4633 m) or Signalkuppe (4553), the type locality of *Buellia leptolepis* [11, 12].

Species richness along altitudinal gradients

In alpine landscapes, lichen diversity has often been reported to follow a unimodal distribution along altitudinal gradients with a midaltitudinal peak. In the Himalayas, the total lichen species richness shows a unimodal relationship with elevation. The maximum modelled total richness occurred at 3100–3400 m [13], i.e., in subalpine landscapes where mountain forests, dwarf shrubs, meadows, and rock formations form complex land cover mosaics, regionally influenced by a long-lasting,

Tab. 2.4.1: Highest records of lichens in African, Asian, and South American Mountain systems according to Körner [6].

Mountain peak	Altitude (m)	Lichen taxa	Habitat	References
Makalu, Nepal	7400	*Carbonea vorticosa*, *Pertusaria bryontha*, *Lecanora polytropa* with parasitic *Cercidospora epiolytropa*	On rock and soil	T. Kunava, cited in [7–9]
K2, Karakorum, Pakistan	7000	*Xanthoria elegans*	On rock	[35]
Cachi, Andes, Argentina	6700			Halloy 1985, cited in [36]
Socompa Volcano, Andes, Argentina, and Chile	6060	Specimens lost in a lab fire: *Stereocaulon* sp., *Cladonia* sp., cf. *Thamnolia*	Warm spots around steam vents	[36]
Kilimanjaro	5896	*Umbilicaria* sp., *X. elegans*	Warm spots near active fumaroles	[37]

extensive human management. At this vegetation belt, epiphytic and lignicolous species reach their upper distribution limit because of the limit of their substrates, i.e., trees, shrubs, and dwarf shrubs with lignified trunks and branches and long-lasting, slowly decomposing lignum (i.e., phanerophytes and chamaephytes) [14]. Light-demanding, rock-inhabiting lichen communities have limited habitat availability in forest landscapes but find sufficient area in the subalpine vegetation belt and above. Therefore, crustose lichens that dominate rock surfaces peaked at a considerably higher altitude (4100–4200m) in the Himalayas, whereas foliose lichens had their maximum richness at consistently lower altitudes between 2400 and 2500 m [13].

In the European Alps, Nimis et al. [12] listed 3017 lichenized taxa (excluding nonlichenized or doubtfully lichenized and particularly dubious records and extremely poorly known taxa). The delimitation of the Alps [15] includes six bioclimatic/altitudinal zones, including the mesomediterranean, submediterranean/colline, montane, subalpine, alpine, and nival belts [12].

For an analysis of the altitudinal distribution of the species of the Alps, we have assigned the presence for each of the 3017 lichenized taxa to the bioclimatic/altitudinal zones and to the major substrate types, including bark, plant debris, calcareous rock, intermediate rock with low carbonate content, siliceous rock, calcareous soil, and noncalcareous soil. Total lichen species richness showed a clear unimodal relationship with elevation. With more than 2000 species, the montane belt includes the highest number of lichen species in the Alps, followed by the subalpine, submediterranean, mesomediterranean, and nival belts (Fig. 2.4.2A).

This montane species peak is characteristic for corticolous, lignicolous, saxicolous (both calcareous, and siliceous) species, whereas saxicolous species growing on rocks with a low calcareous content (saxicolous intermediate), terricolous and bryophilous species, and the growth form of umbilicate species peak at the subalpine belt (Figure 2.4.5.2B–I).

Growth forms of high alpine lichens

Of the 296 taxa occurring in the nival belt of the European Alps, 230 species are crustose (including placodioid, bullate, and endolithic) and only 20 are fruticose, 29 species are foliose, and 17 species are umbilicate. Six species are exclusively associated with cyanobacteria, 13 taxa include cyanobacteria in gall-like structures (cephalodia), and one species (*Solorina crocea*) includes a complex photobiont layer with green algae being concentrated in an upper and cyanobacteria in a lower photobiont layer (data analyzed from Nimis et al. [12], not shown). The vast majority of species include green algae as photobiont, although a considerable number of species is also loosely associated with free-living cyanobacterial microbial communities.

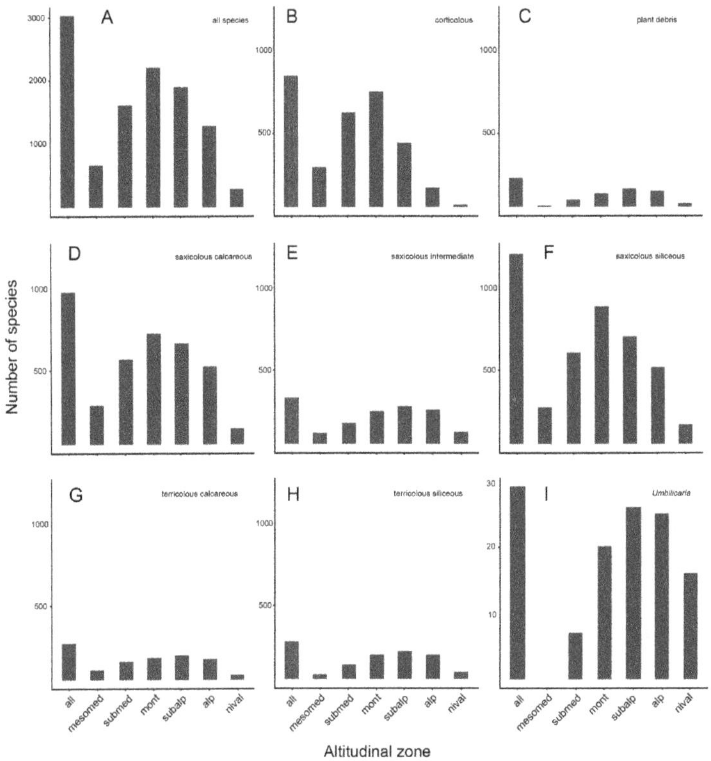

Fig. 2.4.2: Number of all lichen species of the Alps (A) and species growing on the specific substrate types bark (B), plant debris (C), calcareous rock (D), intermediate rock with low carbonate content (E), siliceous rock (F), calcareous soil (G) and noncalcareous soil (H), and umbilicate species as an example of a typically alpine growth form (I). Analyses based on Nimis et al. [12].

The majority of the species known from the nival belt has more often been reported from the alpine and even lower altitudinal zones, but five taxa are hitherto known to be restricted to the nival belt, i.e., *Lecidea leucothallina* var. *discrepans*, *Lecidea nivosa*, *Myriolecis behringii*, *Polysporina limborinella*, and *Umbilicaria virginis* [12].

2.4.3 High alpine lichen ecology and physiology

Although lichen communities up to the alpine belt generally have a high cover percentage, in the nival belt, lichens are generally more scattered and form only rarely dense communities, e.g., on the scattered patches of vegetation debris or on windblown ridges and summits on siliceous rock. Under less favorable conditions such as N-exposed rock slopes, lichen cover is less than 5% and 0% in those areas where snowfields or ice have receded during the last decades. Nevertheless, major parts of rock faces are colonized by a scattered lichen vegetation dominated by a relatively low number of species, often without a visible community structure, because the species pool at this altitude is considerably reduced compared with the alpine zone [16].

Lichens suitable to colonize the nival belt are thus extremophiles adapted to cope with a constantly extreme environment where the most important factors include the following:

- water stress because of desiccation during hours of low water vapor pressure in the atmosphere or because water is frozen,
- subzero temperatures during all seasons,
- intensive solar irradiation with high ultraviolet B index, and
- strong winds with sandblasting effects due to sand and ice crystals.

Nival lichens are poikilohydric organisms that are physiologically [17] and morphologically [18] adapted to frequent and strong desiccation. Switching off physiological activity during unfavorable periods and profiting from beneficial ecological periods give poikilohydric organisms an adaptive advantage over continuously active organisms under a harsh environment such as high alpine and nival areas.

Melting water is an important source for some of the most conspicuous lichen communities at the nival belt, including dense populations of the enigmatic *Umbilicaria virginis* [10, 19, 20]. This is the only macrolichen that is limited to the nival belt where it grows on ridges with frequent melting water supply during late summer (Fig. 2.4.3). With its relatively thick, leathery thallus, this species is able to take up a substantial amount of water and to maintain a high water content also during days with little or no melting water, especially during days where temperature is too low for the melting process.

Water uptake of a desiccated thallus from sources other than liquids is an important ecological strategy of lichens, especially those with green-algal photobionts [21], a process that can lead to nearly 100% of relative net photosynthesis in aerohygrophytic lichen species [18]. Even under subzero temperatures, water vapor uptake under snow can lead to a physiological reactivation of lichen thalli, a process that is especially important for substrate-hygrophytic nival lichens. However, water vapor uptake under snow-free conditions is a slow physical process that takes hours before saturation. Under the rapidly changing weather conditions at the nival belt, the importance of this process remains unclear but there are specifically alpine-nival rehydration processes that deserve further discussion. Under subzero temperature and strong wind, fog frequently forms hoarfrost on some but not all lichen growth forms. During such an event, up to 1 cm long multibranched dendrite ice crystals [22] formed on thalli of *Umbilicaria decussata* (Figure 2.4.4B), whereas several crustose lichens, including *Rhizocarpon* sp. and *Lecanora concolor*, growing close by showed no ice crystals growing on their surface. A second process involves ice condensation on hydrophilic

Fig. 2.4.3: (A) Highly frequented normal route to Mt. Jungfrau (4158 m a.s.l.) with the type locality of *Umbilicaria virginis* at the upper end of the rock ridge. (B) Alpinists pass less than 2 m from the luxuriant population of *U. virginis*. (C) Typical habitat of *U. virginis* at Weissmies (4017 m a.s.l.) with melting water and ice lenses hydrating the thalli during late summer. (D) Upper surface with apothecia and (E) lower surface with rhizines on pink lower cortex.

structures such as the terminal ends, e.g., of *Cladonia arbuscula* during cold nights when the relative air humidity reaches 100% and condensation starts below zero degrees during the minimum temperature at early morning hours. Under these conditions, tubular ice crystals [22] are formed at around -5°C and develop at the very tips of fruticose lichens, e.g., at the pycnidium bearing, hydrophilic ends of the podetia (Figure 2.4.4A). In both the *Umbilicaria* and the *Cladonia* examples, the formation of ice crystals was possible because the thalli were thermally isolated from the still relatively warm ground. In *Cladonia*, thermal isolation is caused by the fruticose growth where the 10-cm or longer thalli limit heat transport from the soil through the thallus to the tips. In *Umbilicaria*, thermal isolation from the massive rock is achieved by the foliose growth form of the lichen where the only contact of the thallus is maintained by the punctiform central holdfast of the umbilicus. In both cases, a subsequent warming above zero degrees will allow the lichen to take up the melting water from the ice crystals and rehydrate the thallus and consequently reactivate physiological activity, including respiration and photosynthesis.

Freezing processes cause various types of injury directly or indirectly associated with the freezing processes of water in tissues [23]. However, in *Umbilicaria aprina*, slow freezing of water-saturated thalli showed neither physiological nor morphological damage [24]. During slow cooling, a specimen in a freezing chamber from +5 to subzero temperatures showed an exotherm reaction at -5.4°C, indicating that this species from Antarctica leads to extracellular but intrathalline freezing of symplastic and apoplastic water, and this translocation of water to the medullary layer inside the lichen leads to a rapid dehydration of the protoplast. Structural changes of the mycobiont and the photobiont occurring during extracellular freezing are similar to processes observed during dehydration at ambient temperatures [24]. The water content of the lichen is maintained during freezing, and during a subsequent rise of the temperature above the melting point, this intrathalline water is taken up into the fungal and photobiont apoplast and symplast, leading to immediate physiological reactivation. We assume that this strategy is widespread among nival lichens of the genus *Umbilicaria* with thick thalli and possibly other lichen genera and plays an important strategy in optimizing water use in a frequently freezing and desiccating environment, especially during sunny periods.

Solar irradiation at high altitudes is strong, especially on mountain ridges and peaks where direct sunlight is available during a major part of the days. Secondary metabolites and pigments concentrated in the cortex reflect or absorb part of the irradiation and, thus, limit irradiation at the algal layer inside the lichen thallus. In addition, the thickness of the cortical and epinecral layer and the pigmentation or the content of calcium oxalate in the epinecral layer are known to vary according to light conditions of the specimen's habitat [25]. In addition to these stable adaptations to a specific solar environment, short-term changes to the activity state are known, which are related to the water content of the lichen thallus. In hydrated thalli, fungal hyphae are turgid, and their symplasts are filled with cytoplasm. During desiccation, the cellular loss of water volume is compensated by collapsing photobiont cells and cavitating fungal hyphae, especially in the cortical layer. Cavitated cortical cells still maintain their shape and size but form an air (or more precisely, a water vapor) bubble in their symplast [18], which includes some 70% to 85% of cell volume. This cavitation bubble substantially increases the light reflectance of the lichen thallus when dry and physiologically inactive and may prevent the (physiologically inactivated) photobionts from light damage [18]. In addition to these adaptations and strategies to intensive light situations, some bullate crustose species have developed a specific adaptation to high-light conditions that was termed Fensterpflanzen [26]. The photobiont layer is protected by a very thick epicortical layer that reflects a major part of light and scatters light that enters the lichen thallus. Photobionts are not concentrated in a typical algal layer below the cortex, but they are arranged laterally along the sidewalls of the thick bullate areoles and receive only indirect light that is scattered in the epinecral layer. *Psorinia conglomerata* is a typical species that refers to this type of window lichen (Fig. 2.4.4C). The borders of the bullate areoles are strongly pigmented and often have a rim of calcium oxalate that reflects sunlight. In the center of the areole, a thick epinecral layer scatters the entering light, and the algal layer is concentrated along the margins of the areole, but not directly under the epinecral layer. Window lichens are not exclusive for the nival belt, and especially in semiarid environments, other examples have been reported [26].

2.4.4 High alpine refugia during last glacial maximum

Alpine landscapes are strongly influenced by glacial oscillations during the Pleistocene and most prominently by the last (Würm) glaciation that had its last glacial maximum (LGM) around 18–20,000 years BP [27]. Biodiversity of the Alps is therefore driven by several cycles of extinction and colonization processes, species often immigrating from distant areas such as other mountain systems several thousand kilometers away [6, 28]. A high number of plant species recolonized the Alps from various peripheral refugia [29], but only a few studies have evidenced the survival of plants on central alpine nunataks during the LGM, e.g., for *Eritrichium nanum* [30–32].

For lichens, no detailed phylogeographic analyses are available that can unambiguously specify the type of their refugia during LGM, although four alternative recolonization processes can be hypothesized, i.e., long-distance migration from remote mountain systems or warmer refugia, upwards migration from periglacial areas or peripheral refugia, and survival on central alpine and nival nunataks, respectively.

Several species, including *Peltigera rufescens*, *Physconia muscigena*, *Psora decipiens*, *Squamarina gypsacea*, and *Xanthoria elegans*, that are currently known from the nival belt colonize a broad altitudinal range from xerothermic lowland sites up to the nival belt, and we assume that their survival during the LGM was possible but did not depend on nunataks or peripheral refugia. It is likely that the majority of these species survived the LGM in periglacial areas or even in the large Mediterranean refugial areas. Other species such as *Bellemerea alpina*, *Cladonia coccifera*, *Frutidella caesioatra*, *Hymenelia coerulea*, *Lecidea confluens*, *Lecidea lapicida*, *Melanelia hepatizon*, *Parmelia omphalodes*, *Rhizoplaca chrysoleuca*, and *Umbilicaria cylindrica* are frequent at montane and subalpine altitudes and likely have survived in peripheral refugia where relatively large habitat areas with suitable climatic conditions were available during the LGM. However, for *Umbilicaria virginis*, the only species that is currently limited to the nival belt and which has been observed relatively frequently in the Alps during the last decades, it is unlikely that suitable habitats were present in the periglacial lowland or on peripheral refugia. Based on the author's field experience, the typical habitat of the species consisting of steep rocks with seasonal water runoffs is relatively frequent also at lower altitudes where infraspecific competition mainly from other taxa of the genus *Umbilicaria* is still low to moderate. Habitat availability and low interspecific competition would allow *U. virginis* to establish at lower altitudes if ecological conditions would allow. However, *U. virginis* is limited to altitudes at or above 3000 m a.s.l. in the Alps, where 13 out of 14 localities studied by E. Frey [20] and the author are known from locations that were alpine nunataks during the LGM [33]. The remaining locality is within 500 m from alpine nunataks on a peak at 2969 m a.s.l. It is likely that temperature dependence of photosynthesis and respiration confines this species distribution to altitudes above 3000 m, making this species a truly nival element of the Alps biodiversity that is physiologically depending on a nival climate. Furthermore, a postglacial recolonization of the Alps from other high mountain systems or from Scandinavian mountains is unlikely, given the sequence differences of the nrITS between alpine specimens and specimens from other regions (Scheidegger, unpublished). We can therefore hypothesize that *U. virginis* is a candidate species that survived the LGM on central alpine nunataks in the Alps and possibly also on ice-free peaks in other mountain systems. Survival on central alpine nunataks is also a likely hypothesis for a number of other alpine lichen species, including the rare saxicolous *Lecanora diaboli* (Figure 2.4.4D) and *Lecanora freyi* on rocks with a low content in carbonates, *Lecanora concolor*, and several lecideoid lichens on siliceous rocks, including infraspecific taxa of *Lecidea atrobrunnea* [10], to name just a few species.

High alpine regions, including the nival belt, thus provide key habitats for extremophile lichen species that depend on cold climates, which are unavailable at lower altitudes. Nival regions of mountain systems are therefore important "museums" for a considerable number of lichens, many of them with boreal arctic-alpine distributions across diverse mountain systems [34]. More studies are however needed to test if high alpine and nival environments also act as "cradles" for lichen diversity, and immediate studies on threats and a possible decline of high alpine and nival lichens are needed to avoid that alpine and nival environments turn into "graves" because of the dramatic retreat of glaciers and permanent snowfields due to climate change.

Fig. 2.4.4: Hoarfrost formation on *Cladonia arbuscular* (A) and *Umbilicaria decussata* (B) is an important water resource in alpine and nival environments. A thick lens-shaped epinectral layer on the center of bullate areoles of *Psorinia conglomerata* (here an almost pigment deficient form of this species) characterize window lichens (C). *Lecanora diaboli* is a rare high alpine species on slightly calcareous rocks (D).

Acknowledgments

The author thanks his mountain companions and alpine guides who patiently accompanied him on several alpine tours in the Alps, the Himalayas, and East African Mountains. Critical feedbacks by Jessica L. Allen on an earlier draft of the manuscript are acknowledged.

References

[1] Spehn EM, Rudmann-Maurer K, Körner C. Mountain biodiversity. *Plant Ecol Divers*. 2011;4(4): 301–2.

[2] Braun-Blanquet J. Die inneralpine Vegetation – von der Provence bis zur Steiermark. Stuttgart: G. Fischer; 1961. 273 p.

[3] Rahbek C, Borregaard MK, Antonelli A, Colwell RK, Holt BG, Nogues-Bravo D, et al. Building mountain biodiversity: geological and evolutionary processes. *Science*. 2019;365(6458):1114.

[4] Rangel TF, Edwards NR, Holden PB, Diniz-Filho JAF, Gosling WD, Coelho MTP, et al. Modeling the ecology and evolution of biodiversity: biogeographical cradles, museums, and graves. *Science*. 2018;361(6399):eaar5452.

[5] Fjeldså J, Bowie RCK, Rahbek C. The role of mountain ranges in the diversification of birds. *Annu Rev Ecol Evol Syst*. 2012;43(1):249–65.

[6] Körner C. Alpine Plant Life. Functional Plant Ecology of High Mountain Ecosystems. Berlin, Heidelberg: Springer; 2003.

[7] Miehe G. Patterns on Mount Everest as influenced by monsoon and föhn. *Vegetation*. 1989;79: 21–32.

[8] Hafellner J. Studien über lichenicole Pilze und Flechten VI. *Herzogia*. 1987;7:353–65.

[9] Hertel H. Gesteinsbewohnende Arten der Sammelgattung Lecidea (Lichenes) aus Zentral-, Ost- und Sudasien. Khumbu Himal, Ergebnisse des. *Forschungsunternehmens Nepal Himalaya*. 1977;6:145–378.

[10] Hertel H, Schuhwerk F. On saxicolous lecideoid lichens growing in the European Alps at high altitudes. *Bibliotheca Lichenologica*. 2010;104:161–239.

[11] Scheidegger C. A revision of European saxicolous species of the genus Buellia De Not. and formerly included genera. *Lichenologist*. 1993;25(4):315–64.

[12] Nimis PL, Hafellner J, Roux C, Clerc P, Mayrhofer H, Martellos S, et al. The lichens of the Alps—an annotated checklist. *MycoKeys*. 2018;31:1–634.

[13] Baniya CB, Solhoy T, Gauslaa Y, Palmer MW. The elevation gradient of lichen species richness in Nepal. *Lichenologist*. 2010;42:83–96.

[14] Ellenberg H, Mueller-Dombois D. A key to Raunkiaer plant life forms with revised subdivisions. *Ber Geobot Inst Eidg Techn Hochschule Stiftung Rubel [Zurich]*. 1967;37:56–73.

[15] Marazzi S. Atlante orografico delle Alpi. SOIUSA. Subdivisione orografica internazionale unificata del Sistema Alpino. Scarmagno (TO). Italy: Priuli & Verlucca; 2005. 416 p.

[16] Frey E. Alpin-nivale Flechten der Tauernketten mit Vergleichen aus den Silikatketten der übrigen Alpen. *Verhandl Zool-Bot Ges Wein*. 1968–1969;108–109:75–98.

[17] Kappen L, Lange OL. Kälteresistenz von Flechten aus verschiedenen Klimagebieten. Deutsche Botanische Gesellschaft. *Neue Folge*. 1970;4:61–5.

[18] Scheidegger C, Schroeter B, Frey B. Structural and functional processes during water vapour uptake and desiccation in selected lichens with green algal photobionts. *Planta*. 1995;197: 399–409.

[19] Schaerer L. Lichenes Helveticorum Spicilegium. Bern; 1823–1846.

[20] Frey E. Cladoniaceae, Umbilicariaceae. Zahlbruckner A, editor. Leipzig: Akademische Verlagsgesellschaft; 1933.

[21] Lange OL, Kilian E, Ziegler H. Water-vapor uptake and photosynthesis of lichens—performance differences in species with green and blue-green-algae as phycobionts. *Oecologia*. 1986; 71(1):104–10.

[22] Libbrecht KG. The physics of snow crystals. *Rep Prog Phys*. 2005;68:855–95.

[23] Burke MJ, Gusta LV, Quamme HA, Weiser CJ, Li PH. Freezing and injury in plants. *Annu Rev Plant Physiol*. 1976;27 507–28.

[24] Schroeter B, Scheidegger C. Water relations in lichens at subzero temperatures—structural changes and carbon-dioxide exchange in the lichen *Umbilicaria aprina* from Continental Antarctica. *New Phytol.* 1995;131(2):273–85.
[25] Dietz S, Büdel B, Lange OL, Bilger W. Transmittance of light through the cortex of lichens from contrasting habitats. In: Schroeter B, Schlensog M, Green TGA, editors. Aspects in Cryptogamic Research Contributions in Honour of Ludger Kappen. Berlin-Stuttgart: Gebrüder Borntraeger; 2000. p. 171–82.
[26] Vogel S. "Niedere Fensterpflanzen" in der südafrikanischen Wüste. Eine ökologische Schilderung. *Beiträge zur Biologie der Pflanzen.* 1955;31:45–135.
[27] Kelly MA, Buoncristiani J-F, Schlüchter C. A reconstruction of the last glacial maximum (LGM) ice-surface geometry in the western Swiss Alps and contiguous Alpine regions in Italy and France. *Eclogae Geol Helv.* 2004;97:57–75.
[28] Leuschner C, Ellenberg H. Vegetation of the Alpine and Nival Belts. In: Leuschner C, Ellenberg H, editors. Ecology of Central European Non-Forest Vegetation: Coastal to Alpine, Natural to Man-Made Habitats; 2017. p. 271–431.
[29] Schönswetter P, Stehlik I, Holderegger R, Tribsch A. Molecular evidence for glacial refugia of mountain plants in the European Alps. *Mol Ecol.* 2005;14(11):3547–55.
[30] Stehlik I. Nunataks and peripheral refugia for alpine plants during quaternary glaciation in the middle part of the Alps. *Bot Helv.* 2000;110(1):25–30.
[31] Stehlik I, Blattner F, Holderegger R, Bachmann K. Nunatak survival of the high alpine *Eritrichium nanum* (L.) Gaudin in the central Alps during the ice ages. *Mol Ecol.* 2002;11:2057–66.
[32] Stehlik I, Schneller JJ, Bachmann K. Resistance or emigration: response of the high-alpine plant *Eritrichium nanum* (L.) Gaudin to the ice age within the Central Alps. *Mol Ecol.* 2001;10(2): 357–70.
[33] Schlüchter C. Die Schweiz während des letzteiszeitlichen Maximums (LGM), 1:500'000. GeoKarten500. Wabern: Bundesamt für Landestopografie swisstopo; 2009.
[34] Galloway DJ. Lichen biogeography. In: Nash III TH, editor. Lichen Biology. Camdridge: Cambridge University Press; 2008. p. 317–77.
[35] Gams H. Die Herkunft der hochalpinen Moose und Flechten. *Jahrbuch des Vereins zum Schutz der Alpenpflanzen und Tiere.* 1963;28:85–95.
[36] Halloy S. Islands of life at 6000 M altitude: the environment of the highest autotrophic communities on earth (Socompa Volcano, Andes). *Arct Alp Res.* 1991;23(3):247–62.
[37] Beck E. Plant life on top of Mt. Kilimanjaro (Tanzania). *Flora.* 1988;181:379–81.

Martin Grube and Lucia Muggia

2.5 Complex lichen communities inhabiting rock surfaces

Bare rocks are the oldest terrestrial substrate, but only few eukaryotic life-forms pioneer growth on their surfaces. Among those lichens are perhaps the most conspicuous and diversified group. Alpine boulders above the tree line have been colonized by lichens since thousands to perhaps even millions of years of substrate availability. In this chapter, we shed light on the formation of complex lichen communities on rock surfaces and reflect on the mechanisms how interactions among the species may influence their composition with selected examples. The common presence of algae shared among different species could facilitate the recolonization of lichen species that compete for space, and specialized parasites, either lichens or non-lichenized fungi, may have specific and varied influence on the fitness of individual host lichens in rock communities. Lichen communities on rocks are complex consortia, which are regulated over long periods of time by a network of interactions. We argue that after initial colonization, these interactions lead to a long-term steady state of species-rich communities with little fluctuation of species composition.

2.5.1 Introduction

In rock habitats with fluctuating water conditions, lichen symbioses are dominant life-forms covering the surfaces. Particularly in alpine or coastal habitats, they may decorate the rocks with characteristic tinges of colors, by which the chemical composition of rocks may be recognized already from a distance. The greenish tinge of the most common map lichens (the genus *Rhizocarpon*) characterizes the presence of siliceous rocks (Fig. 2.5.1A, B). Alternatively, red, orange, and white patches are indicative of lichen species on calcareous rocks (Fig. 2.5.1C, D). However, on these other rocks, lichens are less conspicuous, and their presence is recognized only at a closer inspection. The bare eye may then recognize a mosaic of lichens forming a community of various species and their conspecific individuals. The potential composition is finely determined by the physical and chemical conditions of the rock, and the local microclimate together with the shared presence of species inspired researchers to classify those communities and to create the research branch of lichen sociology. The actual composition of lichens on rocks also depends on the stochastic factors, the variation of weathering, the history of colonization, the competition among lichens, and the relations among species.

Before lichens are able to colonize barren rocks, a sufficient weathering layer needs to be established to develop appropriate conditions for lichen growth. Lichen

https://doi.org/10.1515/9783110646467-007

colonization usually starts to develop after several years of surface weathering of the rocks. Although certain lichens, including *Rhizocarpon geographicum* or *Sporastatia testudinea*, can establish even on barely modified surfaces, e.g., of quartz, other species require more or less weathered surfaces, where hyphae of lichenized fungi can intrude to colonize stable microfissures from which lichenized stages emerge as thallus initials. Many tiny crust lichens can then form simultaneously to spread over the surface. Although endolithic growth is a more common phenomenon on calcareous rocks [1], siliceous rocks are colonized primarily by epilithic species of lichens (Fig. 2.5.1). Lichens with a more conspicuous crustose and epilithic thallus also comprise squamulose, lobate shapes. Crustose lichens themselves induce a range of weathering phenomena [2] because of not only the action of secreted lichen compounds but also, mechanically, the hygroscopic action of their structures.

Fig. 2.5.1: Lichen communities on siliceous (A, B) and calcareous rocks (C, D). Scale bars: A, B = 1 m; C= 10 cm; D = 1cm. A–C, Italy; D, Austria. Photographs by LM.

Photographic documentation for the long-term monitoring of rock lichen communities is scant, and there is still little understanding of the dynamics of whole communities over hundreds of years. An alternative approach has been tried with a mathematical model to stochastically simulate lichen tessellation of rock surface [3]. It has been shown that with the aging of the rock surface, more species with more varied requirements and growth styles settle in the lichen community. Although the modeling approach gives a first impression about the colonization patterns, it tends to neglect the biological interrelationships of species, which go far beyond mere competition for space, but include various levels of parasitism and other strategies of growth success. Richardson [4] compared the biological relations among lichen species with a "war in the world of lichens," and Armstrong [5] also called the dynamics a "two-dimensional warfare in slow motion." However, we think that the term "war" is perhaps too simplistic for these communities. We hypothesize they are tuned for long-term stability in their environment, which may go well beyond the average lifetimes of other terrestrial communities (e.g., plant communities).

2.5.2 Factors influencing the composition of lichens on rock surfaces

The composition of lichen communities on rocks has been studied in great detail in "floristic" works, which also revealed a high correlation with rock chemistry [6, 7]. Beside physical factors, such as the difference in the coherence of chemically different rock types, chemical features are attributed to the availability of nutrients. The availability of calcium (Ca) and pH is higher in limestone than that in acidic rock situations, and additionally differences in pH levels affects solubility of nutrients. Thus, very different lichen communities occur on different types of limestones, on acidic rocks, or on gypsum, for example. In addition, lichens differ in their tolerance to transition metals present in rocks, in particular iron, which can be deposited in intracellular spaces of thalli by specialized species, which are characterized by their rusty colors. Of course, additional nutrient input also originates from sources other than the substrate. In certain habitats, masses of airborne fine particles become incorporated into the surfaces or are deposited in crevices of lichen thalli.

Using the annual growth rate and the thallus size, it has been estimated that individual lichen thalli survive for more than 1000 years [8]. However, it has been shown that growth rates may vary depending on orientation aspect of rocks or due to seasonal variation in correlation with temperature and humidity [9, 10]. Within individuals, growth rates are not constant but appear to be a function of size as they usually decline with size and age [11].

It is beyond the scope of this chapter to revise the lichen flora on various types of rocks or habitats in detail, and we only want to mention general observations here. Variation of lichen composition on the same rock type can be attributed to various other environmental factors besides rock chemistry and large-scale nutrient input. One of the best examples is displayed by coastal rocks, where marine (littoral) and maritime (supralittoral) lichens form beltlike colored arrangements of lichens in temperate and subtemperate regions depending on the salt spray and distance to sea water [12]. In inland habitats, the inclination and exposition explain large-scale spatial patterns on rock faces. In particular, their facing to sunlight may have a significant influence on the species composition and may compensate for macroclimatic variation across latitudes, a phenomenon termed "relative habitat constancy" [13]. In addition, John and Dale [14] pointed out that smaller-scale microhabitats, including cracks and crevices of local water flows, play a role for lichen composition, whereas at smallest scales, the different ecological strategies of lichens, as discussed below, become an important factor in shaping the long-term development of the communities.

2.5.3 Interactions of lichens on rocks

Individuals of many species primarily compete for space, resulting in patterns delimited by their thallus fronts. This behavior does not equal the concept of competition in a narrow sense, which is usually understood as negative effects imposed to another organism by consuming or controlling access to limited resources. Individuals of other species tend to fuse more frequently (e.g., *Dimeleana oreina*; Fig. 2.5.2A). Interestingly, some species seem to develop occasional thalli, which may intrude those of other species. It is not completely clear if these are induced by environmental conditions of whether genotypic variation is responsible for more aggressive invasive strategies. Such forms are frequently observed in, e.g., *Tephromela atra*, *Lecanora polytropa*, or *Protoparmelia badia*, among others. In contrast to the transient lichen-invading behavior, lichenicolous lichens are specialized to live on other lichens, and they are usually very specific for their hosts (Fig. 2.5.2E, F) [15]. Some of the species seem to parasitize other lichens only in juvenile stages (juvenile parasites), until they mature to live independently, without any visible traces of a host lichen.

Apart from the community dynamics involving parasitic behavior among lichens, changing patterns of species may simply involve different growth strategies. As some species tend to develop thicker thalli or grow more superficially, they often may overgrow thalli of neighboring species in an unspecific manner. A good example for this behavior is *Ophioparma ventosa*, a species on siliceous rocks that has rather thick thalli. This behavior may explain the observation of secondary metabolites, which are untypical for this lichen in the medulla [16].

Epilithic lichens with squamulose, lobate, to even more surface-detached growth styles, such as foliose lichens, may represent extreme forms of this strategy

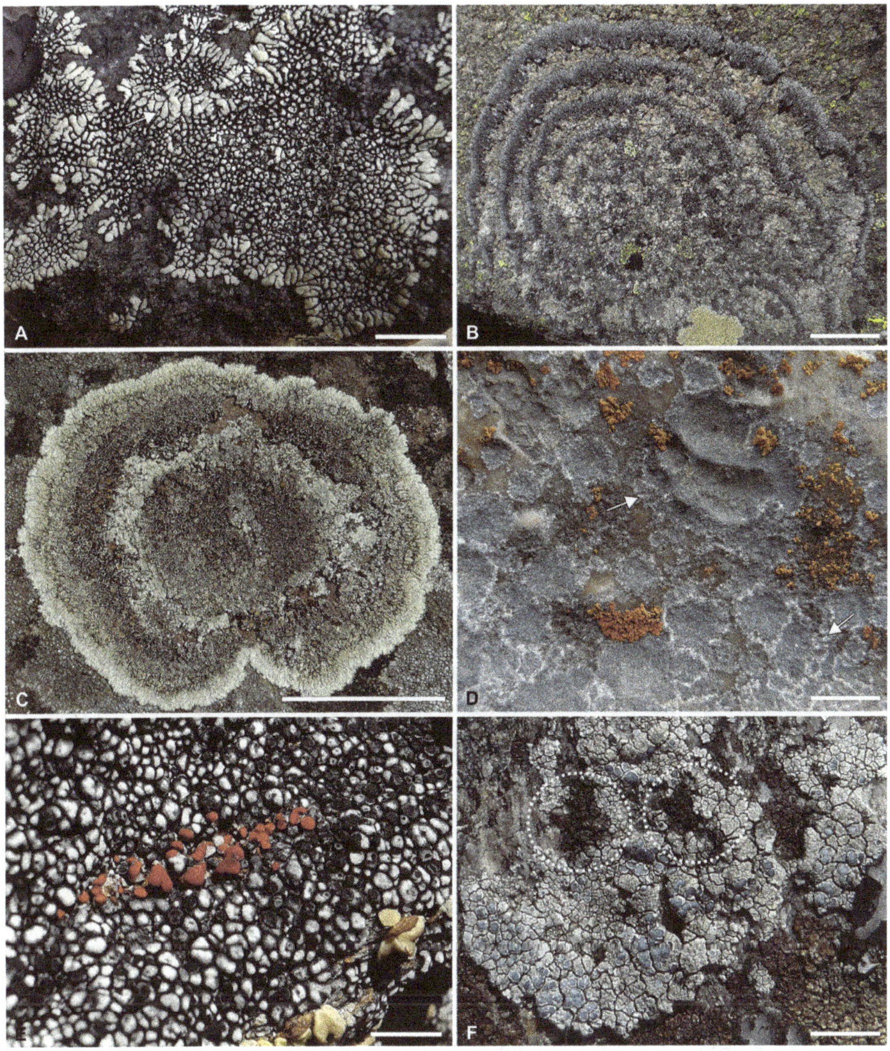

Fig. 2.5.2: Various growth styles of lichens on siliceous and calcareous rocks. (A) The epilithic crustose species *Dimelaena oreina*, with individuals fusing by their growing margins. (B) The foliose species *Brodoa atrofusca*, developing waves of centrifugal thallus fronts as concentric rings. (C) The epilithic *Protoparmeliopsis muralis* growing on concrete, with a regenerative central thallus part. (D) The endolithic *Caloplaca erodens* on calcareous rocks, delimited by a white margin and with younger thalli develop near the margins of the older thalli (arrows). (E) The parasitic lichen *Caloplaca epithallina* on *D. oreina*. E, *Lecanora rupicola* (whitish thallus with bluish-grey apothecia) infected by the lichenicolous lichen *Lambiella insularis* (dotted circle lines around brown thalli). Scale bars: A, D, F = 1 cm; B = 10 cm, C = 2.5 cm, E= 0.5 cm. Photographs A, B, and E by MG; C, D, and F by LM.

to escape competition and instead to literally overgrow neighboring crustose lichens. These lobate to foliose growth forms correlate with the development of an upper eucortex of the thallus that lacks algal cell wall remnants. Such a cortex seemingly acts as a flexible "skeleton," which is presumably a mechanical prerequisite for surface-detached growth forms. By contrast, most crustose lichens develop a phenocortex, which contains gradually decaying algal cell walls. This requires the activity of specific enzymes for algal polysaccharide degradation. Comparative analyses may reveal whether such enzymes or their activities are lost in foliose lichens.

A kind of long-term self-supporting growth strategy is displayed by foliose lichens that may form concentric rings, for example, *Arctoparmelia centrifuga* or species of *Brodoa* (Fig. 2.5.2B). These species with foliose but adnate thalli grow over the surface and can withdraw access to light for those crustose lichens that are overgrown. Such and similarly growing foliose species develop "windows" in the thallus center, which facilitate recolonization, usually by the same species. As the thalli grow centrifugally, this pattern can shape into waves of thalline fronts that develop in the time course of perhaps several decades. To certain extent, the overgrowth of neighboring thalli is also a successful strategy of crustose species with lobate thalli, e.g., *Protoparmeliopsis muralis* (Fig. 2.5.2C).

Another phenomenon to facilitate the same species, called "autoparasitism" [17], has been described for the endolithic *Caloplaca (Pyrenodesmia) erodens* (Fig. 2.5.2D). It has been observed that parts of large thalli of this species are recolonized by juvenile thalli of the same species. However, the biological relations are not completely clear so far, and an alternative interpretation of this phenomenon could as well be rearing care by sharing the algal partners with the progeny. The development of young thalli within larger and centrally aging thalli is perhaps a more widespread phenomenon of many crustose lichens. Clear parasitism often involves relationships between different species, even if these might be closely related (adelphoparasitism). Such forms of parasitism among related species might evolve from a competition for the same type of algae, yet this still needs to be tested by phylogenetic methods.

Fungal parasitism of lichens likely plays an important role in shaping the long-term development of rock lichen communities. The number of accepted lichenicolous fungi, including lichenized parasitic species, is now 2319, with 2000 obligately lichenicolous species, subspecies, or varieties; 257 lichenicolous lichens; and 62 facultatively lichenicolous taxa. These species are found in 10 different classes of Fungi (Ascomycota and Basidiomycota), 55 orders, 115 families, and 397 genera [15].

Lambiella insularis is an example of a lichenicolous lichen that grows on thalli of hosts representing the *Lecanora rupicola* group (Fig. 2.5.2E). It usually invades the older parts of host thalli. In later stages, only few hyphae of the latter remain detectable in the basal part of the parasite's newly formed thallus. De los Ríos et al. [18] showed that the algae of the parasite are apparently captured from the host according to DNA sequencing after direct PCRs from various sections (without prior DNA extraction). Because the algal cells display two morphologically different types, the mycobionts apparently influence the algal morphology differently. PCR and sequencing detected *Lambiella* hyphae also below the host and in the rock below the infected thalli. Whether rock crevices may contain a reservoir of parasitic species merits further study, but we assume that infection by spores may be a more common pathway. Infections of *Lambiella* may partially degrade the thallus of the host by time, occasionally leaving centrally degrading patches.

Many lichenicolous fungi that do not form thallus structures and rather develop infectious mycelia in the host thallus (nonlichenized lichenicolous fungi) are not aggressive parasites. They rather have a commensal to mildly parasitic lifestyle and reproduce without serious damage of the host lichen (Fig. 2.5.3B–D). Some of those impair sexual reproduction of the hosts when they occupy their fruit bodies (Fig. 2.5.3C). Only a lower number of species tend to be more aggressive by producing serious symptoms of damage, which causes the host structures to become brittle and

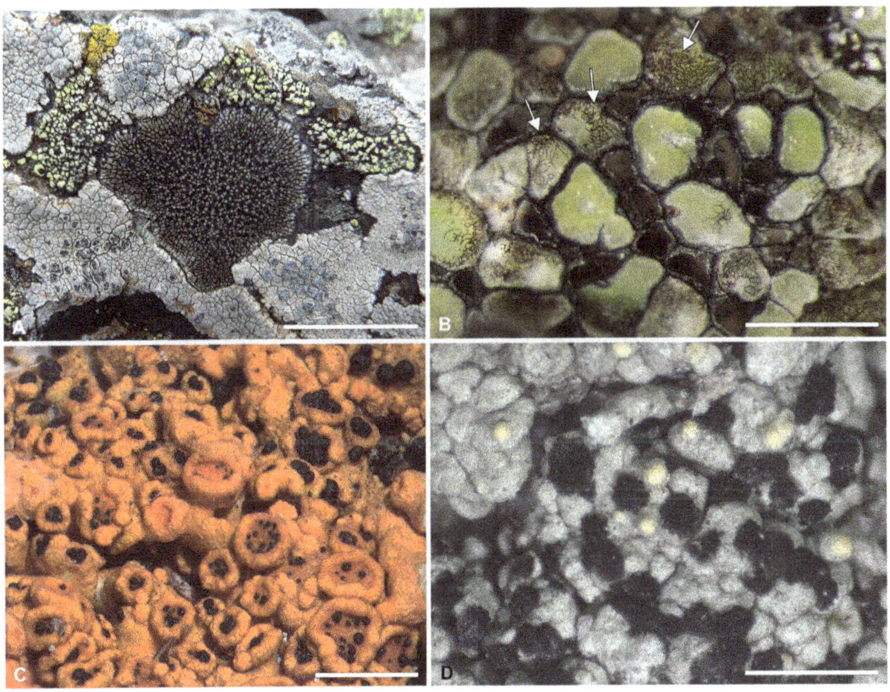

Fig. 2.5.3: Competition for space and lichenicolous fungi. (A) Mature thallus of *Sporastatia testudinea* growing on siliceous rocks between thalli of *Lecanora rupicola* (white), *Aspicilia* sp. (white), and *Rhizocarpon* sp. (yellow). (B) Thallus areoles of *S. testudinea* (immersed in water) with vital areoles having a light green tone, areoles infected by *Sphaerellothecium contextum* and with dark brown old decaying areoles (arrows). (C) Thallus and apothecia of *Xanthoria elegans* (orange red) infected by the lichenicolous fungus *Zwackhiomyces coepulonus* (black perithecia occupying the apothecia). (D) Thallus of *Pertusaria corallina*, infected by the lichenicolous hyphomycete *Sclerococcum sphaerale* (black sporodochia). Scale bars: A = 3 cm; B–D = 1 mm. Photographs A, C, and D by LM; B by MG.

decay. By disruption of the host's integrity, thallus parts tend to break apart more easily and open free rock surface for colonization by other species.

Lichens may even rejuvenate internally with the "help" of lichenicolous fungi. *Sporastatia testudinea* (Fig. 2.5.3A) is frequently colonized by minute mycelia of *Sphaerellotecium contextum*. These melanized mycelia develop in the epinecral layer, a layer of polysaccharides above the upper cortex of the host areoles. Prolonged infection of the areole leads to the formation of denser parasitic mycelia and shading effects. Subsequently, the host areoles tend to degenerate and shrink. The residual spaces are replaced by adjacent faster growing uninfected areoles (Fig. 2.5.3B). A balance seems to exist between infection and rejuvenation of areoles in this species. In other lichens, the regeneration of the same species from central (older) thallus parts appears to be more recurrent phenomenon, which does not require the involvement of lichenicolous fungi.

2.5.4 Other fungi

Typical lichenicolous fungi, which develop reproductive organs on their hosts, seem to live exclusively on lichens. The preferences of diverse other fungi, which have been detected since recent years, is not completely understood. These fungi contribute to the total fungal community of lichens, namely, the lichen mycobiome. The functional significance for the fitness of the host is so far unclear. There is some evidence that the presence of such fungi may be influenced by the surrounding habitats. Also, the technical approach, the culture-dependent isolation with subsequent sequencing, or the culture-independent sequencing approach yields different spectra of lichen-associated fungi [19–22]. Here, we exclusively focus on rock lichen habitats. Culturable black fungi–colonizing lichens were studied in arid habitats by molecular methods [23]. ITS rDNA showed that the genera *Capnobotryella*, *Cladophialophora*, *Coniosporium*, *Mycosphaerella*, and *Rhinocladiella* were repeatedly found in different lichen hosts but lacked any pathogenic symptoms. A specific clade of predominantly lichen-associated strains was found for *Rhinocladiella* strains, whereas samples of the remaining genera were closely related with those from other sources (e.g., *Capnobotryella* and *Coniosporium* strains). The authors concluded that the occurrence of lichens might be opportunistic for these genera, but the high sequence divergence in strains assigned to *Mycosphaerella* does not exclude the possibility of lichen-specific species, which are distantly related to plant inhabitants. A general capacity of rock-inhabiting fungi to form mutualistic interactions with lichen algae was suggested by coculture experiments [24]. A similar result was found by Brunauer et al. [25] for a so far undescribed *Taeniolella*-like fungus isolated from *L. rupicola*. The lichen-inhabitant formed layerlike structures with the hosts photobiont in cultures and appears to be phylogenetically basal in Chaetothyriomycetidae. The genus *Lichenothelia* clearly represents a link between free-living extremotolerant melanized fungi and algal associations [26] because it includes species growing on exposed rocks and is optionally associated with algae or lichens in the natural habitat. Although species diversity becomes better known [27, 28], these and other melanized fungi-forming associations with algae never develop typically stratified thallus morphologies either in nature or under culture conditions [29]. We hypothesize that this lack is due to the limited ability of strongly melanized hyphae to form hygroscopically flexible structures. Whether the presence of melanized hyphae could impair the hygroscopic reactions of the host under fluctuation humidity conditions needs further experimental testing.

Fernandez-Mendoza et al. [19] studied the diversity of lichen-inhabiting fungi (the lichen mycobiome) of neighboring lichens on rocks in an alpine habitat. Lichen mycobiomes reflect an overlap of multiple ecological sets of taxa, which differ in their trophic association with lichen thalli. These comprised pools of generalist environmental, lichenicolous/endolichenic, and transient species. A majority of those belong to the fungal classes Dothideomycetes, Eurotiomycetes, and Tremellomycetes, with close relatives in adjacent ecological niches. There was no evidence that phenotypically recognized lichenicolous fungi influence the occurrence of the other asymptomatic fungi in the host thalli. Fernandez-Mendoza et al. [19] suggest that lichens represent suboptimal habitats for these fungi or act as a complex spore and mycelium bank, which modulate and allow the regeneration of local fungal communities.

2.5.5 Rock lichens as algal metacommunities

Previous works have shown that the development of characteristic thallus structures in lichen-forming fungi requires the association with suitable photoautotrophic partners and that fungi have a specific range of compatible photobionts and selected algal strains also correlate with the habitat conditions [30]. Various studies of rock-inhabiting lichens showed that photobiont species are shared across lichens at the same site. Some lichens have a fairly wide spectrum of partners, whereas others prefer specific types of photobionts. Their selective preferences could be part of the factors determining the habitats of lichens. At the genus level, there seems to be a correlation with growth form and complexity of thallus structure as well. Lower diversity of photobionts has been observed in Verrucariaceae, developing squamulose and foliose structures (e.g., *Placidium* and *Dermatocarpon*), whereas crustose representatives have a higher diversity [31].

Interestingly, species occurring in similar habitats at different latitudes have a shared preference for their local photobionts. Although alpine species of *Tephromela atra*, *L. rupicola*, and *Lecidea fuscoatra* prefer photobionts of the *Trebouxia jamesii* group, Mediterranean or lowland samples of these species are associated with varied photobionts related to *Trebouxia arboricola* [30–33]. The alpine photobionts have never been found in any lichens from similar habitats in the Mediterranean lowlands. We suggest that there is a shared shifting of photobiont selectivity, which works at a community level in ecologically widespread lichen species.

Lichens on rocks were also suggested to represent potential temporary niches for free-living stages of lichen photobionts, which could facilitate the establishment of other lichens in the proximal area. For this purpose, Muggia et al. [34] investigated the rock-inhabiting crust lichen *Protoparmeliopsis muralis*, a cosmopolitan species with high flexibility of algal associations. The authors analyzed the photobiont genotypic diversity within and between thalli and compared the diversity of intrathalline photobionts with that of externally associate algal communities by single-strand conformation polymorphism analyses and ITS sequence data. The results showed that photobiont populations within the lichen thalli are homogeneous, but also that multiple photobiont genotypes occur within single areoles and lobes of individual lichens, whereas the algal communities that superficially colonized the lichen thalli also included taxa known as photobionts in unrelated lichens. The old question of Ahmadjian [35], whether lichen algae exit in free-living stage, has been discussed controversially. The work of Muggia et al. suggests that lichen algae can have a habitat on thalli (particularly between areoles) of other lichens. In lichen communities, this reservoir of algae could facilitate continuous recolonization with species, thus contributing to maintenance of high diversity.

2.5.6 Bacterial communities of lichens

Bjelland et al. [36] studied bacterial metacommunities of several rock-inhabiting lichens, comprising the major lineages *Acidobacteria*, *Actinobacteria*, *Alpha-*, *Beta-*, and *Gammaproteobacteria*, *Bacteriodetes*, *Chloroflexi*, *Deinococcus*, *Firmicutes*, *Planctomycete*, *Tenericutes*, and *Cyanobacteria*. Also, *Crenarchaeota* were documented to be associated with rock-inhabiting lichens, although at low abundances. A higher bacterial diversity was generally observed inside the rocks than within the epilithic lichen thalli directly above. By contrast, the abundance of bacteria was higher in the lichen thalli compared with within the rock. The microbial communities of *Ophioparma ventosa*, *Pertusaria corallina*, and *Rhizocarpon geographicum* were more similar to each other, in terms of both the number and the types of different sequences, than to *Hydropunctaria maura*. This observation suggests a significant influence of habitat ecology on the prokaryotic community of rock lichens. These results demonstrated that the lichen-rock interfaces are complex habitats, and that the macroscopic lichens are specific habitats of bacterial metacommunities. The habitat influence was also demonstrated by West et al. [37]. In contrast to inlandic or other rock-inhabiting species higher up at maritime rock coasts where Alphaproteobacteria are often dominating the bacterial community, intertidal *Lichina* species are characterized by a high proportion of *Bacteriodetes*. The long-persisting (fungal) structures of lichen thalli provide a stable organic habitat for bacterial colonization, but their growth seems to be strictly controlled, e.g., by secondary metabolites or poikilohydric conditions, which generally limit bacterial growth.

Omics approaches such as metagenomics, metatranscriptomics, and metaproteomics suggest the involvement of bacteria in multiple processes, including provision of nutrients, vitamins, and hormones [38]. Further study is needed to show to what extent living or dead bacteria could contribute to nutrient flows in the lichen system. *In situ* hybridization experiments showed a high number of bacteria in lichen structures, including lichens from rock habitats [39], which corroborates that they represent a significant factor in lichen ecology and in nutrition under otherwise nutrient-poor conditions. The total number of bacteria in lichen thalli varies to some extent, which to our observation depends on abiotic conditions and on intrinsic factors of the lichens. We observed lower abundances of bacteria on surfaces of thalli in dry habitats than in humid habitats (in contrast to fungi). The hydrophilic nature of thalline surfaces also seems to be important for bacterial colonization. Hydrophobic surfaces or melanized parts of the lichen appear to have lower and certainly different communities than hydrophilic surfaces, where biofilms of bacteria can develop. A common niche for bacterial communities, including cyanobacteria, is the spaces between areoles of crustose lichens, which tend to stay humid for longer periods than the exposed surfaces.

2.5.7 Conclusions

In accordance with the growth speeds of lichens, their community dynamics is very slow, certainly too slow for comprehensive documentation in a single scientific career. If we could use a time lapse camera producing pictures at intervals of 5 years over a period of several hundred to thousands of years, it would probably show considerable dynamics of thallus development on rock faces, but with overall maintenance of species composition. There is so far little evidence that climate change would lead to rapid and massive changes of rock-inhabiting lichen communities because the diversity of microhabitats and variable aspects of rocks will locally relocate species to suitable niches but less likely cause their complete local disappearing. Apart from competition for space, the community regulation also includes effects of lichen parasitism by fungi and overgrowth by faster growing lichens. Lichen thalli themselves or perhaps cryptic stages in rock fissures are reservoirs for the recolonization of surfaces, which all contribute to an intricately weaved regulatory network of communities designed for persisting throughout extremely long periods of times.

References

[1] Pinna D, Salvadori O, Tretiach M. An anatomical investigation of calcicolous endolithic lichens from the Trieste karst (NE Italy). *Plant Biosyst.* 1998;132:183–95.

[2] Chen J, Blume HP, Beyer L. Weathering of rocks induced by lichen colonization—a review. *Catena*. 2000;39:121–46.
[3] Jettestuen E, Nermoen A, Hestmark G, Timdal E, Mathiesen J. Competition on the rocks: community growth and tessellation. *PLOS One*. 2010;5:1–5.
[4] Richardson DHS. War in the world of lichens: parasitism and symbiosis as exemplified by lichens and lichenicolous fungi. *Mycol Res*.1999;103:641–50.
[5] Armstrong RA. Lichen competition: two-dimensional warfare in slow motion. *Microbiologist*. 2008;9:27–31.
[6] Roux C. Etude Ecologique et Phytosociologique des Peuplements Licheniques Saxicoles-Calicoles du Sud-Est de la France; 1981. p. 1–537.
[7] Wirth V. Die Silikatflechten-Gemeinschaften im außeralpinen Zentraleuropa. *Diss Bot*. 1972;17: 1–306.
[8] Beschel RE. Dating rock surfaces by lichen growth and its application to glaciology and physiography (lichenometry). In: Raasch GO, editor. Geology of the Arctic, Vol. 2. Toronto: University of Toronto Press; 1961. p. 1044–62.
[9] Armstrong RA. Radial growth of *Rhizocarpon* section *Rhizocarpon* lichen thalli over six years at Snoqualmie Pass in the Cascade Range, Washington State. *Arct Antarct Alp Res*. 2005;37:411–5.
[10] Armstrong RA. Seasonal growth of the crustose lichen *Rhizocarpon geographicum* (L.) DC. in South Gwynedd, Wales. *Symbiosis*. 2006;41:97–102.
[11] Bradwell T, Armstrong RA. Growth rates of *Rhizocarpon geographicum* lichens: a review with new data from Iceland. *J Quaternary Sci*, 2007;22:311–20.
[12] Fletcher A. Marine and maritime lichens of rocky shores: their ecology, physiology and biological interactions. In: Price JH, Irvine DEG, Farnham WF, editors. The Shore Environment. Volume 2: Ecosystems. The Systematics Association Special Volume No. 17B. London: Academic Press; 1980. p. 789–842.
[13] Poelt J. Das Gesetz der relativen Standortkonstanz bei den Flechten. *Bot Jahrb Syst*. 1987;108: 363–71.
[14] John E, Dale MRT. Determinants of spatial pattern in saxicolous lichen communities. *Lichenologist*. 1991;23:227–36.
[15] Diederich P, Lawrey JD, Ertz D. The 2018 classification and checklist of lichenicolous fungi, with 2000 non-lichenized, obligately lichenicolous taxa. *Bryologist*. 2018;121:340–425.
[16] Le Pogam P, Legouin B, Geairon A, Rogniaux H, Lohézic-Le Dévéhat F, Obermayer W, Boustie J, Le Lamer A-C. Spatial mapping of lichen specialized metabolites using LDI-MSI: chemical ecology issues for *Ophioparma ventosa*. *Sci Rep*. 2016;6:37807.
[17] Hafellner J, Muggia L. Über Vorkommen von *Caloplaca erodens* in der Steiermark (Österreich). *Mitt naturwiss Ver Steiermark*. 2006;135:33–49.
[18] De los Ríos A, Ascaso C, Grube M. An ultrastructural, anatomical and molecular study of the lichenicolous lichen *Rimularia insularis*. *Mycol Res*. 2002;8:946–53.
[19] Fernández-Mendoza F, Fleischhacker A, Kopun T, Grube M, Muggia L. ITS1 metabarcoding highlights low specificity of lichen mycobiomes at a local scale. *Mol Ecol*. 2017;26:4811–30.
[20] Banchi E, Stanković D, Fernandez-Mendoza F, Gionechetti F, Palavicini A, Muggia L. ITS2 metabarcoding analysis complements lichen mycobiome diversity data. *Mycol Prog*. 2018;17:1049–66.
[21] U'Ren JM, Lutzoni F, Miadlikowska J, Arnold EA. Community analysis reveals close affinities between endophytic and endolichenic fungi in mosses and lichens. *Microb Ecol*. 2010;60: 340–53.
[22] U'Ren JM, Riddle JM, Monacell JT, Miadlikowska J, Arnold EA. Tissue storage and primer selection influence pyrosequencing-based inferences of diversity and community composition of endolichenic and endophytic fungi. *Mol Ecol Resour*. 2014;14:1032–48.

[23] Harutyunyan S., Muggia, L., Grube, M. Black fungi in lichens from seasonally arid habitats. *Stud Mycol.* 2008;61:83–90.

[24] Gorbushina AA, Beck A, Schulte A. Microcolonial rock inhabiting fungi and lichen photobionts: evidence for mutualistic interactions. *Mycol Res.* 2005;109:1288–96.

[25] Brunauer G, Blaha J, Hager A, Türk R, Stocker-Wörgötter E, Grube M. An isolated lichenicolous fungus forms lichenoid structures when co-cultured with various coccoid algae. *Symbiosis.* 2007;44:127–36.

[26] Muggia L, Gueidan C, Knudsen K, Perlmutter G, Grube M. The lichen connections of black fungi. *Mycopathologia.* 2013;175:523–35.

[27] Muggia L, Kocourková J, Knudsen K. Disentangling the complex of *Lichenothelia* species from rock communities in the desert. *Mycologia.* 2015;107:1233–53.

[28] Ametrano CG, Knudsen K, Kocourkova J, Grube M, Selbmann L, Muggia L. Phylogenetic relationships of rock-inhabiting black fungi belonging to the widespread genera *Lichenothelia* and *Saxomyces*. *Mycologia.* 2019;111:127–60.

[29] Ametrano CG, Selbmann L, Muggia L. A standardized approach for co-culturing Dothidealean rock-inhabiting fungi and lichen photobionts *in vitro*. *Symbiosis.* 2017;73:35–44.

[30] Blaha J, Baloch E, Grube M. High photobiont diversity associated with the euryoecious lichen-forming ascomycete *Lecanora rupicola* (Lecanoraceae, Ascomycota). *Biol J Linn Soc.* 2006;99:283–93.

[31] Thüs H, Muggia L, Pérez-Ortega S, Favero-Longo SE, Joneson S, O'Brien H, Nelsen MP, Duque-Thüs R, Grube M, Friedl T, Brodie J, Andrew CJ, Lücking R, Lutzoni F, Gueidan C. Revisiting photobiont diversity in the lichen family Verrucariaceae (Ascomycota). *Eur J Phycol.* 2011;46:399–415.

[32] Muggia L, Zellnig G, Rabensteiner J, Grube M. Morphological and phylogenetic study of algal partners associated with the lichen-forming fungus *Tephromela atra* from the Mediterranean region. *Symbiosis.* 2010;51:149–60.

[33] Muggia L, Pérez-Ortega S, Kopun T, Zellnig G, Grube G. Photobiont selectivity leads to ecological tolerance and evolutionary divergence in a polymorphic complex of lichenized fungi. *Ann Bot.* 2014;114:463–75.

[34] Muggia L, Vancurova L, Škaloud P, Peksa O, Wedin M, Grube M. The symbiotic playground of lichen thalli—a highly flexible photobiont association in rock-inhabiting lichens. *FEMS Microbiol Ecol.* 2013;85:313–23.

[35] Ahmadjian V. The lichen alga *Trebouxia*: does it occur free-living? *Plant Syst Evol.* 1988;158: 243–7.

[36] Bjelland T, Grube M, Hoem S, Jorgensen SL, Daae FL, Thorseth IH, Øvreås L. Microbial metacommunities in the lichen–rock habitat. *Environ Microbiol Rep.* 2011;3:434–42.

[37] West N, Parrot D, Fayet C, Grube M, Tomasi S, Suzuki M. Marine cyanolichens from different littoral zones are associated with distinct bacterial communities. *Peer J.* 2018;6:e5208

[38] Grube M, Cernava T, Soh J, Fuchs S, Aschenbrenner I, Lassek C, Wegner U, Becher D, Riedel K, W. Sensen C, Berg G. Exploring functional contexts of symbiotic sustain within lichen-associated bacteria by comparative omics. *ISME J.* 2015;9:412–24.

[39] Grube M, Cardinale M, de Castro JV, Müller H, Berg G. Species-specific structural and functional diversity of bacterial communities in lichen symbioses. *ISME J.* 2009;3:1105–15.

Tatyana Darienko and Thomas Friedl

2.6 Eukaryotic algal communities of rock surfaces

Terrestrial eukaryotic algae are an ecological group of photoautotrophic protists that is phylogenetically enormously diverse. They conquered land many times independently. Rock surfaces are generally an unfavorable substrate for the colonization by algae because of their low nutrient content and the little water holding capacity. The rock substrate's hardness and passing light inside the stone are further limiting factors. The rock surface habitats connect the eukaryotic algae and cyanobacteria, other nonphototroph protists, and bacteria to functional assemblages of changeable community structure. Algae and bacteria in biofilms build up the first organic structures on exposed rock surfaces in arid environments. Microbial mats can cause forming cracks and fissures on the rock surface, leading to the rocks' biodeterioration. On dark hard limestone surfaces, the rock-colonizing or lithophytic algae can form either drilling or tuff-forming communities. Granite is the most challenging substrate for colonization by algae because of its density. Light cannot pass much below the surface, and it has practically no water holding capacities. By contrast, on sandstone, all ecological groups of lithophytic algae can occur because it provides a good water holding capacity, and light can pass well through the substrate beneath the surface. Lithophytic algae are mostly of coccoid organization, forming unicells or cell packages or form short filaments that can quickly become fragmented. In terms of the number of lithophytic algal species, green algae (Chlorophyta and Streptophyta), particularly of the class Trebouxiophyceae, and diatoms (Stramenopiles) are most dominant. The lithophytic algae form distinct communities dependent on whether they occur on limestone (e.g., in caves), sandstone, granite, or man-made artificial substrates such as monument surfaces. There is even a variety of formally described algal communities with different characteristic species.

2.6.1 Introduction

Algae comprise all eukaryotic photoautotrophic microorganisms, except the embryophytes. They conquered land many times independently from which only a single line, the Streptophyta, developed further into the land plants. Algae are phylogenetically enormously diverse. They originated several times by primary and secondary endosymbiosis in almost all major lineages of the eukaryotes. Intimately connected to the eukaryotic algae are the cyanobacteria because they gave rise to the earliest algae plastids. The terrestrial habitats, such as rock surfaces, connect eukaryotic algae and cyanobacteria, other nonphototroph protists, and bacteria to functional assemblages of changeable community structure. Because of their frequent co-occurrence with nonphototrophic microeukaryotes, the microscopic eukaryotic

https://doi.org/10.1515/9783110646467-008

algae are often summarized under the term protists and called the photoautotrophic protists. The algae and cyanobacteria are crucial for photoautotrophic energy and C (and O_2) input into young and developing terrestrial habitats like soils or biofilms on rock surfaces. There may be numerous beneficial effects of the photoautotrophic protists and cyanobacteria to their terrestrial habitats. The algae's impacts are caused by their broad biochemical diversity of pigments, photosynthetic storage products, cell walls and mucilage, fatty acids and lipids, oils, sterols and hydrocarbons, and bioactive compounds.

Terrestrial algae are an ecological group of eukaryotic algae that spend their whole life cycle in terrestrial habitats such as soil, rocks, caves, snow, and ice and can also live on animals and plants [1–4]. They occur worldwide and also colonize many extreme habitats, where often they are the only primary producers. Lithophytic algae are a specialized ecological group of algae that live on or within rocks or other hard substrate types, e.g., man-made substrates and materials and solid mineral surfaces. [2, 5–11].

2.6.2 The lithophytic habitats and ecological factors

Types of lithophytic habitats

Lithophytic algae colonize those bare rock surfaces provided the availability of sufficient moisture [12, 13]. Where often, they are associated with lichens and mosses [14]. They can survive under extreme climate conditions such as hot and cold deserts, high alpine regions, and cliffy rock faces. For lithophytic algae, microhabitats and microclimatic conditions often play more important roles than the macroclimate [15–18]. Therefore, the widely accepted classification of lithophytic or lithobiontic habitats has been defined for the colonization of microniches proposed by Golubic et al. [5] (see also Chapter 2.2. in this volume). The classification distinguishes three major groups of lithophytic habitats: epilithic, hypolithic, and endolithic habitats. Epilithic algae are colonizing exposed rock surfaces. Hypolithic algae live underneath the surface of translucent rocks that are partly embedded in the soil. Endolithic algae are inhabiting the interior of rocks. There are subgroups of the latter. The chasmoendolithic algae colonize fissures and cracks open to the rock surface. Cryptoendolithic algae colonize structural cavities within porous rocks (e.g., in sandstone). The euendolithic algae can actively penetrate the rock's interior and form tunnels that conform with the shape of their body. Those rock-boring algae are common on limestone but are also dwelling on other kinds of stone ([6, 9, 13, 19] and references therein).

Often, the algal communities of lithophytic habitats are difficult to separate from those of other terrestrial habitats. For example, algal species found in soil often can also dwell on rock surfaces (Figs. 2.6.1–2.6.7). Hollerbakh and

Shtina [1] have presented a terminology for defining the various types of terrestrial algal habitats. Aerophilic algal communities are those living in habitats exposed to air, and the air is their life cycles' dominating environment.

Typical habitats for aerophilic algae are hard substrates where they live on the interface between substrate and air. The hard substrate, e.g., granite, has (almost) no physical-chemical impact on the algae. Examples may be the different kinds of rocks and stone, tree bark, and artificial man-made hard substrates. Limestone, however, can easily be influenced by interactions of algae and other organisms with their substrate. Two groups of aerophilic algal communities are recognized. Aerial algal communities receive water only in the form of rain or dew, which means unstable humidity, i.e., fluctuations in the water supply. Aquatic-aerial or pseudaerial algal communities live at complete moisturization. However, in both cases, the algal communities are exposed to full aeration.

Lithophytic algal communities on the surfaces of dark hard limestone can either be drilling or tuff-forming communities. The drilling algae are actively dissolving their surrounding substrate. The algae live within small caverns and pores, formed by the algal excretion of substances that dissolve the limestone. The caverns and pores connect to the air. Tuff-forming algae actively produce (precipitate) carbonates around their cells. Typically, the algae live in the substrate's peripheral layers, which provide sufficient water and light, whereas the algae are successively dying in the middle. The drilling and tuff-forming algae are less exposed to the air in comparison to aerial communities.

Ecological factors determining lithophytic algal communities

The most important ecological factors for the growth of lithophytic algae are light, water, temperature, and the rock surfaces' chemicophysical properties. Because the rock surfaces are the habitats, where the environmental factors are frequently changing not only seasonally but also within the day, the algae must have adapted their metabolism to relatively short vegetation periods. According to Friedmann et al. [15], it could be restricted up to a total of a few hundred hours per year in Antarctic cold deserts.

Light is a limiting factor, especially for epilithic and endolithic algae. Intense PAR illumination inhibits the growth of epilithic algae [20–22]. By contrast, endolithic algae have to cope with low light intensities. Porosity and translucence of the rock substrate are the most important properties that determine the colonization of endolithic algae [23, 24].

The rock's interior surfaces are characterized by a steep light gradient along with depth due to light attenuation by rock substrates [20]. The porosity of rock surfaces can be up to 50%. Many factors such as rock type, mineral content, and grain particle size determine it [24]. In the polar regions, the colonization of opaque rocks by the hypolithic organisms may be promoted by local periglacial activities (the freezing and thawing of groundwater), which create the openings around the margins of quartz rocks and enable the penetration of light to the undersides [25, 26]. The decrease of light by rocks also helps organisms circumvent UVR stress in refuge lithic niches [22, 27, 28]. In hot deserts, algae cannot colonize the epilithic habitat besides the limited water availability because of inhibition by high light intensities and great temperature changes of the substrate surface with relatively cold overnight and high temperatures during the day. In hot deserts, algae on rock surfaces usually live as endoliths and hypoliths [13, 29]. Epilithic algae are reported mostly from the temperate zone or caves, primarily in limestone or sandstone [30–33].

Water availability is one of the essential requirements for the algal development in the hot as well in the cold deserts [27, 34]. The algal communities in the hot desert depend on dew and local rainfall. Water is present in solid form as ice and snow in cold deserts and, thus, only available when melted [16–18].

Temperature also plays an essential role for algal communities on rock surfaces. In hot deserts, the temperature, together with the illumination, makes the colonization of rock surfaces (almost) impossible and restricts algal life to the endolithic and hypolithic habitats [20, 35]. The colonization of the chasmolithic and cryptoendolithic habitats is an efficient way to escape overheating at the surfaces. In cold deserts, however, cryptoendolithic growth of algae prevails because the microhabitats inside the rock surface are less exposed to high temperatures than the surface ("greenhouse effect") [12, 27, 36, 37].

Rock as a substrate for the colonization by algae

Rock surfaces are generally an unfavorable substrate for the colonization by algae due to their low nutrient content (especially with nitrogen being somewhat limited) and the little water holding capacity. Besides the water availability, the hardness of the rock substrate and the ability to pass light inside the stone are significantly limiting factors. Therefore, the most favorable rocky substrates for the colonization by algae are limestone and sandstone. Limestone is porous and has a good water holding capacity, and it is soft and rich in calcium. On limestone, endoliths and epiliths can be frequently found. However, other lithophytic habitats are not accessible to algae because light cannot pass into deeper limestone layers. Sandstone provides a good water holding capacity as well. Because sandstone is rich in silicium, light can pass through the substrate up to several millimeters or even a few centimeters beneath the surface. It is porous and middle hard [7, 9]. Therefore, all ecological groups of lithophytic algae may occur on sandstone. Granite is the most challenging substrate for colonization by algae. It is very dense, i.e., light cannot pass much below the surface, very hard, and has practically no water holding capacities. Chasmoendolithic growth is most common on granite, but a few epiliths may also colonize the granite surface if it is shadowed [30].

Biological interactions of algae with the lithic substrate

Eukaryotic algae, together with bacteria, and fungi develop subaerial biofilms, which cause biodeterioration or bioweathering, i.e., the substrate's damage by different biological agents [38–46]. The interaction with the lithic substrate and the excretion of substantial amounts of microbial extracellular polymeric substances (EPS) create a biofilm organization [47, 48]. Typically, biofilms represent a consortium of different organisms embedded in a mucilage layer formed by various EPS. EPS plays an important role in protecting microbial cells, in community organization, and in mitigating microenvironmental conditions [48]. Biofilms are known to build up the first organic structures on exposed rock surfaces in arid environments [14, 18, 49, 50]. Algae in the complex with bacteria and fungi can also colonize man-made substrates and cause different types of destruction by forming cracks and fissures. Microbial mats can cause them on the rock surface, which produces organic acids or EPS, finally leading to the rocks' biodeterioration [42, 51–59]. Eukaryotic algae being active in biodeterioration have become a popular topic for the last 30 years with various studies from biodiversity assessment to studies of structural damages in stones and other man-made substrates (see also Chapter 2.2. in this volume).

2.6.3 Taxonomic composition of algal communities growing on rock habitats

Lithophytic algae are mostly of coccoid organization, forming unicells or cell packages, or form short filaments that can easily become fragmented (Fig. 2.6.1). Fortunately, most lithophytic algae are easy to isolate and then maintain under laboratory conditions, which enable the study of their morphology. The identification of the lithophytic algae traditionally uses only morphological features. Sequence analyses of DNA marker molecules, like the SSU rRNA genes by metabarcoding approaches, are crucial for the correct identification (e.g., see [60–63]). However, the molecular approach is dependent on the availability of proper reference sequence. Because of the scarcity of morphological features, many lithophytic algae are difficult to identify by morphology alone, and there are frequent cases of parallelisms and convergent evolution. Methods for the cultivation of terrestrial algae, including lithophytes, have been described in Büdel et al. [50].

There are two major groups of algae in terms of the number of species known from lithophytic habitats: the green algae, which are distributed on the two divisions, Chlorophyta (Figs. 2.6.2–2.6.6) and Streptophyta (Fig. 2.6.7), and the diatoms (Diatomeae and Stramenopiles; Fig. 2.6.1). More than 80 species of diatoms may live on rock surfaces (Tab. 2.6.1) [4, 64–66]. Less rich in species are the classes Xanthophyceae and Eustigmatophyceae (Stramenopiles; Tab. 2.6.1; see Fig. 2.6.1 for overview). Besides, unicellular red algae (Rhodophyta) are known as rarely occurring lithophytes [4] (Tab. 2.6.1). They seem to prevail on rock surfaces at higher humidity and low light intensities only. Reports of flagellated algae as lithophytes, such as dinoflagellates (Dinoflagellata and Alveolates) [32, 67] or members of Haptista (*Ruttnera*) [4], on wet rock surfaces still need confirmation.

We will not treat diatoms in detail here but simply list diatom species most commonly reported in Tab. 2.6.1. Examples of lithophytic members of Xanthophyceae and Eustigmatophyceae are shown in Fig. 2.6.1.

Chlorophyta

The systematics and phylogeny of green algae are still in progress. The traditional systematics based on morphology is artificial, and many new lineages were described. Most challenging is the scarcity of characteristic morphological features. Besides, there is a large amount of plasticity, i.e., variation of morphology dependent on various environmental conditions. Most members of Trebouxiophyceae (Tab. 2.6.2)

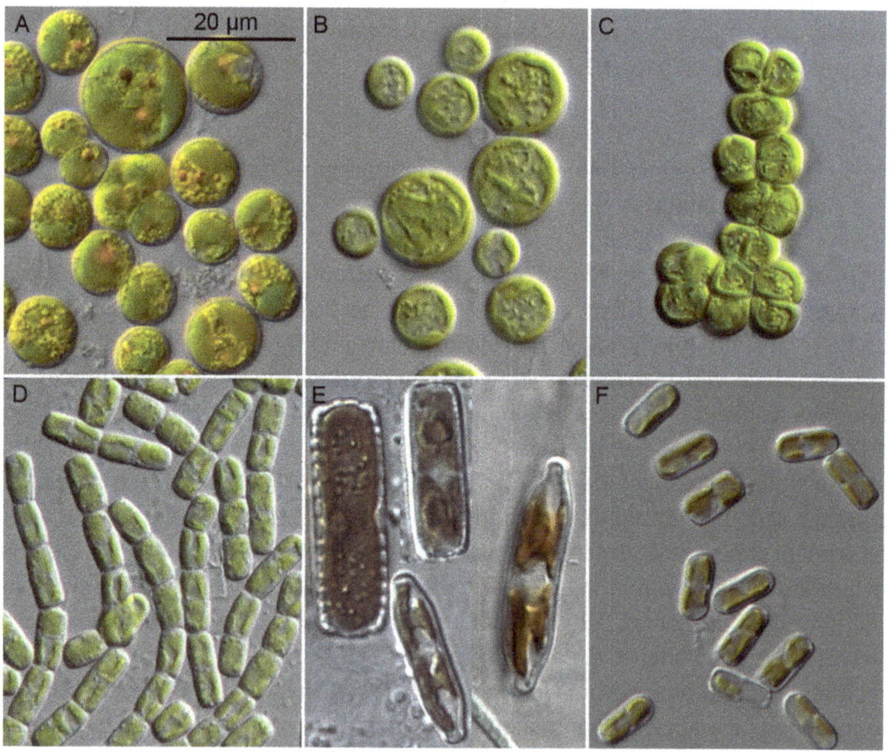

Fig. 2.6.1: Examples of lithophytic Stramenopiles algae in culture. (A) *Vischeria magna* (Eustigmatophyceae) isolated from quartz, La Campana, Chile. (B) *Botrydiopsis* sp. (Xanthophyceae) isolated from sandstone, Trakhtemyriv Natural Park, Ukraine. (C) *Heterococcus leptosiroides* SAG 2572 (Xanthophyceae). (D) *Xanthonema exile* strain SAG 2573 (Xanthophyceae). (E) *Hantzschia amphioxys* (Diatomeae), isolated from soil, Trakhtemyriv Natural Park, Ukraine. (F) *Humidophila contenta* (Diatomea), isolated from granite walls of Kilmartin, Scotland.

are described from cultured material, which makes it one of the best-studied groups of green algae. The terrestrial representatives of the green algal classes Chlorophyceae (Tab. 2.6.3) and Ulvophyceae (Tab. 2.6.4) can also easily be detected using cultural methods. However, many genera have been described based on morphological observations from environmental samples, and often they have been found only once by their descriptions.

A prominent example is the genus *Chlamydomonas*. More than 500 species have been described, but less than 20% of these species are known from cultures [68]. Almost nothing is known about the molecular phylogeny of terrestrial diatoms. Molecular phylogenetic studies are available for their freshwater relatives. Therefore, there is no genotypic (molecular) evidence of whether the freshwater and terrestrial populations are identical or not.

A considerable number of green algae (Chlorophyta) has adapted to life on land and rock surfaces. Representatives of three classes of Chlorophyta are typical on rock surfaces, i.e., Trebouxiophyceae (Figs. 2.6.2–2.6.4; Tab. 2.6.2), Chlorophyceae (Fig. 2.6.5; Tab. 2.6.3), and Ulvophyceae (Fig. 2.6.6; Tab. 2.6.4). The most extensive green algal diversity known from rock surfaces belongs to the class Trebouxiophyceae (Figs. 2.6.2–2.6.4; Tab. 2.6.2). Members of Trebouxiophyceae are typical for all kinds of rocks.

Members of the class Chlorophyceae (Fig. 2.6.5; Tab. 2.6.3) are less common on rock surfaces substrates than in the soil. About 21 species of lithophytic Chlorophyceae have been reported up to now. They are distributed on two major phylogenetic lineages of the class, i.e., the Sphaeropleales (or DO group) and the Volvocales (or CW group) (Tab. 2.6.3). Like with the Trebouxiophyceae, the genera are distributed on several monophyletic and well-resolved clades within the class (Tab. 2.6.3) [69–71]. The most common genera are *Bracteacoccus*, "*Chlamydomonas*"-like algae, i.e., representatives of the *Chloromonas* and *Oogamochlamys* clades, *Chlorosarcinopsis*, *Chlorococcum*, *Deasonia*, and *Scenedesmus*-like algae (Fig. 2.6.5).

Although less common in lithophytic habitats than those of the other two green algal classes, species of class Ulvophyceae (Fig. 2.6.6) are widely distributed on sandstone surfaces. About nine genera of Ulvophyceae have been reported from rock surfaces yet (Tab. 2.6.4). They were often overlooked and misidentified as a member of the Chlorophyceae [60]. The systematics of lithophytic Ulvophyceae is still difficult because of the phenotypic plasticity of morphological features caused by different environmental conditions. Molecular phylogenies of the lithophytic Ulvophyceae disagree with the classification based on morphology (e.g., [60]). Examples of lithophytic species of Ulvophyceae are shown in Fig. 2.6.6.

Streptophyta

The Streptophyta is the lineage of green plants (Chlorobionta), which is sister to the Chlorophyta but developed the embryophytes and features the terrestrialization of green algae. The diversity of streptophyte green algae (Fig. 2.6.7; Tab. 2.6.5) is low in terrestrial

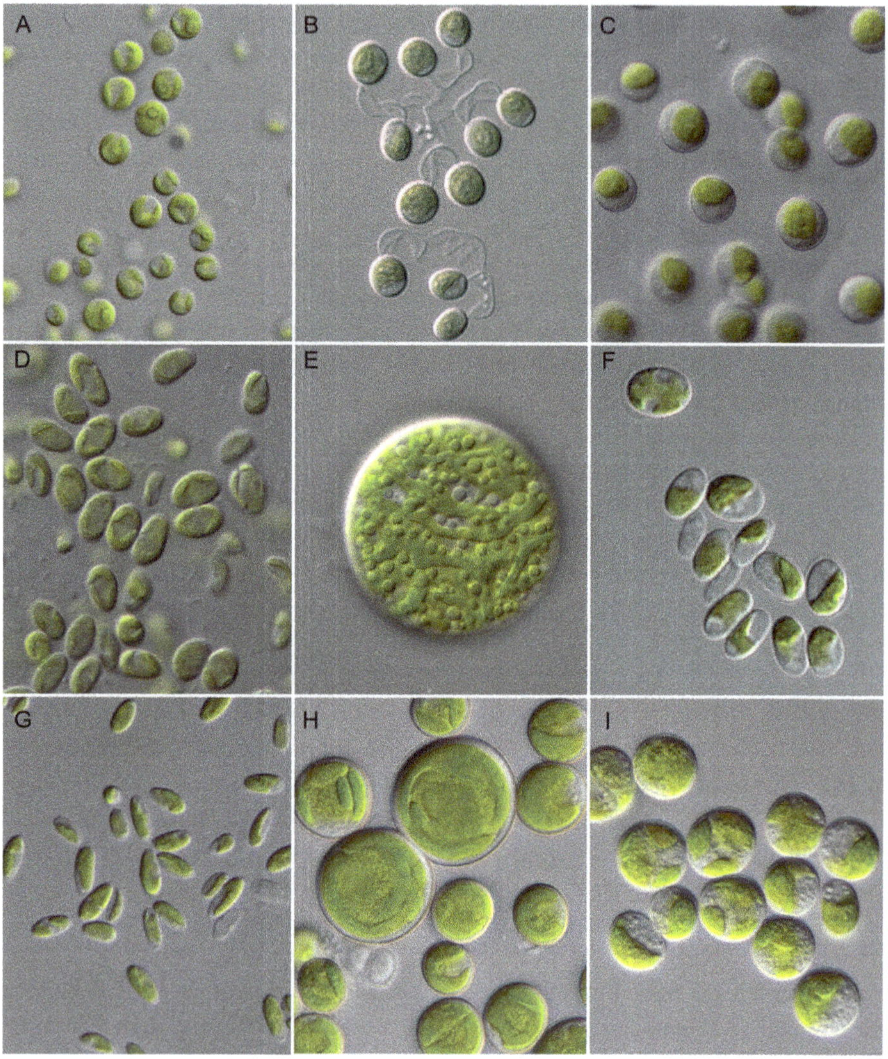

Fig. 2.6.2: Examples of lithophytic green algae of the Trebouxiophyceae (Chlorophyta) in culture, part 1. (A) *Chlorella vulgaris*, isolated from a sandstone, Obelisk, castle Schönbrunn, Vienna, Austria. (B) *Dictyosphaerium ehrenbergianum*, isolated from sandstone, Trakhtemyriv Natural Park, Ukraine. (C) *Coenochloris* sp., isolated from granite, La Campana, Chile. (D) *Neocystis brevis*, isolated from granite, La Campana, Chile. (E) *Dictyochloropsis splendida*, isolated from soil, Trakhtemyriv Natural Park, Ukraine. (F) *Coccomyxa viridis*, isolated from sandstone, Obelisk, castle Schönbrunn, Vienna, Austria. (G) *Coccomyxa simplex*, isolated from sandstone, Obelisk, castle Schönbrunn, Vienna, Austria. (H) *Elliptochloris bilobata*, isolated from granite, Nahuelbuta, Chile. (I) *Elliptochloris subsphaerica*, isolated from granite, Nahuelbuta, Chile.

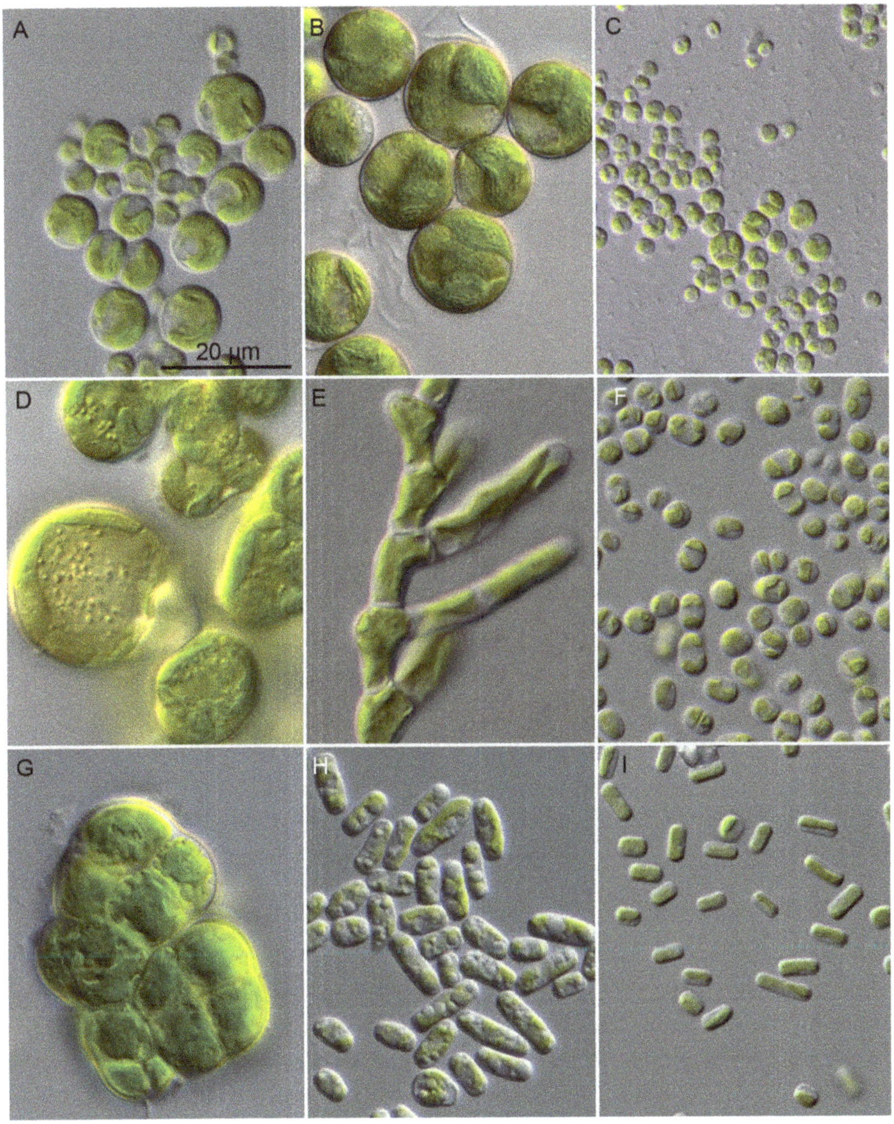

Fig. 2.6.3: Examples of lithophytic green algae of the Trebouxiophyceae (Chlorophyta) in culture, part 2. (A) *Lobosphaera tirolensis* strain SAG 2007. (B) *Parietochloris alveolaris*, isolated from Scythian statues, granite, Khomutovsky Steppe National Natural Park, Ukraine. (C) *Muriella terrestris* strain SAG 2435. (D) *Myrmecia biatorellae*, isolated from granite, La Campana, Chile. (E) *Desmococcus olivaceus*, isolated from granite walls of Kilmartin, Scotland. (F) *Diplosphaera chodatii* strain SAG 49.86. (G) *Prasiolopsis ramosa* strain SAG 26.83. (H) *Pseudostichococcus monallantoides*, isolated from granite, Nahuelbuta, Chile. (I) *Stichococcus* sp., isolated from granite, La Campana, Chile.

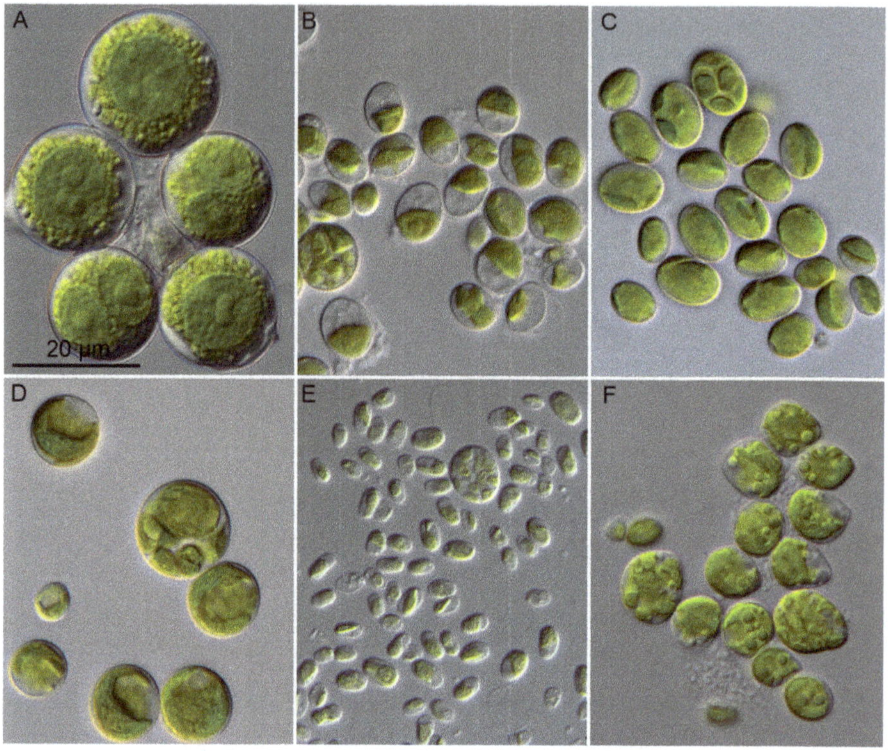

Fig. 2.6.4: Examples of lithophytic green algae of the Trebouxiophyceae (Chlorophyta) in culture, part 3. (A) *Trebouxia* sp., isolated from granite, Nahuelbuta, Chile. (B) *Chloroidium ellipsoideum* strain SAG 3.95. (C) *Chloroidium saccharophilum*, isolated from sandstone, Obelisk, castle Schönbrunn, Vienna, Austria. (D) *Jaagichlorella luteoviridis* strain SAG 211-2a. (E) *Watanabea borysthenica* strain SAG 2550. (F) *Watanabea alpicola*, isolated from granite, Nahuelbuta, Chile.

habitats compared with it in freshwaters. Thus, they prevail in relatively humid and shadowed microniches of the lithophytic habitats. About 10 genera of streptophyte green algae have been reported from rock surfaces so far, with members of the classes Klebsormidiophyceae and Zygnematophyceae being most common (Tab. 2.6.5). Examples of lithophytic species of the streptophyte green algae are shown in Fig. 2.6.7.

2.6.4 Rock substrates and their algal communities

Limestone

The most prominent limestone habitats of lithophytic algae are caves, the conspicuous black crusts called "Tintenstriche", and man-made hard surfaces like walls of buildings. Characteristics for cave habitats are high humidity levels and low

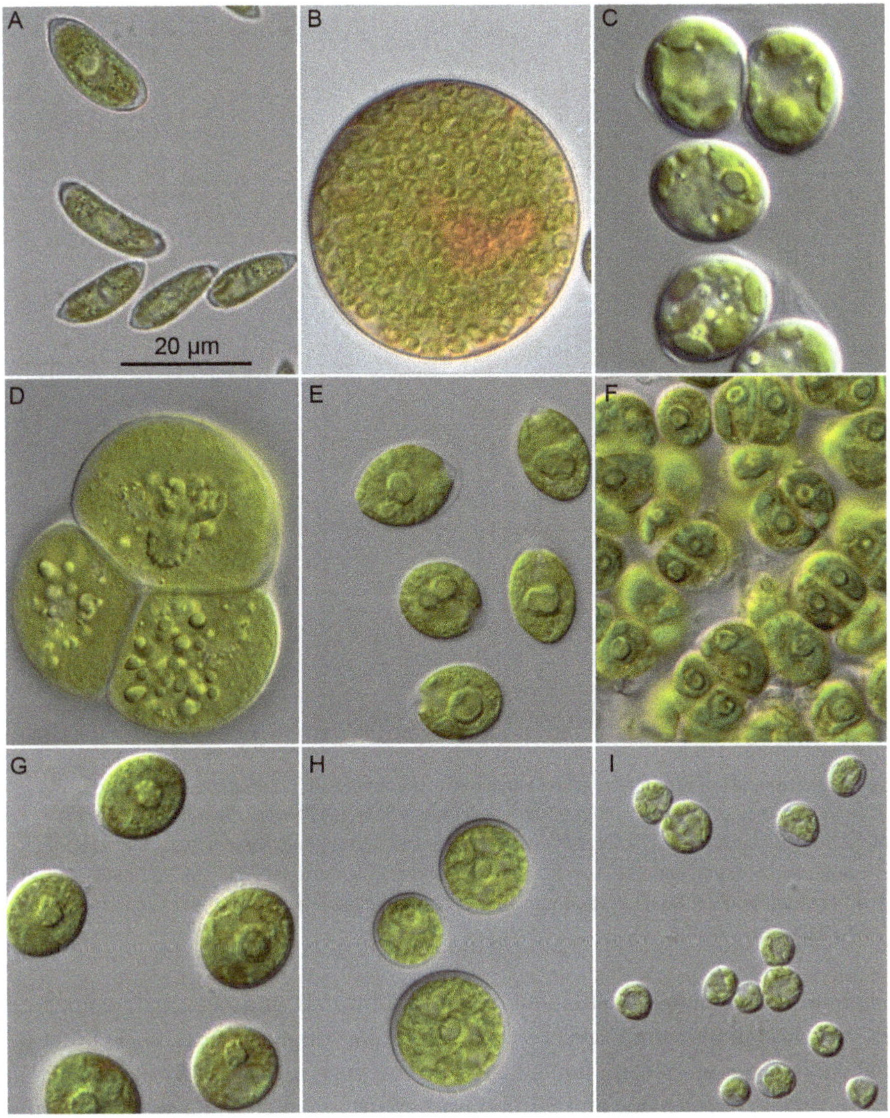

Fig. 2.6.5: Examples of lithophytic green algae of the Chlorophyceae (Chlorophyta) in culture. (A) *Chlorolobion lunulatum*, isolated from soil crusts of Namibia, Africa. (B) *Bracteacoccus grandis*, isolated from soil crusts of Namibia, Africa. (C) *Coelastrella oocystiformis* strain SAG 277-1. (D) *Eubrownia isobilateralis* strain SAG 95.80. (E) *Lobochlamys culleus* strain SAG 17.73. (F) *Chlorosarcinopsis arenicola*, isolated from soil, Azov-Syvash National Nature Park, Ukraine. (G) A young cell of *Deasonia* sp., isolated from soil crusts of Namibia, Africa. (H) *Spongiochloris minor*, isolated from soil, Trakhtemyriv Natural Park, Ukraine. (I) *Jenufa aeroterrestrica* strain SAG 2383.

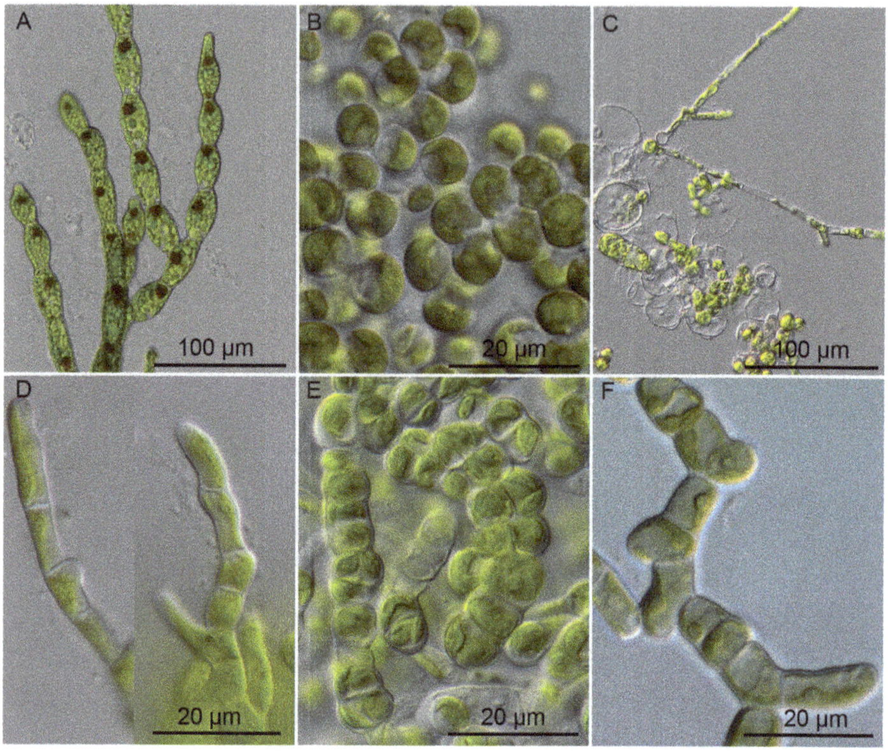

Fig. 2.6.6: Examples of lithophytic green algae of the Ulvophyceae (Chlorophyta) in culture. (A) *Trentepohlia* sp., isolated from granite, Nahuelbuta, Chile. (B) *Planophila laetevirens* strain SAG 465-1. (C) *Ctenocladus circinnatus*, isolated from sandstone, Bock Casemates, Luxembourg. (D) *Halofilum ramosum*, isolated from archaeological remains, Carthage, Tunisia. (E) *Paulbroadya prostrata* strain SAG 23.92. (F) *Pseudendoclonium commune*, isolated from quartz cliffs, Snake Island, Black Sea, Ukraine.

temperatures, minimal fluctuations of daily (diurnal) and seasonal temperatures, and low illumination levels. Most cave algae are cyanobacteria. Reports on eukaryotic algae are mostly limited to very few green algal genera, i.e., *Chlorella*, *Chlorococcum*, *Coccomyxa*, and *Stichococcus*. Many green algae have been left unidentified as "unicellular green algae." Diatoms are the most abundant eukaryotic cave algae. Falasco et al. [66] provided a detailed review, which focused on the diatom flora colonizing different caves, summarizing the information published from a period of more than 100 years up to date. The most frequent diatom species reported are *Diadesmis biceps*, *Hantzschia amphioxys*, *Humidophila contenta*, *Luticola mutica*, *Luticola nivalis*, *Orthoseira roeseana*, and *Pinnularia borealis* (Tab. 2.6.1). Unfortunately, there are no phylogenetic studies of the lithophytic diatoms yet. Czerwik-Marcinkowska and coauthors studied the algae of European caves [32, 64, 65]. Using morphological and ultrastructural analyses, they recognized that epilithic

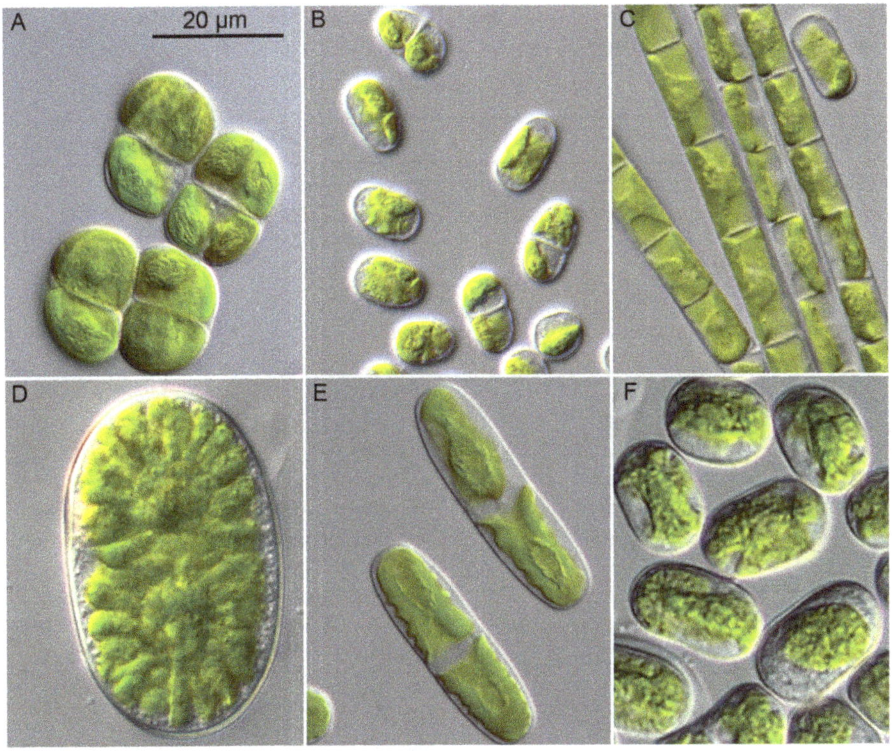

Fig. 2.6.7: Examples of lithophytic green algae of the Streptophyta in culture. (A) *Chlorokybus atmophyticus* strain SAG 34.98 (Chlorokybophyceae). (B) *Interfilum terricola* (Klebsormidiophyceae), isolated from granite, Nahuelbuta, Chile. (C) *Klebsormidium nitens* (Klebsormidiophyceae), isolated from granite, Nahuelbuta, Chile. (D) *Cylindrocystis crassa* strain SAG 23.97 (Zygnematophyceae). (E) *Mesotaenium endlicherianum* strain SAG 12.97 (Zygnematophyceae). (F) *Mesotaenium* sp. (Zygnematophyceae), isolated from granite, Nahuelbuta, Chile.

cyanobacteria and green algae (Chlorophyta) were almost the only components of the cave microflora.

The so-called "Tintenstriche" are often present on bare limestone rock surfaces [72]. In correlation with light intensity and humidity, Golubic [73] distinguished eight different associations dominated by either unicellular or filamentous cyanobacteria, which dominate over green algae and diatoms [2]. Jaag [72] investigated different types of limestone, dolomite, and calcareous schist in the Alps. Most common were cyanobacteria, followed by green algae especially members of the Zygnematophyceae (Streptophyta), which represented around 20% of diversity. Coccoid green algae made up about only 12% of the reported diversity. Most common were members of the Ulvophyceae, i.e., *Trentepohlia aurea*, *T. umbrina*, and *T. iolithus*, as well as Trebouxiophyceae, i.e., *Stichococcus bacillaris*, *Apatococcus* sp. (as "*Protococcus*"), *Chloroidium lichinum*, and species of *Coccomyxa*. Recent studies

Tab. 2.6.1: List of genera and species of red algae (Rhodophyta), diatoms, and other algae of Stramenopiles known from lithophytic habitats so far. Names in bold indicate algae shown in Fig. 2.6.1.

Division or class	Genus / species	Genus / species	Genus / species
Rhodophyta	*Chroothece rupestris*	*Phragmonema sordidum*	
	Cyanidium calcaridum	*Rhodospora sordida*	
Diatomeae (Stramenopiles)	*Achnanthes coarctata*	*Fragilariforma*	*Nitzschia*
	Angusticopula	*Frustulia*	*Nupela*
	Aulacoseira	*Gomphonema*	*Orthoseira roeseana*
	Caloneis	*Grunowia* spp.	*Pinnularia borealis*
	Cavinula	*H. gallica*	*Placoneis*
	Cymbella spp.	*Halamphora*	*Platessa*
	Cymbopleura	*Hantzschia*	*Psammothidium*
	Diadesmis bicep	***Hantzschia amphioxys*** (Fig. 2.6.1 E)	*Pseudostaurosira*
	Diatoma spp.	***Humidophila contenta*** (Fig. 2.6.1 F)	*Reimeria*
	Diploneis spp.	*Hygropetra*	*Rhopalodia*
	Ellerbeckia	*Kurtkrammeria*	*Sellaphora*
	Encyonopsis	*Luticola nivalis, L. mutica*	*Simonsenia*
	Epithemia	*Mayamaea*	*Stauroneis* spp.
	Eunotia spp.	*Meridion* spp.	*Staurosirella*
	Fallacia	*Microcostatus*	*Tryblionella*
	Fistulifera	*Navicula* spp.	
	Fragilaria spp.	*Neidium* spp.	
Xanthophyceae (Stramenopiles)	***Botrydiopsis*** (Fig. 2.6.1 B)	***Heterococcus*** (Fig. 2.6.1 C)	***Xanthonema*** (Fig. 2.6.1 D)
	Botryochloris	*Pleurochloris*	
	Ellipsoidion	*Tribonema*	
Eustigmatophyceae (Stramenopiles)	*Chlorobotrys*	*Ellipsoidion*	***Vischeria*** (Fig. 2.6.1 A)

based on molecular cloning of lithophytes colonizing limestones in the Alps (Austria and Switzerland) confirmed cyanobacteria's dominance with green algae found only occasionally [74–78].

Available information about algae on limestone walls of buildings in open space has been summarized by John [38] and Macedo et al. [57]. Green algae mostly from the Trebouxiophyceae, e.g., *Chloroidium*, *Coccomyxa*, and *Desmococcus*, as well as the streptophyte *Klebsormidium* were reported. In general, the diversity of green algae on man-made limestone substrates is still poorly known [38, 43].

Tab. 2.6.2: List of genera and species of Trebouxiophyceae (Chlorophyta) from lithophytic habitats, their taxonomic positions within the class. Additional genera mean those genera which have not been approved by molecular phylogeny, or whose phylogenetic position is still unclear. Names in bold indicate algae shown in Figs. 2.6.2–2.6.4.

Trebouxiophyceae (Chlorophyta)			
clade	**genus**	**Figure**	**genus, additional**
Apatococcus-clade	*Apatococcus*		*Coleochlamys*
Chlorella-clade	***Chlorella***	2.6.2 A	*Keriochlamys*
	Dictyosphaerium	2.6.2 B	*Pleurastrosarcina*
Coenocystis-clade	***Coenochloris***	2.6.2 C	*Podohedra*
	Neocystis	2.6.2 D	*Radiococcus*
Dictyochloropsis-clade	***Dictyochloropsis***	2.6.2 E	*Sporotetras*
Elliptochloris-clade	***Coccomyxa***	2.6.2 F, G	*Symbiochloris*
	Elliptochloris	2.6.2 H, I	*Thelesphaera*
Lobosphaera-clade	***Lobosphaera***	2.6.3 A	
	Parietochloris	2.6.3 B	
Muriella-clade	***Muriella***	2.6.3 C	
Myrmecia-clade	***Myrmecia***	2.6.3 D	
Prasiola-clade	***Desmococcus***	2.6.3 E	
	Deuterostichococcus		
	Diplosphaera	2.6.3 F	
	Edaphochlorella		
	Prasiococcus		
	Prasiola		
	Prasiolopsis	2.6.3 G	
	Pseudochlorella		
	Pseudostichococcus	2.6.3 H	
	Stichococcus s.l.	2.6.3 I	
Trebouxia-clade	***Trebouxia***	2.6.4 A	
Watanabea-clade	***Chloroidium***	2.6.4 B, C	
	Jaagichlorella	2.6.4 D	
	Watanabea	2.6.4 E, F	
Choricystis-clade	*Choricystis*		
Leptosira-clade	*Leptosira*		

Sandstone and granite

Natural sandstone rocks have so far been studied only occasionally [79, 80]. A high diversity of green algae has been reported from sandstone surfaces in the Carpathian Mountains [80]. The most abundant have been members of Trebouxiophyceae (e.g., *Myrmecia bisecta*, *Stichococcus bacillaris*), but also some Ulvophyceae (e.g., *Ctenocladus printzii*), and the streptophyte *Klebsormidium nitens* were present. In contrast to the natural substrates, cultural monuments have actively been studied Europe [42, 51, 54, 58, 59, 81]. Green algae of the Trebouxiophyceae like *Chlorella*, *Chloroidium*, *Elliptochloris*, and *Stichococcus* have been common on monument surfaces,

Tab. 2.6.3: List of genera and species of Chlorophyceae (Chlorophyta) from lithophytic habitats, their taxonomic positions within the class. Additional genera mean those genera which have not been approved by molecular phylogeny, or whose phylogenetic position is still unclear. Names in bold indicate algae shown in Fig. 2.6.5.

Chlorophyceae (Chlorophyta)					
lineage	**clade**	**Figure**	**Figure**	**genus, additional**	**Figure**
Sphaeropleales (DO group)	*Ankistrodesmus*-clade	***Chlorolobion***	2.6.5 A	*Characium*	
		Monoraphidium			
	Bracteacoccus-clade	***Bracteacoccus***	2.6.5 B	*Dictyochloris*	
	Chaetopeltis-clade	*Hormotilopsis*		*Hormotila*	
	Mychonastes-clade	*Mychonastes*		*Inoderma*	
	Pseudomuriella-clade	*Pseudomuriella*		***Jenufa***	2.6.5 I
	Scenedesmus-clade	***Coelastrella***	2.6.5 C	*Keratococcus*	
		Desmodesmus		*Macrochloris*	
		Halochlorella		*Phaseolaria*	
Volvocales (CW group)	*Chloromonas*-clade	*Chloromonas*			
		Ostravamonas		*Pseudodictyochloris*	
	Moewusii-clade	*Chlorococcum*			
		Eubrownia	2.6.5 D		
	Oogamochlamys-clade	*Desmotetra*			
		Lobochlamys	2.6.5 E		
	Reinhardtii-clade	*Radiosphaera*			
	Stephanosphaera-clade	*'Chlorococcum'*		*Protosiphon*	
		Chlorosarcinopsis	2.6.5 F		
		Deasonia	2.6.5 G		
		Palmellopsis			
		Spongiochloris	2.6.5 H		

Tab. 2.6.4: List of genera and species of Ulvophyceae (Chlorophyta) from lithophytic habitats, their taxonomic positions within the class. Additional genera mean those genera which have not been approved by molecular phylogeny, or whose phylogenetic position is still unclear. Names in bold indicate algae shown in Fig. 2.6.6.

Ulvophyceae (Chlorophyta)			
order	**clade**	**genus**	**Figure**
Cladophorales		*Wittrockiella*	
Scotinosphaeriales		*Kentrosphaera*	
Trentepohliales	*Trentepohlia*-clade	***Trentepohlia***	2.6.6 A
Ulotrichales	*Planophila*-clade	***Planophila***	2.6.6 B
		Rhexinema	
Ulvales	*Ctenocladus*-clade	***Ctenocladus***	2.6.6 C
	Halofilum-clade	***Halofilum***	2.6.6 D
		Paulbroadya	2.6.6 E
	Pseudendoclonium-clade	***Pseudendoclonium***	2.6.6 F

Tab. 2.6.5: List of genera and species of Streptophyta (Chlorophyta) from lithophytic habitats, their taxonomic positions within the class. Additional genera mean those genera which have not been approved by molecular phylogeny, or whose phylogenetic position is still unclear. Names in bold indicate algae shown in Fig. 2.6.7.

Streptophyta		
class	**genus**	**Figure**
Chlorokybophyceae	***Chlorokybus***	2.6.7 A
Klebsormidiophyceae	***Interfilum***	2.6.7 B
	Klebsormidium	2.6.7 C
Zygnematophyceae	*Cosmarium*	
	Cylindrocystis	2.6.7 D
	Euastrum	
	Mesotaenium	2.6.7 F
	Penium	
	Spirotaenia	
	Staurastrum	

particularly in the Mediterranean [52]. The occurrence of *Trebouxia* and *Trentepohlia* indicated their putative involvement in lichen symbioses, initiating the colonization of stone surfaces by lichens. Endolithic growth of algae was under a black patina comprising *Chlorella*, *Stichococcus*, *Trentepohlia*, and *Klebsormidium* [52, 57].

Investigations of algal communities on the granitic substrate are scarce. Previously, the walls of monuments and castles in Spain have been studied [55, 56]. Other studies on granite surfaces have been carried out in the Alpine region by Jaag [72], and with a focus on cyanobacteria and natural granite canyons in the southern part of Ukraine by Mikhailyuk et al. [30]. Jaag [72] investigated granite outcrops in the Swiss Alps, located at an altitude up to 2700 m a.s.l. Besides cyanobacteria and diatoms, he listed species of green algae (Chlorophyta), i.e., *Coccomyxa* and *Trentepohlia*, and the streptophyte *Cylindrocystis* (Zygnematophyceae). Mikhailyuk et al. [30] have studied algae from the epilithic, chasmoendolithic, and epiphytic habitats. Members of Chlorophyta were rather diverse in the chasmoendolithic habitat. Sun-exposed granite surfaces appeared to be an extreme habitat, and the development of algae occurred only in areas protected from direct sunlight and excessive dehydration.

Artificial man-made hard substrates

Different man-made artificial hard substrates in urban and suburban areas have been studied using rRNA gene sequence analyses in conjunction with establishing clone libraries and culture strains [82–85]. A variety of green algae, all from the Trebouxiophyceae, were found on various man-made substrates such as plastic switch boxes, roof tiles, and facades of buildings. The most common were species of *Apatococcus*, *Chloroidium*, *Coccomyxa*, *Coenochloris*, *Jaagichlorella*, and *Pseudochlorella*. The diversity of detected algae from crude cultures was higher than from environmental samples. There was a shift in the abundance of specific operational taxonomic units (OTUs) between environmental samples and crude cultures. The authors assumed that members of Trebouxiophyceae might be more tolerant towards desiccation than members of the Chlorophyceae. Generally, epilithic algae's species composition on the man-made substrates was similar to those of natural stones but characterized by a lower diversity. The low water availability due to the somewhat limited water holding capacity of the artificial substrates with smooth surfaces from which the water quickly runs off may explain this. Generally, the investigations of Hallmann et al. [82–85] showed a good correlation with results obtained from traditional morphology and cultivation methods. It confirms that studies of lithophytic algae may have only a marginal bias because they are easy to culture. Cultural techniques are well developed and appropriate for biodiversity assessment, ecology, and taxonomical studies of this group of organisms.

2.6.5 Formally described algal communities

Eight different aerophilic algal communities have been described in the literature. Barkman [86] described several algal communities such as the *Apatococcetum*, *Trentepohlietum*, and *Prasioletum*.

The *Apatococcetum* was initially described by Barkman [86] under the name *Pleurococcetum* and later on renamed into *Apatococcetum* by Gärtner and Ingolic [87]. In the temperate and subalpine zones, the community is the most abundant. It is typical for shaded, humid sites. Characteristic community members are species of *Apatococcus*, *Desmococcus*, *Coenochloris*, and *Myrmecia*. Such communities were also investigated using molecular methods (rRNA gene clone libraries) by Hallmann and coauthors [85] as well as cultivation methods [88, 89]. The *Apatococcetum* is regarded as a toxic- and nitrate-tolerant community [90, 91]. It has also been reported from granite outcrops in Ukraine [30].

The *Trentepohlietum* occurs on tree bark of shaded and humid places in mountainous and subalpine Europe [86]. Generally, *Trentepohlia* species are distributed in the tropical regions where they are very abundant [92]. There they may cause the progressive mechanical degradation of buildings (biodeterioration). In tropical areas with high humidity conditions, the damage to concrete structures caused by *Trentepohlia* spp. is considered a serious problem. In West Europe, the distribution of these communities was studied by Rindi et al. [88].

The *Prasioletum* is typical for the humid and cool living conditions of the Arctic and Antarctica [86, 89, 92, 93]. The community is also known as nitrophilic and nitrotolerant because it is common in places strongly polluted with bird feces [4]. The typical representatives are *Prasiola crispa*, *Prasiolopsis ramosa*, *Prasiococcus calcarius*, and *Schizogonium murale* Kützing. Rindi et al. [93] also reported Prasiolales from urban walls at Galway (Ireland). Some representatives of the community also were found in cracks in granite canyons in Ukraine [30].

Several types of lithophytic communities from sandstones' surfaces in Europe have been described ([2] and references therein). The *Bacillarietum* can be found on wet sandstones in the humid seasons and disappears during the dry seasons. The most common species are the diatoms *Fragilariforma virescens*, *Pinnularia borealis*, *Pinnularia appendiculata*, *Orthoseira roseana*, and *Frustulia saxonica*. The *Mesotaenietum* and the *Gloeocystetum* have been reported from a mountain valley in Central Europe [72] ([2] and references therein). They have been found at places with high humidity and light intensities. The characteristic species the green algae (Zygnematophyceae) *Mesotaenium macrococcum*, *Mesotaenium chlamydosporum*, and *Cylindrocystis crassa*. The distinct members of the *Gloeocystetum* are species of *Gloeocystis* (Chlorophyceae), *Klebsormidium* (Klebsormidiophyceae), and the dinoflagellate *Rufusiella insignis*. Species of those communities can produce massive, violet pigmented mucilage.

The *Klebsormidium* community has initially been a part of the *Prasioletum*, but later it was demonstrated that in temperate zones where air humidity is lower and temperature higher, the species of Prasiolales were exchanged by *Klebsormidium* [88]. On granite canyons in the Ukraine, the community was present on the north-west exposed shaded sampling sites [30].

The *Dilabifilum* community has been proposed by Mikhailyuk et al. [30] for wet granite sites in canyons in Ukraine, where it covered the north exposed sites of granite outcrops. The ulvophycean filamentous algae are difficult for morphological identification, and the main morphological features are not supported by molecular phylogeny [60]. The *Dilabifilum* sp. as reported by Mikhailyuk et al. [30] likely belongs to *Ctenocladus circinnatus* but needs to be investigated by rRNA gene sequence analyses.

The *Elliptochloris* community typically lives on granite outcrops [30]. There, species of *Elliptochloris* made up biofilms in the cracks on granites and were spread on dry sampling sites with high illumination. The most common species were *Elliptochloris subsphaerica* and *Elliptochloris bilobata* (probably representing *Elliptochloris perforata*), whereas *Elliptochloris reniformis* was less frequently encountered.

References

[1] Hollerbakh MM, Shtina EA. Pochvenniye vodorosli (Soil algae). Leningrad, Nauka; 1969 (In Russian).

[2] Hoffmann L. Algae of terrestrial habitats. *Bot Rev*. 1989;55:77–105.

[3] Büdel B, Darienko T, Deutschewitz K, et al. Southern African biological soil crusts are ubiquitous and highly diverse in drylands, being restricted by rainfall frequency. *Microb Ecol*. 2009;57:229–47.

[4] Ettl H, Gärtner G. Syllabus der Boden-, Luft- und Flechtenalgen. Berlin, Germany: Springer; 2014.

[5] Golubic S, Friedmann I, Schneider J. The lithobiontic ecological niche, with special reference to microorganisms. *J Sediment Petrol*. 1981;51:475–8.

[6] Hoham RW, Remias D. Snow and glacial algae: a review. *J Phycol*. 2020;56:264–82.

[7] Bell RA. Cryptoendolithic algae of hot semiarid lands and deserts. *J Phycol*. 1993;29:133–9.

[8] Chan Y, Lacap DC, Lau MC, et al. Hypolithic microbial communities: between a rock and a hard place. *Environ Microbiol*. 2012;14:2272–82.

[9] Nienow JA, Friedmann EI. Terrestrial lithophytic (rock) communities. In: Friedmann EI, editor. Antarctic Microbiology. New York: Wiley-Liss; 1997. p. 343–412.
[10] Vestal JR. Cryptoendolithic communities from hot and cold deserts: speculation on microbial colonization and succession. In: Primary Succession on Land. Vol. 12. Blackwell Scientific; BES Special Publication; 1993. p. 5–16.
[11] Villar SE, Edwards HGM, Benning LG. Raman spectroscopic and scanning electron microscopic analysis of a novel biological colonisation of volcanic rocks. *Icarus*. 2006;184:158–69.
[12] Wynn-Williams D. Cyanobacteria in deserts—life at the limit? In: Whitton BA, Potts M, editors. The Ecology of Cyanobacteria. Boca Raton: Kluwer Academic Publishers; 2002. p. 341–366.
[13] Makhalanyane TP, Pointing SB, Cowan DA. Lithobionts: Cryptic and Refuge Niches. In: Cowan D, editor. Antarctic Terrestrial Microbiology. Berlin, Germany: Springer, 2014. p. 163–179.
[14] Pointing SB, Belnap J. Microbial colonization and controls in dryland systems. *Nat Rev Microbiol.* 2012;10:551–62.
[15] Friedmann EI, Weed R. Microbial trace-fossil formation, biogenous, and abiotic weathering in the Antarctic cold desert. *Science*. 1987;236:703–5.
[16] Komarek J, Nedbalova L. Green cryosestic algae. In: Seckbach J, editor. Algae and cyanobacteria in extreme environments. Dordrecht, Netherlands: Springer; 2007. p. 323–42.
[17] Hoham RW, Duval B. Microbial ecology of snow and freshwater ice with emphasis on snow algae. In: Jones HG, Pomeroy JW, Walker DA, Hoham RW, editors. Snow Ecology. An Interdisciplinary Examination of Snow-Covered Ecosystems. Cambridge: Cambridge University Press; 2001. p. 168–228.
[18] Flechtner V. North American desert microbiotic soil crust communities: diversity despite challenge. In: Seckbach J, editor. Algae and Cyanobacteria in Extreme Environments. Dordrecht, Netherlands: Springer; 2007. p. 539–54.
[19] Garty J. Lithobionts in the Eastern Mediterranean. In: Seckbach J, editor. Enigmatic Microorganisms and Life in Extreme Environments. Dordrecht, Netherlands: Kluwer Academic Publishers; 1999. p. 257–76.
[20] Friedmann EI. Endolithic microbial life in hot and cold deserts. *Limit Life*. 1980;4:33–45.
[21] Broady PA. Ecological and taxonomic observations on subaerial epilithic algae from Princess Elizabeth Land and Mac. Robertson Land, Antarctica. *Br Phycol J*. 1981;16:257–66.
[22] Broady PA. The ecology of sublithic terrestrial algae at the Vestfold Hills, Antarctica. *Br Phycol J* 1981;16:231–40.
[23] Cowan DA, Ah Tow L. Endangered Antarctic environments. *Ann Rev Microbiol*. 2004;58:649–90.
[24] Šperl J, Trčková J. Permeability and porosity of rocks and their relationship based on laboratory testing. *Acta Geodyn Geomater.* 2008;5:41–7.
[25] Cockell CS, McKay CP, Omelon C. Polar endoliths an anti-correlation of climatic extremes and microbial biodiversity. *Int J Astrobiol.* 2002;1:305–10.
[26] Cockell CS, Stokes MD. Widespread colonization by polar hypoliths. *Nature*. 2004;431:414–4.
[27] Cowan DA, Pointing SB, Stevens MI, Craig Cary S, Stomeo F, Tuffin IM. Distribution and abiotic influences on hypolithic microbial communities in an Antarctic Dry Valley. *Polar Biol.* 2010;34:307–11.
[28] Wong FK, Lacap DC, Lau MC, Aitchison JC, Cowan DA, Pointing SB. Hypolithic microbial community of quartz pavement in the high-altitude tundra of central Tibet. *Microbiol Ecol.* 2010;60:730–9.
[29] Büdel B, Weber B, Kühl M, Pfanz H, Sültemeyer D, Wessels D. Reshaping of sandstone surfaces by cryptoendolithic cyanobacteria: bioalkalization causes chemical weathering in arid landscapes. *Geobiology*. 2004;2:261–8.
[30] Mikhailyuk TI, Kondratyuk SY, Nyporko SO et al. Lichen-forming fungi, bryophytes and terrestrial algae of granitic canyons of Ukraine. Kyiv, Ukraine: Alterpress; 2011 (in Ukrainian).
[31] Vinogradova O, Mikhailyuk T. Algal flora of the caves and grottoes of the National Nature Park "Podilsky Tovtry" (Ukraine). *Int J Algae*. 2008;11:289–304.

[32] Czerwik-Marcinkowska J, Mrozińska T. Algae and cyanobacteria in caves of the Polish Jura. *Polish Bot J.* 2011;56:203–43.

[33] Škaloud P. Species composition and diversity of aero-terrestrial algae and cyanobacteria of the Boreč Hill ventaroles. *Fottea.* 2009;9:65–80.

[34] Broady PA. Survey of algae and other terrestrial biota at Edward VII Peninsula, Marie Byrd Land. *Antarct Sci.* 1989;1:215–24.

[35] Friedmann EI, Galun M. Desert algae, lichens and fungi. In: Brown GWJ, editor. Desert Biology. New York, USA: Academic Press; 1974. p. 165–212.

[36] Friedmann EI. Endolithic microorganisms in the Antarctic cold desert. *Science.* 1982;215: 1045–53.

[37] Ziolkowski LA, Mykytczuk NCS, Omelon CR, Johnson H, Whyte LG, Slater GF. Arctic gypsum endoliths: a bioheochemical characterization of a viable and active microbial community. *Biogeosci Discuss.* 2013;10:2269–304.

[38] John DM. Algal growths on buildings: a general review and methods of treatment. *Biodeterior Abstr.* 1988;2:81–102.

[39] Krumbein WE. Microbial interactions with mineral materials. In: Houghton DR, Smith RN, Eggins HOW, editors. Biodeterioration. Vol. 7. London: Elsevier Applied Science; 1988. p. 78–100.

[40] Ascaso C, Wierzchos J. Study of biodeterioration zone between the lichen thallus and the substrate. *Cryptogam Bot.* 1995;5:270–81.

[41] Edwards HGM, Russell NC, Seaward MR, Slarke D. Lichen biodeterioration under different microclimates: an FT Raman spectroscopic study. *Spectrochim Acta.* 1995;51A, 2091–100.

[42] Altieri A, Pietrini AM, Ricci S, Roccardi A. The temples of the archaeological area of Paestum (Italy): a case study on biodeterioration. In: Fassina V, editor. Proceedings of the 9th International Congress on Deterioration and Conservation of Stone, Amsterdam, Elsevier; 2000. p. 433–443.

[43] Kovacik L. Cyanobacteria and algae as agents of biodeterioration of stone substrata of historical buildings and other cultural monuments. In: Choi S, Suh M, editors. Proceedings of the New Millennium International Forum on Conservation of Cultural Property, Kongju, Korea, Kongju National University; 2000. p. 44–58.

[44] Huek HJ. The biodeterioration of materials—an appraisal. *Int Biodeterior Biodegradation.* 2001;48:5–11.

[45] De los Ríos A, Grube M, Sancho LG, Ascaso C. Ultrastructural and genetic characteristics of endolithic cyanobacterial biofilms colonizing Antarctic granite rocks. *FEMS Microbiol Ecol.* 2007;59:386–95.

[46] Gorbushina A. Life on the rocks. *Environ Microbiol.* 2007;9:1613–31.

[47] De los Ríos A, Wierzchos J, Sancho LG, Ascaso C. Acid microenvironments in microbial biofilms of Antarctic endolithic microecosystems. *Environ Microbiol.* 2003;5:231–7.

[48] De los Ríos A, Cary C, Cowan D. The spatial structures of hypolithic communities in the Dry Valleys of East Antarctica. *Polar Biol.* 2014;37:1823–33.

[49] Büdel B, Wessels DC. Rock inhabiting blue-green algae/cyanobacteria from hot arid regions. *Algol Stud.* 1991;64:385–98.

[50] Büdel B, Dulic T, Darienko T, Rybalka N, Friedl T. Cyanobacteria and algae in biological soil crusts. In: Weber B, Büdel B, Belnap J, editors. Biological Soil Crusts: An Organizing Principle in Drylands. Ecological Studies (Analysis and Synthesis), 226. Cham, Germany: Springer; 2016. p. 55–80.

[51] Ortega-Calvo JJ, Sanchez-Castillo PM, Hernandez-Marine M, Saiz-Jimenez C. Isolation and characterization of epilithic chlorophyta and cyanobacteria from two Spanish cathedrals (Salamanca and Toledo). *Nova Hedwigia.* 1993;57:239–53.

[52] Ortega-Calvo JJ, Arino X, Hernandez-Marine M, Seiz-Jiminez C. Factors affecting the weathering and colonization of monuments by phototrophic microorganisms. *Sci Total Environ.* 1995;167:329–41.

[53] Krumbein W E, Urzi CE. Microbiological impacts on cultural heritage with special reference to rock materials. In: Lefevre R, editor. La pietra dei Monumenti nel suo Ambiente fisico. Rome, Italy: Instituto Poligrafico Roma; 1995. p. 151–78.

[54] Flores M, Lorenzo J, Gómez-Alarcón G. Algae and bacteria on historic monuments at Alcala de Henares, Spain. *Int Biodeterior Biodegradation*. 1997;40:241–6.

[55] Noguerol-Seoane A, Rifon-Lastra A. Epilithic phycoflora on monuments. A survey of San Esteban de Ribas de Sil monastery (Ourense, NW Spain). *Cryptogamie Algol*. 1997;18:351–61.

[56] Rifón-Lastra A, Noguerol-Seoane Á. Green algae associated with the granite walls of monuments in Galicia (NW Spain). *Cryptogamie Algol*. 2001;22:305–26.

[57] Macedo MF, Miller AZ, Dionísio A, Saiz-Jimenez C. Biodiversity of cyanobacteria and green algae on monuments in the Mediterranean Basin: an overview. *Microbiology*. 2009;155:3476–90.

[58] Miller AZ, Macedo MF. Mapping and characterization of a green biofilm inside of Vilar de Frades Church (Portugal). In: Fort R, Alvarez de Buergo M, Gomez-Heras, M, Vazquez-Calvo C, editors. Heritage, Weathering and Conservation. London, UK: Taylor & Francis; 2006. p. 329–35.

[59] Miller AZ, Laiz L, Dionısio A, Macedo MF, Saiz-Jimenez C. Growth of phototrophic biofilms from limestone monuments under laboratory conditions. *Int Biodeterior Biodegradation*. 2009;63, doi: 10.1016/j.ibiod.2009.04.004.

[60] Darienko T, Pröschold T. Toward a monograph of non-marine Ulvophyceae using an integrative approach (molecular phylogeny and systematics of terrestrial Ulvophyceae II). *Phytotaxa*. 2017;324:1–41.

[61] Darienko T, Pröschold T. The genus *Jaagichlorella* Reisigl (Trebouxiophyceae, Chlorophyta) and its close relatives: an evolutionary puzzle. *Phytotaxa*. 2019;388:47–68.

[62] Rybalka N, Mikhailyuk T, Darienko T, Dultz S, Blanke M, Friedl T. Genotypic and phylogenetic diversity of new isolates of terrestrial Xanthophyceae (Stramenopiles) from maritime sandy habitats. *Phycologia*. 2020;59:506–514.

[63] Kryvenda A, Rybalka N, Wolf M, Friedl T. Species distinctions among closely related strains of Eustigmatophyceae (Stramenopiles) emphasizing ITS2 sequence-structure data: *Eustigmatos* and *Vischeria*. *Eur J Phycol*. 2018;53:471–91.

[64] Czerwik-Marcinkowska J. Observations on aerophytic cyanobacteria and algae from ten caves in the Ojców National Park. *Acta Agrobotanica*. 2013;66:39–52.

[65] Czerwik-Marcinkowska J, Wojciechowska A, Massalski A. Biodiversity of limestone caves: aggregations of aerophytic algae and cyanobacteria in relation to site factors. *Polish J Ecol*. 2015;63:481–99.

[66] Falasco E, Ector L, Isaia M, Wetzel, CE, Hoffmann L, Bona F. Diatom flora in subterranean ecosystems: a review. *Int J Speleol*. 2014;43:231–51.

[67] Starmach K. New species of the order Dinococcales (Pyrrhophyta) from the Twardowski Cave in Kraków. *Acta Hydrobiologica*. 1963;5:337–42.

[68] Pröschold T, Marin B, Schlösser UG, Melkonian M. Molecular Phylogeny and taxonomic revision of *Chlamydomonas* (Chlorophyta). I. Emendation of *Chlamydomonas* Ehrenberg and *Chloromonas* Gobi, and description of *Oogamochlamys* gen. nov. and *Lobochlamys* gen. nov. *Protist*. 2001;152:265–300.

[69] Pröschold T, Leliaert F. Systematics of the green algae: conflict of classic and modern approaches. In: Brodie J, Lewis J, editors. Unravelling the Algae, the Past, Present, and Future of Algal Systematics. Boca Raton: CRC Press, 2007;123–53.

[70] Friedl T, Rybalka N. Systematics of the green algae: a brief introduction to the current status. *Prog Bot*. 2012;73:252–80.

[71] Leliaert F, Smith DR, Moreau H, Herron MD, Verbruggen H, Delwiche CF, De Clerck O. Phylogeny and molecular evolution of the green algae. *Crit Rev Plant Sci*. 2012;31:1–46.

[72] Jaag O. Untersuchungen über die Vegetation und Biologie der Algen des nackten Gesteins in den Alpen, im Jura und im schweizerischen Mittelland. *Beitr Kryptogamenfl Schweiz*. 1945;9:1–560.

[73] Golubic S. Algenvegetation der Felsen. Eine ökologische Algenstudie im dinarischen Karstgebiet. In: Elster HJ, Ohle W, editors. Die Binnengewässer 23, Stuttgart, Germany: E. Schweizerbart'sche Verlagsbuchhandlung; 1967. p. 1–183.
[74] Pohl W. Wechselwirkungen zwischen endolithischen Biofilmen and Karbonatgesteinen in alpinen Gebieten Mitteleuropas. PhD dissertation, University of Göttingen; 2000. http://webdoc.gwdg.de/diss/ 2000/pohl/.
[75] Pohl W, Schneider J. Impact of endolithic biofilms on carbonate rock surfaces. In: Siegesmund S, Weiss T, Vollbrecht A, editors. Natural Stone, Weathering Phenomena, Conservation Strategies and Case Studies. London, UK: Geological Society; 2002. p. 177–194.
[76] Horath TD, Neu, TR, Bachofen R. An endolithic microbial community in dolomite rock in central Switzerland: characterization by reflection spectroscopy, pigment analyses, scanning electron microscopy, and laser scanning microscopy. *Microb Ecol.* 2006;51:353–64.
[77] Horath TD. Endolithic microbial communities in dolomite. PhD dissertation, University of Zürich; 2010. https://doi.org/10.5167/uzh-163992.
[78] Horath T, Bachofen R. Molecular characterization of an endolithic microbial community in dolomite rock in the Central Alps (Switzerland). *Microb Ecol.* 2009;58:290–306.
[79] Garbacki N, Ector L, Kostikov I, Hoffmann L. Contribution àl'étude de la flore des grottes de Belgique. *Belgian J Bot.* 1999;132:43–76.
[80] Mikhailyuk TI, Darienko TM. Algae of terrestrial habitats. In: Prorochuk VV, Stefurak YP, Derzhypilskyi LM, editors. National Natural Park "Hutzulshchyna," Lviv, Ukraine, Karty i atlasy; 2013. p. 129–130 (in Ukrainian).
[81] Bellinzoni A M, Caneva G, Ricci S. Ecological trends in travertine colonization by pioneer algae and plant communities. *Int Biodeterior Biodegradation.* 2003;51:203–10.
[82] Hallmann C, Rüdrich J, Enseleit M, Friedl T, Hoppert M. Microbial diversity on a marble monument—a case study. *Environ Earth Sci.* 2011;63:1701–11.
[83] Hallmann C, Stannek L, Fritzlar D, Hause-Reitner D, Friedl T, Hoppert M. Molecular diversity of phototrophic biofilms on building stone. *FEMS Microbiol Ecol.* 2013;84:355–72.
[84] Hallmann C, Wedekind W, Hause-Reitner D, Hoppert M. Cryptogam covers on sepulchral monuments and re-colonization of a marble surface after cleaning. *Environ Earth Sci.* 2013;69: 1149–60.
[85] Hallmann C, Hoppert M, Mudimu O, Friedl T. Biodiversity of green algae covering artificial hard substrate surfaces in a suburban environment: a case study using molecular approaches. *J Phycol.* 2016;52:732–44.
[86] Barkman JJ. Phytosociology and ecology of cryptogamic epiphytes. Assen, Netherlands; 1969.
[87] Gärtner G, Ingolic E. Ein Beitrag zur Kenntnis von *Apatococcus lobatus* (Chlorophyta, Chaetophorales, Leptosiroidea). *Plant Syst Evol.* 1989;164:133–43.
[88] Rindi F, Guiry MD. Composition and spatial variability of terrestrial algal assemblages occurring at the bases of urban walls in Europe. *Phycologia.* 2004;43:225–35.
[89] Rindi F. Diversity, distribution and ecology of green algae and cyanobacteria in urban habitats. In Seckbach J, editor. Algae and Cyanobacteria in Extreme Environments. Dordrecht, Netherlands: Springer; 2007. p. 571–82.
[90] Lüttge U, Büdel B. Resurrection kinetics of photosynthesis in desiccation-tolerant terrestrial green algae (Chlorophyta) on tree bark. *Plant Biol.* 2010;12:437–44.
[91] Freystein K, Salisch M, Reisser W. Algal biofilms on tree bark to monitor airborne pollutants. *Biologia.* 2008;*63*:866–72.
[92] Rindi F, Lam DW, López-Bautista JM. Phylogenetic relationships and species circumscription in *Trentepohlia* and *Printzina* (Trentepohliales, Chlorophyta). *Mol Phylogen Evol.* 2009;52: 329–39.
[93] Rindi F, Guiry MD, Barbiero RP, Cinelli F. The marine and terrestrial Prasiolales (Chlorophyta) of Galway City, Ireland: a morphological and ecological study. *J Phycol.* 1999;35:469–82.

Stefan Porembski

2.7 Desiccation-tolerant vascular plants: systematic distribution, ecology, and biogeography

The vast majority of vascular plants does not survive large variations in the water content of their vegetative tissues. There is, however, a number of ferns and angiosperms that developed vegetative desiccation tolerance (DT) independently several times. The exact number of DT species is still debatable but could reach approximately 1500. Most DT plants occur in the tropics where they colonize rock outcrops (e.g., inselbergs) and as epiphytes in the canopy of forests. Within ferns and fern allies, filmy ferns (Hymenophyllaceae) make up the largest group. Among angiosperms, monocotyledons outnumber the dicotyledons, with Cyperaceae, Poaceae, and Veloziaceae being the most important. Within dicotyledons, Gesneriaceae and Linderniaceae comprise most DT plants. Despite the lack of detailed information, it can be stated that southeastern Brazil, East Africa, and central Madagascar form centers of diversity for DT species. Because of the widespread destruction of their natural growth sites by various human activities, numerous DT plants are endangered.

2.7.1 Introduction

Over the last years, the interest in poikilohydric vascular plants has increased steadily. This trend is driven by the expectation that a detailed understanding of how plants survive extreme fluctuations of dehydration and rehydration could help in improving the desiccation tolerance (DT) of agricultural crops. Based on the study of Walter [1], the water content of poikilohydrous plants closely follows fluctuations of humidity in their environment. Surveys about different ecological, physiological, and molecular attributes of vascular DT species were provided by various authors [e.g., 2–11].

Many resurrection plants occur on exposed slopes of tropical rock outcrops and in particular on inselbergs (Fig. 2.7.1A). The latter often consist of granitic and gneissic rocks and form terrestrial habitat islands [12]. They are characterized by extreme environmental conditions, i.e., high temperatures, lack of soil, and unpredictable droughts, which render them microenvironmental deserts even when located in humid regions.

First remarks on DT vascular plants from rock outcrops were provided by, e.g., Dinter [13] and Heil [14]. Later, Hambler [15] and Gaff [16] emphasized the important role of rock outcrops as growth sites for resurrection plants. Own research over decades in many parts of the tropics combined with laboratory work and experiments in botanical gardens clearly showed that tropical inselbergs form centers of diversity for poikilohydrous vascular plants [17, 18]. Interestingly, the canopy of moist tropical forests contains many DT vascular plants that has also been largely

https://doi.org/10.1515/9783110646467-009

Fig. 2.7.1: (A) Inselbergs are widespread on old crystalline shields (Minas Gerais, Brazil). (B) *Selaginella sellowii* is a pioneer that commonly occurs on inselbergs in southeastern Brazil. (C) *Selaginella nivea* in the dry state. The species has a mosslike habit and grows on Malagasy inselbergs. (D) On tropical inselbergs, numerous quillwort species can be found in seasonally wet localities (*Isoetes* sp., Angola). (E) It is still debatable to what extent species of *Isoetes* are DT (*Isoetes* sp., Côte d'Ivoire). (F) Epiphytic member of Hymenophyllaceae with desiccated fronds (Cameroon).

neglected hitherto. In general, the number of vascular plants that show vegetative DT is relatively small and could reach approximately 1500 species.

2.7.2 Systematic overview

There are still many open questions concerning the problem of which species qualify as resurrection plants. For example, there are controversial views concerning the status of Velloziaceae. Based on observations in the field, Porembski [19] advocates that all members of this family are DT, whereas Alcantara et al. [20] state that a number of Brazilian members of Velloziaceae are non-DT.

Vegetative DT occurs among ferns and fern allies and angiosperms but is completely lacking within gymnosperms. The nonexistence of poikilohydrous gymnosperms is due to certain ecophysiological constraints that prohibit trees from being DT.

It has been estimated that approximately 300 species of DT vascular plants occur on rock outcrops [17]. In an updated calculation, Porembski [18] estimated the number of poikilohydric vascular plant species to reach approximately 1300. In the following sections, the current state of knowledge about the diversity and systematic distribution of vascular resurrection plants is presented. This account is based on recently published articles and on own fieldwork on rock outcrops in Brazil, India, Madagascar, and East and West Africa.

Ferns and fern allies

DT species occur both within lycophytes and monilophytes sensu [21]. Most poikilohydric ferns and fern allies occur in the tropics, but they are present in temperate regions too. Preferentially, they colonize xeric habitats such as rock outcrops or tree canopies. For many ferns, it can only be suspected that they are DT, and more observations in the field and in the laboratory are needed.

Spike mosses (*Selaginella*, Selaginellaceae) form an ancient group of lycopods and date back to the carboniferous period (330–350 million years ago). The genus *Selaginella* comprises an unclear number of DT species that belong to the subgenera *Tetragonostachys* (mosslike with small leaves that possess thick cuticles, distributed throughout the tropics) and *Stachygynandrum* (distributed throughout the tropics), forming rosettes such as the "Rose of Jericho" (*Selaginella lepidophylla*) as well as a clade that is still unnamed [22]. Obviously, poikilohydric spike mosses evolved at least three times in different clades of the genus.

Published literature and own fieldwork indicate that approximately 50 species of *Selaginella* are DT. The vast majority of them grows fully sun-exposed on tropical inselbergs (Fig. 2.7.1B, C). On inselbergs in southeastern United States, poikilohydric species of *Selaginella* (e.g., *Selaginella rupestris*) survive freezing temperatures [23].

Most quillworts (*Isoetes*, Isoetaceae) possess a grasslike habit and likewise belong to the lycopods and comprise mostly aquatic or semiaquatic species. A large number of *Isoetes* species colonize seasonally water-filled rock pools on inselbergs in tropical and temperate regions (Fig. 2.7.1D, E). These rock pools usually dry out completely during the dry season. It is conceivable that *Isoetes* species that grow in seasonal rock pools are DT, but further observations are needed.

DT species independently evolved several times within the monilophytes. The most speciose order containing resurrection plants is Hymenophyllales with filmy ferns (Hymenophyllaceae) containing more than 600 species that grow epiphytically and epilithically in very damp places. It has been shown by Kornás [24] and Nitta [25] as well as by own personal observations in different parts of tropical Africa, tropical Asia, and New Zealand that filmy ferns survive dry spells in a desiccated state (Fig. 2.7.1F). Members of Hymenophyllaceae possess particular adaptive traits (e.g., lamina only one cell-layer thick, no stomata, cuticle absent or largely reduced) to survive shortages in water supply. They are DT and have thus reduced investments into leaf tissues to a minimum.

Other fern orders that contain resurrection plants are Polypodiales and Schizaeales. Within Polypodiales, Aspleniaceae (*Asplenium* s.l.), Polypodiaceae (e.g., *Drynaria*, *Phymatosorus*, *Platycerium*, and *Polypodium*; Fig. 2.7.2A), and Pteridaceae (e.g., *Actiniopteris*, *Cheilanthes*, *Doryopteris*, *Hemionitis*, *Notholaena*, and *Pellaea*; Fig. 2.7.2B) have DT members. Within Schizaeales, Anemiaceae (*Anemia*) and Schizaeaceae (*Schizaea*) comprise poikilohydric species. The genera mentioned above mainly consist of species that grow as epiphytes or lithophytes. For a number of species, observations at their natural growth sites have shown that their leaves are gray (i.e., they are poikilochlorophyllous such as is the case with *Actiniopteris radiata*) in the desiccated state and regain their green coloration immediately after rewatering. In total, the number of DT ferns might reach more than 1200 species.

Angiosperms

Within angiosperms, the number of DT monocotyledons is much higher than those of dicotyledons. DT evolved several times independently within different angiosperm lineages. Although the most basal lineages do not contain any resurrection plants, this adaptive trait appeared in rather derived herbaceous lineages.

DT developed in the following orders of monocotyledons: Alismatales (Aponogetonaceae, *Aponogeton desertorum* from southern Africa seems to be poikilohydric), Asparagales (Boryaceae, several species of *Borya* in Australia), Pandanales (Velloziaceae), and Poales (Cyperaceae, Poaceae, and possibly Bromeliaceae). An overview on DT genera is shown in Tab. 2.7.1.

Velloziaceae comprise approximately 250 species that show a Gondwanan distribution with a center of diversity on various types of rock outcrops in Brazil [26], whereas eastern Africa and Madagascar form a secondary center. In parts of tropical Africa and Madagascar, Velloziaceae form typical elements of rock outcrop vegetation too. Many members of this family have a shrubby or treelet-like habit (Fig. 2.7.2C, D) with stems reaching a height of more than 5 m.

Based on observations at many growth sites both in the neotropics (Brazil) and paleotropics (East Africa, Madagascar), I hypothesize that all members of this family are DT because nearly all of them occur on azonal sites where water supply is lacking periodically. This view is opposed by Alcantara et al. [20], who stated that there are DT and non-DT species of *Vellozia* in the Brazilian campo rupestre vegetation where they are growing side by side.

Within Cyperaceae, all members of the tribe Trilepideae seem to be DT plants. Genera belonging to this tribe (*Afrotrilepis*, *Coleochloa*, *Microdracoides*, and

Fig. 2.7.2: (A) The epiphytic fern *Platycerium stemaria* in the desiccated state (Cameroon). (B) DT species of *Actiniopteris* are typical elements on paleotropical inselbergs. (C) *Vellozia plicata* is a mat-forming DT species on Brazilian inselbergs. (D) Numerous species of *Xerophyta* occur on rock outcrops in Madagascar (dry season aspect). (E) Stems of *Afrotrilepis pilosa* can reach more than 1 m in height (Nigeria). (F) *Coleochloa setifera* forms dense mats on inselbergs in East Africa and Madagascar.

Trilepis; Figs. 2.7.2E, F and 2.7.3A, B) show a disjunct distribution between the neotropics and the paleotropics. Whether other members of Cyperaceae are poikilohydric is debatable (e.g., species of *Bulbostylis*).

Poaceae comprise several genera belonging to different subfamilies that include DT species [27]. Based on own observations, it can be stated that certain genera

Tab. 2.7.1: Genera of vascular plants containing desiccation-tolerant taxa.

Genus	Family	Distribution	Growth sites
*Acanthochlamys**	Velloziaceae	SW China	Rock outcrops
Actiniopteris	Pteridaceae	Paleotropics	Rock outcrops
Afrotrilepis	Cyperaceae	W. Africa	Rock outcrops
Anemia	Schizaeaceae	S. America	Rock outcrops
*Aponogeton**	Aponogetonaceae	Paleotropics	Rock outcrops
Asplenium	Aspleniaceae	Subcosmopolitan	Rock outcrops
Barbacenia	Velloziaceae	S. America	Rock outcrops
Barbaceniopsis	Velloziaceae	S. America	Rock outcrops
Boea	Gesneriaceae	Paleotropics	Rock outcrops
Borya	Boryaceae	Australia	Rock outcrops
Bulbostylis	Cyperaceae	S. America	Rock outcrops
Chamaegigas	Linderniaceae	Namibia	Rock outcrops
Cheilanthes	Pteridaceae	Subcosmopolitan	Rock outcrops
Coleochloa	Cyperaceae	W./E. Africa, Madagascar	Rock outcrops Rarely epiphytic
*Corallodiscus**	Gesneriaceae	Tropical Asia	Rock outcrops
Craterostigma	Linderniaceae	W./E. Africa	Rock outcrops
Doryopteris	Pteridaceae	S. America	Rock outcrops
Drynaria	Polypodiaceae	Subcosmopolitan	Canopy
Eragrostiella	Poaceae	Australia	Rock outcrops
Eragrostis	Poaceae	E./S. Africa	Rock outcrops
Fimbristylis	Cyperaceae	Tropical Africa, Australia	Rock outcrops
*Guzmania**	Bromeliaceae	Neotropics	Canopy
Haberlea	Gesneriaceae	S. Europe	Rock outcrops
Hemionitis	Pteridaceae	S. America	Rock outcrops
*Henckelia**	Gesneriaceae	Tropical Asia	Rock outcrops
Hymenophyllum	Hymenophyllaceae	Subcosmopolitan	Canopy
Jancaea	Gesneriaceae	Greece	Rock outcrops
Lindernia	Linderniaceae	Tropical Africa	Rock outcrops
Linderniella	Linderniaceae	E./S. Africa, Madagascar	Rock outcrops
Micrairia	Poaceae	Australia	Rock outcrops
Microchloa	Poaceae	mostly Paleotropics	Rock outcrops
Microdracoides	Cyperaceae	W. Africa	Rock outcrops
Microsorum	Polypodiaceae	Paleotropics to Fr. Polynesia	Canopy Rock outcrops
Myrothamnus	Myrothamnaceae	E./S. Africa, Madagascar	Rock outcrops
Notholaena	Pteridaceae	N./S. America	Rock outcrops
*Oreocharis**	Gesneriaceae	Tropical Asia	Rock outcrops
Oropetium	Poaceae	Paleotropics	Rock outcrops
Paraboea	Gesneriaceae	Tropical Asia	Rock outcrops
Paraisometrum	Gesneriaceae	Tropical Asia	Rock outcrops
Pellaea	Pteridaceae	Tropical Africa	Rock outcrops
*Petrocosmea**	Gesneriaceae	Tropical Asia	Rock outcrops
*Pitcairnia**	Bromeliaceae	S. America	Rock outcrops
Platycerium	Polypodiaceae	mainly Paleotropics	Canopy Rock outcrops

Tab. 2.7.1 (continued)

Genus	Family	Distribution	Growth sites
Pleopeltis	Polypodiaceae	Subcosmopolitan	Canopy
Polypodium	Polypodiaceae	Subcosmopolitan	Rock outcrops Canopy
*Primulina**	Gesneriaceae	Tropical Asia	Rock outcrops
Ramonda	Gesneriaceae	S. Europe	Rock outcrops
Satureja	Lamiaceae	S. America	Rock outcrops
Schizaea	Schizaeaceae	E. Africa, Seychelles	Rock outcrops
Selaginella	Selaginellaceae	Pantrop., N. America	Rock outcrops
Sporobolus	Poaceae	Paleo-/Neotropics	Rock outcrops
Streptocarpus	Gesneriaceae	E. Africa, Madagascar	Rock outcrops
Styppeiochloa	Poaceae	Madagascar, tropical Africa	Rock outcrops
Trichomanes	Hymenophyllaceae	Subcosmopolitan	Canopy
Trilepis	Cyperaceae	S. America	Rock outcrops
Tripogon	Poaceae	Subcosmopolitan	Rock outcrops
*Trisepalum**	Gesneriaceae	Tropical Asia	Rock outcrops
Vellozia	Velloziaceae	S. America (mainly Brazil)	Rock outcrops
Xerophyta	Velloziaceae	Tropical Africa, Madagascar	Rock outcrops

Information on geographic distribution and growth sites is based on literature sources and own personal observations. For genera marked with asterisk (*), direct proof of DT is still lacking.

probably exclusively comprise resurrection species such as *Micraira*, *Microchloa*, *Oropetium* (Figs. 2.7.3C, D), *Styppeiochloa* (Fig. 2.7.3E), and *Tripogon*. Other genera (e.g., *Eragrostis* and *Sporobolus*) contain both resurrection and nonresurrection species. It can be estimated that approximately 40 species of grasses are DT.

The small family Boryaceae comprises several woody resurrection species within the genus *Borya* (Fig. 2.7.3F), which occur on rock outcrops in Australia. They possess needlelike leaves, which attain a bright orange color during desiccation.

There are hints that certain members of Bromeliaceae are DT. Based on the study of Zotz and Andrade [28], the epiphytic *Guzmania monostachya* loses more than 90% of the water present in full turgor and behaves like a DT plant. It has been argued that *Pitcairnia burchellii* qualifies as a DT species [29] too. This species grows on rock outcrops in the Brazilian Cerrado, and during long dry spells, its xeromorphic leaves lose most of their tissue water and the plants survive with an underground rhizome. It seems to be unclear whether dry leaves are able to regreen after rainfall. Viera et al. [30] mentioned that rehydration results in the production of new leaves and roots after leaf senescence. In addition, the genera *Stigmatodon* and *Tillandsia* might contain resurrection species that occur in perpendicular rock walls where they grow on bare rock and are able to survive weeks and months without precipitation (own observations).

Within dicotyledons, DT members occur in Gunnerales (Myrothamnaceae) and Lamiales (Gesneriaceae, Linderniaceae, and Plantaginaceae). Myrothamnaceae

Fig. 2.7.3: (A) Small (approximately 10 cm high) individual of *Microdracoides squamosus* during desiccation (Cameroon). (B) *Trilepis ihotzkiana* is a common mat-former on Brazilian inselbergs. (C) The DT annual *Oropetium thomaeum* is widespread on rock outcrops in southern India. (D) In West Africa, the DT annual *Oropetium aristatum* grows on inselbergs and ferricretes (Burkina Faso). (E) *Styppeiochloa hitchcockii* belongs to the most important mat formers on Malagasy inselbergs. (F) Several species of Borya ("pincushions") occur with DT members on Australian rock outcrops

comprise only the genus *Myrothamnus* with two species (*Myrothamnus flabellifolius* in eastern and southern Africa and *Myrothamnus moschatus* in Madagascar; Fig. 2.7.4A, B), which are typical elements of inselberg plant communities. It has to be emphasized that the *Myrothamnus* species are the only shrubby dicotyledons resurrection plants. Although there is detailed information on various morphological and biochemical aspects of *M. flabellifolius* (see [31]), almost no data are available concerning these aspects for the Malagasy species.

Lamiales comprise most resurrection plants within the dicotyledons. Many Gesneriaceae that occur on rock outcrops seem to be DT as has been shown, for

Fig. 2.7.4: (A) The shrub *Myrothamnus flabellifolius* is widespread on inselbergs in eastern and southern Africa (Angola). (B) Folded dry leaves of *Myrothamnus moschatus* (Madagascar). (C) *Haberlea rhodopensis* grows in clefts in rocky habitats in Bulgaria and Greece. (D) *Craterostigma yaundense* is endemic to inselbergs around Yaoundé (Cameroon). (E) *Afrotrilepis pilosa* individuals on inselberg with grayish, yellow, and green leaves depending on water availability (Côte d'Ivoire). (F) *Coleochloa setifera* can form monodominant mats on open rocky slopes (Madagascar).

example, for endemics in Mediterranean mountains (*Haberlea*, *Jankaea*, and *Ramonda*; Fig. 2.7.4C). From tropical outcrops, it is also known that Gesneriaceae are DT, such as species of *Boea*, *Paraboea*, *Paraisometrum*, and *Streptocarpus*. However, there are other promising candidates growing on rocks in tropical Asia such as *Corallodiscus*, *Hemiboea*, *Henkelia*, *Oreocharis*, *Petrocosmea*, *Primulina*, and *Trisepalum* that should be tested concerning their DT.

Linderniaceae comprise several genera that include DT species: *Chamaegigas*, *Craterostigma* (Fig. 2.7.4D), and *Linderniella*, which mainly occur on rock outcrops in tropical Africa and Madagascar.

2.7.3 Ecological aspects

In this section, a short overview on certain adaptive traits is provided mainly from a morphological-anatomical and field ecological viewpoint. Based on own observations, it can be stated that there are differences in drought hardiness between individual DT species. Our still fragmentary knowledge shows that species of *Selaginella* belong to the most drought-tolerant colonizers of rock outcrops that behave similar to certain mosses. In particular, *Selaginella* species that have a mosslike habit (e.g., *Selaginella nivea*, Madagascar), which grow fully sun-exposed directly on the underlying rock, regreen shortly (i.e., within a few hours) after rain and return to dormancy more or less immediately after the onset of a drier period.

Dehydration of resurrection plants is often accompanied by a drastic change in leaf color. During the cycle of drying and regreening, the leaves of most poikilohydric angiosperms are characterized by the following behavior: desiccated grayish leaves unfold and obtain at first a yellow coloration within approximately 48 hours after rain and subsequently turn green (Fig. 2.7.4E). After prolonged periods of drought, the leaves fold and turn yellow first and subsequently gray (own observations in the field and in the laboratory). Occasionally (e.g., in some Linderniaceae and Velloziaceae), dry leaves have a reddish-violet coloration. In general, it can be distinguished between poikilochlorophyllous (i.e., losing most or all of their chlorophyll while drying) and homoiochlorophyllous (i.e., chlorophyll content will be preserved during drying) plants. The majority of monocotyledons are poikilochlorophyllous, whereas most DT dicotyledons are homoiochlorophyllous. In nature, poikilohydric vascular plants on rock outcrops persist for prolonged periods in a dormant state such as *Afrotrilepis pilosa* (3–8 months), depending on whether individuals grow on inselbergs located in rainforests or savannas. During the rainy season, leaves can stay continuously green for weeks or months, but even then they react rapidly when water supply is not sufficient and dry out. Thus, they are able to go several times through the cycle of drying and regreening within a couple of weeks.

Widespread in many groups of vascular resurrection plants is the folding or curling of leaves during the process of desiccation, which seems to be important for avoiding photoinhibitory damage. The typical parallel nervature of leaves of monocotyledons seems to offer mechanical advantages with regard to leaf folding motions during dehydration-rehydration cycles compared with the netlike nervature of dicot leaves. This might explain why poikilohydric monocots largely outnumber dicots. Remarkably, many DT dicots have modified the typical leaf nervature of dicots and possess a parallel leaf nervature what could allow better leaf folding. For example, this is the case within Myrothamnaceae and Linderniaceae. A remarkable exception is provided by DT members of Gesneriaceae, which possess a netlike venation pattern. According to Kampowski et al. [32], *Ramonda myconi* leaves undergo complex morphological and anatomical alterations that result in drastic changes of its mechanical properties, making the leaves more flexible.

Poikilohydric arborescent (i.e., possessing woody trunks) monocotyledons form a remarkable example of convergent evolution. They occur in both tropical and temperate regions on rock outcrops. Typical elements are Boryaceae (*Borya*), Cyperaceae (*Afrotrilepis*, *Microdracoides*, *Trilepis*), and Velloziaceae (e.g., *Barbacenia*, *Vellozia*, and *Xerophyta*). Interestingly, their trunks consist mainly of adventitious roots with a velamen radicum, which is useful for rapid water uptake [33]. Certain Velloziaceae from nutrient-poor Brazilian campo rupestres possess roots ("Velloziod roots") that are able to mobilize phosphorus via root exudates [34]. Usually, DT mat-forming monocots possess the ability to form large clonal populations of considerable age (i.e., hundreds of years) by means of stolons, which allows the long-lasting occupation of suitable sites [35]. Trunks of Cyperaceae and Velloziaceae can attain a considerable diameter and are regularly colonized by epiphytic orchids (e.g., *Polystachya microbambusa* in West Africa and *Pseudolaelia vellozicola* in Brazil), which show a high degree of phorophyte specificity [36].

Ecologically, very remarkable is the rock pool specialist *Chamaegigas intrepidus*, which is endemic to Namibia. The species occurs on inselbergs in semiarid areas, where it colonizes seasonally water-filled rock pools [14, 37]. The species has contractile, DT-submerged leaves and develops nonpoikilohydric floating leaves after rainfall. During desiccation, the submerged leaves shrink by 75–80%, mainly due to contraction of xylem vessels [38].

2.7.4 Habitats

Poikilohydric vascular plants have their center of diversity on tropical rock outcrops and in forest canopies (ferns), but some occur in temperate regions too. The altitudinal spectrum of their distribution reaches from sea level to heights of nearly 3000 m a.s.l. (e.g., on Mt. Mulanje, Malawi [39]). It can be speculated that in southwestern China, possible DT species (e.g., *Acanthochlamys bracteata*, Velloziaceae; *Tripogon* spp., Poaceae) colonize dry rocky slopes at altitudes even higher than 4000 m a.s.l. where they are exposed to deep freezing temperatures.

DT angiosperms nearly exclusively grow on rock outcrops such as inselbergs, which are characterized by very harsh environmental conditions (e.g., lack of water and nutrients, high temperatures). Throughout the tropics, among the most dry-adapted resurrection species are those which form carpetlike mats (mainly monocotyledons such as *A. pilosa* in West Africa or *Coleochloa setifera* in Madagascar; Fig. 2.7.4F) on open, sometimes steep rocky slopes. Mat-forming species are literally glued to the underlying rock via their roots, which allow them to colonize "bare rock." Frequently, mat communities are colonized by DT dicotyledons such as Gesneriaceae (e.g., *Streptocarpus*), Linderniaceae (e.g., *Craterostigma*), and *Myrothamnus*. These taxa can also be found in shallow depressions and seasonally water-filled rock pools (e.g., the famous *Chamaegigas intrepidus* in Namibia). Both on tropical and temperate rock outcrops, species of *Selaginella* form mats on sun-exposed rocks (Fig. 2.7.5A), and they occur at the periphery of mats formed by DT monocotyledons.

Still understudied are lateritic plateaus (e.g., ferricretes), which are widespread in many parts of the tropics, but it has been shown that they harbor numerous resurrection plants. Because of their extreme seasonality and thus extreme seasonal dryness for several months, they are colonized by various grasses, Velloziaceae, and ferns (in Africa, e.g., *Actiniopteris* spp.). Among Poaceae, the genera *Microchloa* (mostly *Microchloa indica*) and *Oropetium* are prominent on lateritic plateaus in the paleotropics.

Steep, cliff-like, basaltic mountain ranges are likewise important growth sites for poikilohydric species. This is particularly true for the Western Ghats in southern India, which extend for approximately 1500 km. In parts of this mountain range, annual rainfall exceeds 5000 mm, which falls within only 4–5 months. The cliff-like, nearly perpendicular rocky slopes form the largest ecosystem on earth that is characterized by large stands of DT species (mainly species of *Tripogon*), which form extensive mats (Fig. 2.7.5B).

As far as is known, resurrection species among vascular epiphytes are nearly exclusively ferns. Particularly, speciose are filmy ferns (Hymenophyllaceae) that sometimes grow lithophytically too. According to Kornás [24] and Nitta [25], filmy ferns prefer shady conditions where they can build up large stands. They form the largest family of vascular resurrection plants (approximately 700 spp.). Moreover, species of Aspleniaceae (e.g., *Asplenium*) and Polypodiaceae (e.g., *Microsorum*, *Platycerium*, and *Polypodium*) occur with DT members in tropical and temperate regions. Only very rarely specialized resurrection species of Cyperaceae (i.e., *Coleochloa* ssp. in tropical Africa) and Poaceae (e.g., *Tripogon* spp. in India) grow as epiphytes. A rather limited number of resurrection plants have managed to colonize urban environments. For example, in Europe several species of *Asplenium* (e.g., *A. ceterach*, *A. trichomanes*) are common colonizers of crevices of buildings (Fig. 2.7.5C). They are typical elements of particular plant communities (e.g., *Asplenietea trichomanis*) that can fall completely dry for weeks or months and which survive freezing in Europe. Villages and cities in humid regions of tropical Africa are characterized by the staghorn fern *Platycerium stemaria*, which is common on roofs and walls of buildings.

Fig. 2.7.5: (A) *Selaginella* sp. (gray) growing together with *Aloe* sp. (Madagascar). (B) Near vertical cliffs in the Western Ghats in southern India are covered by dense stands of DT species of *Tripogon* (here *Tripogon lisboae*). (C) In the Mediterranean region, the DT fern *Asplenium ceterach* commonly colonizes urban habitats (southern France). (D) A species of *Selaginella* on sale in a market in India. (E) Twigs of *Myrothamnus moschatus* in a market for medicinal plants in Antananarivo (Madagascar).

2.7.5 Geographic distribution

Most vascular resurrection plants occur in the tropics where they colonize tree canopies and rock outcrops. It is well documented that the tropical parts of Africa and South America are rich in poikilohydric species. Tropical Asia seems to be less speciose, but this could be a result of rather limited fieldwork conducted there. It can be expected that better exploration of tropical Asia (in particular, tower karst formations in China and Vietnam seem to be rich in candidates) for resurrection species will demonstrate that in particular for Gesneriaceae (e.g., within *Boea*, *Corallodiscus*, *Paraboea*), and Poaceae (within *Tripogon*), the number of taxa will increase. Interestingly, certain candidates that should be tested for desiccation tolerance occur at high altitudes (i.e., higher than 4000 m a.s.l.) in the Himalayas, such as *Corallodiscus kingianus* and *Acanthochlamys bracteata* (the only Velloziaceae in this region which seems to be the most basal member of the family).

In tropical Africa, the Sudano-Zambezian Region (mainly eastern and southern Africa) harbors many poikilohydric species. Deserts and semideserts in Africa are almost devoid of resurrection plants except the small grass *Tripogon minimus*, which can be found up to the southern fringes of the Sahara. Rock outcrops on the central highlands in Madagascar form global centers of diversity for resurrection species [40]. Particularly rich in species are Gesneriaceae (*Streptocarpus*), Linderniaceae (*Craterostigma*, *Linderniella*), and Velloziaceae (*Xerophyta*). Cyperaceae (*Afrotrilepis*, *Coleochloa*, and *Microdracoides*) and Poaceae (*Microchloa*, *Sporobolus*, *Styppeiochloa*, and *Tripogon*) contain fewer species but can be dominant as mat formers (in particular, *Styppeiochloa*). Most important mat formers are *A. pilosa* (West Africa [41]), *C. setifera* (East Africa, Madagascar), *Microdracoides squamosus* (West Africa), *Styppeiochloa hitchcockii* (Madagascar), and various species of *Xerophyta* (mainly East Africa and Madagascar). The shrubby Myrothamnaceae are endemic to eastern and southern Africa (*M. flabellifolius*) and Madagascar (*M. moschatus*). Fern and fern allies occur in Africa and Madagascar with rock outcrop dwellers (e.g., *Actiniopteris*, *Asplenium*, *Pellaea*, and *Selaginella*) and epiphytes (e.g., *Hymenophyllum*, *Microsorum*, and *Platycerium*).

Rock outcrops in North and Central America are colonized by poikilohydric ferns and grasses (*Sporobolus atrovirens*). Among ferns, the genera *Cheilanthes*, *Notholaena*, *Pellaea*, and *Polypodium* (with the lithophytic/epiphytic *Polypodium polypodioides* and *Polypodium virginianum*) are important [42]. Well known is the so-called Rose of Jericho (*S. lepidophylla*), a typical element of the Chihuahuan Desert.

In South America, rock outcrops in Brazil (mostly inselbergs, lateritic plateaus, and campo rupestre) form centers of diversity for DT plants [43]. Campo rupestre vegetation (mainly in Minas Gerais and Bahia) is a hot spot for Velloziaceae of which a large number might be DT (but see [20]). In general, Velloziaceae occur on all types of rock outcrops in Brazil from sea level up to high altitude grasslands at approximately 2000 m a.s.l. (e.g., Itatiaia Mts., Serra do Brigadeiro [44]). Shrubby to treelike Velloziaceae have their center of diversity in the quartzitic Serra do Espinhaço, where they occur with numerous local endemics (mainly *Barbacenia* spp. and *Vellozia* spp.). Ferns and fern allies (e.g., *Anemia*, *Doryopteris*, and *Selaginella*) are likewise important, whereas Cyperaceae (mainly *Trilepis*) are of minor importance and DT Poaceae and dicotyledons are almost absent. A secondary center of diversity for poikilohydric species is formed by Sugar Loaf Land, which comprises inselberg landscapes in southeastern Brazil [45]. Information on poikilohydrous epiphytic ferns is sparse, but it seems species of *Pleopeltis* and *Trichomanes* belong to this group (cf. [46]). Whether certain Bromeliaceae that occur on rocks or as epiphytes are poikilohydric still has to be tested.

On various types of rock outcrops in Europe, poikilohydric ferns occur (e.g., *Asplenium ceterach*). Well known is the occurrence of DT Gesneriaceae (*Haberlea* with two species, *Jankaea heldreichii*, *Ramonda* with three species), which form small rosettes. These gesneriads are relicts from the tertiary period and grow in mountain ranges in the Mediterranean region.

All resurrection species in Australia are endemic to this continent. Prominent are woody, cushion-forming species of *Borya*, which are widespread on rock outcrops. In addition, ferns (e.g., *Cheilanthes* spp.), grasses (*Micraira* spp.), and the gesneriad *Boea hygroscopica* are present [47].

2.7.6 Economic use

Over the last decades, a number of vascular resurrection plants have been studied to understand the strategies of these plants to cope with drought [48, 49]. It is conceivable that these efforts might be of interest for the future development of extremely drought-tolerant crop species [50].

It should not be underestimated that a number of DT plants are widely used economically as ornamentals. Among the better known cases is the Rose of Jericho *S. lepidophylla*, which is commonly sold; this is also the case with other species of *Selaginella* (Fig. 2.7.5D). Another poikilohydric fern that is sold as ornamental is the epiphytic *Platycerium stemaria* [51]. Increasingly popular as ornamentals are poikilohydric Gesneriaceae such as species of *Boea*, *Ramonda*, and *Streptocarpus*.

Certain resurrection species are sold in large quantities as medicinal plants. This is well documented for the two species of *Myrothamnus* (*M. flabellifolius* and *M. moschatus*), which are offered in markets in parts of Africa and Madagascar (Fig. 2.7.5E). Because these species are not cultivated, their twigs are collected in the wild, which possibly has negative consequences for them.

2.7.7 Conservation

Most vascular resurrection plants are not protected by law because in many cases their growth sites have escaped the attention of conservationists. However, their growth sites (especially rock outcrops) are highly endangered at present due to various human activities such as quarrying, fire, grazing, and illegal collection of plants [52]. A number of resurrection plants are endangered by the collection of large amounts of individuals for ornamental and medicinal purposes. This is embodied by

S. lepidophylla and other species of *Selaginella*, which are collected in large quantities in the wild.

2.7.8 Research perspectives

Our knowledge about the diversity of vascular DT plants and their natural growth sites is still far from being complete. In particular, rock outcrops in the tropics are still underexplored with regard to DT plants. Inventorying work is an old fashioned but still necessary task and is particularly needed in parts of tropical Asia, eastern Africa, and Madagascar. Moreover, in many cases, it is still not clear whether certain taxa are poikilohydric or not, which is particularly true for many members of Velloziaceae. Here, more direct observations in the field and laboratory analyses are necessary.

Research on the molecular level hitherto concentrated on a limited number of DT species (e.g., *Craterostigma plantagineum* and *Xerophyta elegans*), which acted as model organisms. If our knowledge concerning the cultivation of other DT taxa could be enhanced, other rather neglected DT species could be tested with regard to their ability to cope with desiccation.

Acknowledgments

This research was financially supported by the Deutsche Forschungsgemeinschaft, the Akademie der Wissenschaften und der Literatur Mainz, and the Andrew W. Mellon Foundation. Numerous colleagues are thanked for discussions and common fieldwork in different parts of the world: S. Andriambololonera (Antananarivo), L. O. Azevedo (Belo Horizonte), D. Bartels (Bonn), W. Barthlott (Bonn), L. Bondi (Rio de Janeiro), M. N. Datar (Pune), J. M. Farrant (Cape Town), S. D. Hopper (Albany), J. Köhler (Rostock), L. P. Mauad (Rio de Janeiro), R. Mello-Silva (Sao Paulo), L F. A. de Paula (Belo Horizonte), M. Oliver (St. Louis), M. Rabarimanarivo (Antananarivo), J. Rexroth (Rostock), Elycée Tindano (Ouagadougou), and A. Watve (Pune).

References

[1] Walter H. Die Hydratur der Pflanze und ihre physiologisch-ökologische Bedeutung. Jena: Gustav Fischer; 1931.

[2] Gaff DF. The biology of resurrection plants. In: Pate JS, McComb AJ, editors. The Biology of Australian plants. Perth, Australia: University of Western Australia Press, 1981. p. 114–46.

[3] Gaff DF. Responses of desiccation tolerant "resurrection" plants to water stress. In: Kreeb KH, Richter H, Hinckley TM, editors. Structural and Functional Responses to Environmental Stresses: Water Shortages. The Hague: SPB Academic Publishing; 1989. p. 264–311.

[4] Gaff DF, Oliver MJ. The evolution of desiccation tolerance in angiosperm plants: a rare yet common phenomenon. *Funct Plant Biol.* 2013;40:315–28. doi: 10.1071/FP12321

[5] Bewley JD. Physiological aspects of desiccation tolerance—a retrospect. *Int J Plant Sci.* 1995;156:393–403. doi: 10.1086/297261
[6] Hartung W, Schiller P, Dietz K-J. The physiology of poikilohydric plants. *Prog Bot.* 1998;59: 299–327.
[7] Tuba Z, Proctor MCF, Csintalan Z. Ecophysiological responses of homoiochlorophyllous desiccation tolerant plants: a comparison and an ecological perspective. *Plant Growth Regul.* 1998;24:211–17. doi: 10.1023/A:1005951908229
[8] Kluge M, Brulfert J. Ecophysiology of vascular plants on inselbergs. In: Porembski S, Barthlott W, editors. Inselbergs: Biotic Diversity of Isolated Rock Outcrops in Tropical and Temperate Regions. Ecological Studies, 146. Berlin: Springer; 2000. p. 143–74.
[9] Walters C, Farrant JM, Pammenter NW, Berjak P. Desiccation stress and damage. In: Black M, Pritchard H, editors. Desiccation and Survival in Plants: Drying without Dying. Wallingford: CAB International; 2002. p. 263–91.
[10] Kappen L, Valladares F. Opportunistic growth and desiccation tolerance: the ecological success of poikilohydrous autotrophs. In: Pugnaire F, Valladares F, editors. Functional Plant Ecology. 2nd ed. Boca Raton: CRC/Taylor and Francis Group; 2007. p. 7–65.
[11] Oliver MJ, Farrant JM, Hilhorst HWM, Mundree S, Williams B, Bewley JD. Desiccation tolerance: avoiding cellular damage during drying and rehaydration. *Annu Rev Plant Biol.* 2020;71:435–60. doi.org/10.1146/annurev-arplant-071219-105542
[12] Porembski S, Barthlott W, editors. Inselbergs: Biotic Diversity of Isolated Rock outcrops in tropical and temperate regions. In: Ecological Studies. Vol. 146. Berlin, Heidelberg: Springer; 2000. doi: 10.1007/978-3-642-59773-2
[13] Dinter K. Botanische Reisen in Deutsch-Südwest-Afrika. *Feddes Rep Beih.* 1918;3:1–169.
[14] Heil H. *Chamaegigas intrepidus* Dtr., eine neue Auferstehungspflanze. *Beitr Bot Zentralbl* 1924;41:41–50.
[15] Hambler DJ. A poikilohydrous, poikilochlrophyllous angiosperm from Africa. *Nature.* 1961;191:1415–16.
[16] Gaff DF. Desiccation tolerant vascular plants of southern Africa. *Oecologia.* 1977;31:95–109. doi: 10.1007/BF00348713
[17] Porembski S, Barthlott W. Granitic and gneissic outcrops (inselbergs) as centres of diversity for desiccation-tolerant vascular plants. *Plant Ecol.* 2000;151:19–28. doi.org/10.1023/A:1026565817218
[18] Porembski S. Evolution, diversity, and habitats of poikilohydrous vascular plants. In: Lüttge U, Beck E, Bartels D, editors. Plant Desiccation Tolerance. Vol. 215. Ecological Studies. Berlin, Heidelberg: Springer; 2011. p. 139–56. doi.org/10.1007/978-3-642-19106-0_8
[19] Porembski S. 2007 Tropical inselbergs: habitat types, adaptive strategies and diversity patterns *Brazil J Bot.* 2007;30:579–86. doi: 10.1590/S0100-84042007000400004
[20] Alcantara S, Mello-Silva R, Teodoro GS, Drequeceler K, Ackerly DD, Oliveira RS. Carbon assimilation and habitat segregation in resurrection plants: a comparison between desiccation- and non-desiccation-tolerant species of neotropical Velloziaceae (Pandanales). *Funct Ecol.* 2015;29:1499–512. doi: 10.1111/1365-2435.12462
[21] Pryer KM, Schuettpelz E, Wolf PG, Schneider H, Smith AR, Cranfill R. Phylogeny and evolution of ferns (monilophytes) with a focus on the early leptosporangiate divergences. *Am J Bot.* 2004;91:1582–98. doi: 10.3732/ajb.91.10.1582
[22] Korall P, Kenrick P. Phylogenetic relationships in Selaginellaceae based on rbcL sequences. *Am J Bot.* 2002;89:506–17. doi: 10.3732/ajb.89.3.506
[23] Leslie KA, Burbanck MP. Vegetation of granitic outcroppings at Kennesaw Mountain, Cobb County, Georgia. *Castanea.* 1979;44:80–7.
[24] Kornás J. Life-forms and seasonal patterns in the pteridophytes of Zambia. *Acta Soc Bot Pol.* 1977;46:669–90. doi: 10.5586/asbp.1977.055

[25] Nitta JH. Distribution, ecology and systematics of the filmy ferns (Hymenophyllaceae) of Moorea (French Polynesia). Berkeley: University of California, Department of Integrative Biology; 2006.
[26] Mello-Silva R, Santos DYAC, Salatino MLF, Motta LB, Cattai MB, Sasaki D, Lovo J, Pita PB, Rocini C, Rodrigues CDN, Zarrei M, Chase MW. Five vicarious genera from Gondwana: the Velloziaceae as shown by molecules and morphology. *Ann Bot.* 2011;108:87–102. doi: 10.1093/aob/mcr107
[27] Pardo J, Wai CM, Chay H, Madden CF, Hilhorst HWM, Farrant JM, VanBuren R. Intertwined signatures of desiccation and drought tolerance in grasses. *Proc Natl Acad Sci.* 2020;117:10079–88. doi: 10.1073/pnas.2001928117/-/DCSupplemental
[28] Zotz G, Andrade J-L. Water relations of two co-occurring epiphytic bromeliads. *J Plant Physiol.* 1998;152:545–54. doi: 10.1016/S0176-1617(98)80276-9
[29] Viera EA, Silva KR, Oriani A, Moro CF, Braga MR. Mechanisms of desiccation tolerance in the bromeliad *Pitcairnia burchellii* Mez: biochemical adjustments and structural changes. *Plant Physiol Biochem.* 2017;121:21–30. doi: 10.1016/j.plaphy.2017.10.002
[30] Viera EA, Centeno D, Freschi L, da Silva EA, Braga MR. The dual strategy of the bromeliad *Pitcairnia burchellii* Mez to cope with desiccation. *Environ Exper Bot.* 2017;143:135–48. doi: 10.1016/j.envexpbot.2017.09.003
[31] Moore JP, Lindsey GG, Farrant JM, Brandt, WF. An overview of the biology of the desiccation-tolerant resurrection plant *Myrothamnus flabellifolia. Ann Bot.* 2007;99:211–17. doi: 10.1093/aob/mcl269
[32] Kampowski T, Demandt S, Poppinga S, Speck T. Kinematical, structural and mechanical adaptations to desiccation in poikilohydric *Ramonda myconi* (Gesneriaceae). *Front Plant Sci.* 2018;9:1701. doi: 10.3389/fpls.2018.01701
[33] Porembski S, Barthlott W. On the occurrence of a velamen radicum in tree-like Cyperaceae and Velloziaceae. *Nord J Bot.* 1995;15:625–29. doi: 10.1111/j.1756-1051.1995.tb02131.x
[34] Teodoro GS, Lambers H, Nascimento DL, de Britto Costa P, Flores-Borges DNA, Abrahao A, Mayer JLS, Sawaya ACHF, Ladeira FSB, Abdala DB, Pérez CA, Oliveira RS. Specialized roots of Velloziaceae weather quartzite rock while mobilizing phosphorus using carboxylates. *Funct Ecol.* 2019;33:1–12. doi: 10.1111/1365-2435.13324
[35] Porembski S. Vegetative architecture of desiccation-tolerant arborescent monocotyledons. *Aliso.* 2006;22:129–34. doi: 10.5642/aliso.20062201.10
[36] Porembski S. Epiphytic orchids on arborescent Velloziaceae and Cyperaceae: extremes of phorophyte specialisation. *Nord J Bot.* 2005;23:505–13. doi: 10.1111/j.1756-1051.2003.tb00424.x
[37] Heilmeier H, Durka W, Woitke M, Hartung W. Ephemeral pools as stressful and isolated habitats for the endemic aquatic resurrection plant *Chamaegigas intrepidus. Phytocoenologia.* 2005;35:449–68. doi: 10.1127/0340-269X/2005/0035-0449
[38] Schiller P, Wolf R, Hartung W. A scanning electron microscopical study of hydrated and desiccated submerged leaves of the aquatic resurrection plant *Chamaegigas intrepidus. Flora.* 1999;194: 97–102. doi: 10.1016/S0367-2530(17)30882-4
[39] Porembski S. Notes on the vegetation of inselbergs in Malawi. *Flora.* 1996;191:1–8. doi: 10.1016/S0367-2530(17)30685-0
[40] Rabarimanarivo MN, Ramandimbisoa B, Rakotoarivelo NH, Phillipson PB, Andriambololonera S, Callamander MW, Porembski S. The extraordinary botanical diversity of inselbergs in Madagascar. *Candollea.* 2019;74:65–84. doi: 10.15553/c2019v741a8
[41] Porembski S, Brown G, Barthlott W. A species-poor tropical sedge community: *Afrotrilepis pilosa* mats on inselbergs in West Africa. *Nord J Bot.* 1996;16:239–245. doi:10.1111/j.1756-1051.1996.tb00222.x
[42] Iturriaga G, Gaff DF, Zentella R. New desiccation-tolerant plants, including a grass, in the central highlands of Mexico, accumulate trehalose. *Austr J Bot.* 2000;48:153–58. doi: 10.1071/BT98062

[43] Porembski S, Martinelli G, Ohlemüller R, Barthlott W. Diversity and ecology of saxicolous vegetation mats on inselbergs in the Brazilian Atlantic rainforest. *Divers Distrib.* 1998;4:107–119. doi.org./10.1046/j.1365-2699.1998.00013.x

[44] Campos PV, Villa PM, Schaefer CEGR, Nunes JA, Porembski S, Neri AV. Beta diversity, community composition and structure of high altitude grasslands along an altitudinal gradient in southeastern Brazil. *Rev Biol Trop.* 2020;68:977–86. doi: 10.15517/rbt.v68i3.37704

[45] de Paula LFA, Forzza RC, Neri AV, Bueno ML, Porembski S. Sugar Loaf Land in south-eastern Brazil: a centre of diversity for mat-forming bromeliads on inselbergs. *Bot J Linn Soc.* 2016;181:495–76. doi: 10.1111/boj.12383

[46] Hietz P, Briones O. Correlation between water relations and within-canopy distribution of epiphytic ferns in a Mexican cloud forest. *Oecologia.* 1998;114:305–16. doi: 10.1007/s004420050452

[47] Gaff DF, Latz PK. The occurrence of resurrection plants in the Australian flora. *Austr J Bot.* 1978;26:485–92. doi: 10.1071/BT9780485

[48] Alam A, Dwivedi A, Emmanuel I. Resurrection plants: imperative resources in developing strategies to drought and desiccation pressure. *Plant Sci Today.* 2019;6:333–41. doi: 10.14719/pst.2019.6.3.542

[49] Farrant JM, Moore JP, Hilhorst HWM. Editorial: unifying insights into the desiccation tolerance mechanisms of resurrection plants and seeds. *Front Plant Sci.* 2020;11:1089. doi: 10.3389/fpls.2020.01089

[50] Hilhorst HWM, Farrant JM. Plant desiccation tolerance: a survival strategy with exceptional prospects for climate smart agriculture. *Ann Plant Rev Online.* 2018;1:1–27. doi: 10.1002/9781119312994.apr0637

[51] Porembski S, Biedinger N. Epiphytic ferns for sale: influence of commercial plant collection on the frequency of *Platycerium stemaria* (Polypodiaceae) in coconut plantations on the southeastern Ivory Coast. *Plant Biol.* 2001;3:72–6. doi: 10.1055/s-2001-11752

[52] Porembski S, Silveira FAO, Fiedler PL, Watve A, Rabarimanarivo, M, Kouame, F, Hopper SD. Worldwide destruction of inselbergs and related rock outcrops threatens a unique ecosystem. *Biodivers Conserv.* 2016;25:2827–30. doi: 10.1007/s10531-016-1171-1

Index

https://doi.org/10.1515/9783110646467-010

www.ingramcontent.com/pod-product-compliance
Lightning Source LLC
Chambersburg PA
CBHW081207260926
51849CB00009B/479
* 9 7 8 3 1 1 0 6 4 2 6 1 2 *